Bernoulli   Bayes   Laplace
1713   *1763*   *1813*

Anniversary Volume

Proceedings of an International Research Seminar
Statistical Laboratory
University of California, Berkeley 1963

Edited by
Jerzy Neyman and Lucien M. Le Cam

Springer-Verlag Berlin Heidelberg GmbH

ISBN 978-3-642-49466-6   ISBN 978-3-642-49749-0 (eBook)
DOI 10.1007/978-3-642-49749-0

Library of Congress Catalog Card Number 65-26 236

Title No. 1286

Foreword

The present volume represents the Proceedings of an International Research Seminar organized in 1963 by the Statistical Laboratory, University of California, Berkeley, on the occasion of a remarkable triple anniversary:

the 250th anniversary of JACOB BERNOULLI's *"Ars Conjectandi"*,

the 200th anniversary of THOMAS BAYES' *"Essay towards solving a problem in doctrine of chance"*, and

the 150th anniversary of the PIERRE-SIMON DE LAPLACE's *"Essai philosophique sur les probabilités"*.

Financial assistance of the National Science Foundation, without which the Seminar could not have been held, is gratefully acknowledged.

The publication of *Ars Conjectandi*, in 1713, was a milestone in the history of probability theory. Here, for the first time, appeared a careful description of the now well-known combinatorial methods which give solutions of many problems on simple games of chance. Also, *Ars Conjectandi* contains the Bernoulli numbers, theorems relating to the duration of games, and to the ruin of gamblers and, above all, the statement and proof of the famous Bernoulli weak law of large numbers.

Even though the original Latin edition of *Ars Conjectandi* was followed by several in modern languages, currently the book is not easily accessible. Apparently the last re-publication, in German, occurred in 1899, in two issues, No. 107 and No. 108, of the series "Ostwald's Klassiker der exakten Wissenschaften", Wilhelm Engelman, Leipzig. The two books are difficult to locate.

In 1763, exactly 50 years after *Ars Conjectandi*, THOMAS BAYES' *"Essay"* was posthumously published by Richard Price (*Philosophical Transactions, Royal Society*, London, Vol. 53, 1763, pp. 376—398). This paper has been the focus of what may be the most heated controversy in the history of probability and statistics, a controversy that extends to the present time, The contents of Bayes' paper are limited and mathematically unsophisticated, and the most surprising thing about the paper is that it could have become the center of frequently bitter and prolonged debate.

Bayes' „*Essay*" is readily accessible. Sometime in the 1930's it was photographically republished by the Graduate School of the U.S. Department of Agriculture, Washington, D.C., with commentaries by W. EDWARDS DEMING and EDWARD C. MOLINA. More recently it was again re-published, with commentaries by G. A. BARNARD, in *Biometrika*, Vol. 45 (1958).

In 1812, 49 years after the appearance of Bayes' paper, the French Academy published the memorable book *"Théorie analytique des probabilités"* by PIERRE-SIMON DE LAPLACE. In spite of the then developing Napoleon's debacle in Russia, the book must have sold well, presumably not only in France, because the second edition appeared in 1814, only two years later. In addition to the original text of almost 500 pages, this second edition contains several supplements and a 153 page *"Introduction"*. This *"Introduction"*, then, must have been written in 1813, 150 years before the Berkeley Seminar of 1963. It appeared also as a separate publication, under the title *"Essai philosophique sur les probabilités"*.

"Théorie analytique", including the *Introduction*, has again been republished in 1820 and several times thereafter and is currently accessible in many university libraries. An English version of the *"Essai philosophique"* was issued in 1951 by Dover Publications.

The interest that a contemporary reader may find in the three famous publications must be conditioned by two factors: the character of contents and the time interval dividing us from the epoch when the given work was completed. These two factors combine heavily to favor LAPLACE. In fact, we found *"Théorie analytique"* not only readable, but highly interesting and thoroughly enjoyable, both because of its contents and because of the elegance of LAPLACE's style. Regretfully, this elegance is easily lost in translations.

"Essai philosophique" is described by LAPLACE as an extended text of a lecture he delivered in the Ecoles Normales in 1795, and contains no mathematics. Essentially, it may be considered as a summary, of LAPLACE's knowledge in the various domains of science and of his thinking over the period between the French Revolution and the year of disaster marking the decline of the Napoleonic era. This by itself makes *"Essai philosophique"* very interesting.

The leading idea of the book is that each particular phonomenon in Nature, including social and political developments, is governed by forces of two distinct kinds, the permanent forces and the accidental forces. In each particular phenomenon, the effect of accidental forces may appear stronger than that of permanent forces, with the result that such phenomena become appropriate subjects for probabilistic studies. On the other hand, in a long series of similar occurrences, the accidental forces average out and the permanent forces prevail. This is considered by LAPLACE as a consequence of Bernoulli's law of large numbers and LAPLACE is emphatic in praising BERNOULLI. Considerations of the above kind are, of course, quite usual with reference to lotteries, games of dice, insurance, and so on. However, LAPLACE's musings go much farther. Here is an illustration.

"This theorem (the weak law of large numbers) implies also that, in the long run the action of regular and constant forces must prevail upon that of irregular forces. It is this circumstance that makes the earnings of lotteries as certain as those of agriculture: the chances reserved for the lottery insure its gains in the total of a large number of games. Similarly, since numerous favorable chances are tied with the eternal principles of reason, justice and humanity, the principles that are the foundation of societies and their mainstays, there is a great advantage in adhering to these principles and serious inconveniences in departing from them. Both history and personal experiences support this theoretical result. Consider the benefits to the nations from institutions based on reason and on the natural rights of man, the nations who knew how to establish such institutions and how to maintain them. Consider also the advantages that good faith earns governments whose policies are based on good faith, and how richly these governments are repaid for the sacrifices incurred in scrupulous observance of their commitments. What immense internal credit! What authority abroad! Consider, on the contrary, the abyss of miseries into which the peoples are frequently thrown by the ambitions and treachery of their rulers. Whenever a great power, intoxicated by lust for conquests, aspires to world domination, the spirit of independence among the menaced nations leads to a coalition, to which the aggressor power almost invariably succumbs. Similarly, the natural boundaries of a State, acting as constant causes, must eventually prevail over the variable causes that alternatively expand or compress the given State. Thus, it is important for stability, as well as for the happiness of empires, not to extend them beyond the limits into which they are repeatedly thrown by the constant causes, just as ocean waves whipped up by violent tempest fall back into their basin due to gravitation. This is another example of a probability theorem being confirmed by disastrous experiences."

Clarity of the concept of probability and of its relation to physical phenomena was reached early in the 20th century mainly through the works of Kolmogorov, on the one hand, and of von Mises, on the other. Thus, Laplace's interpretation of probability is far from consistent and unambiguous. Many of his writings indicate that, for him, probability is a measure of confidence or diffidence, independent of any frequency connotations. If there is no reason to believe that one of the contemplated events is more likely to occur than any other, then, for Laplace, these events are equiprobable. Here then, the intensity of expectation appears as the decisive moment in assigning probabilities. On the other hand, in many other passages, the decisive role is given to frequencies. For example, in discussing the familiar incident with Chevalier de Méré, Laplace appears to consider that the disagreement between DE MÉRÉ's experiments with dice and DE MÉRÉ's solution of the corresponding probability problem merely confirms the fact, established by PASCAL and FERMAT, that this solution is wrong. Also, a very substantial section of the "Essai philosophique" is given to "illusions" leading to mistakes in assigning probabilities to events. Here, then, probability appears as something independent of subjective emotions of particular individuals. Incidentally, this section on "illusions" includes LAPLACE's discussion of physiology of the nervous system and of the brain, for which LAPLACE proposes the summary term "Psychology". We were unable to verify whether this now commonly adopted term was actually introduced by LAPLACE.

Frequency interpretation of probability is also apparent in LA-PLACE's studies of a number of applied problems. These studies, described in both the "*Essai philosophique*" and in the "*Théorie analytique*" proper, and also in several sections of "*Mécanique céleste*", are very much in the spirit of the present day applications of probability and statistics to the various domains of science and we found them very interesting and stimulating. In the 19th and in the early years of the present century, when LAPLACE's writings were read more frequently than they now are, these particular studies exercised a very considerable influence both on theoretical and on applied research. Undoubtedly LAPLACE's discussion of the sex ratio, customarily indicating a prevalence of male births, influenced LEXIS and later BORTKIEWICZ. Also, LAPLACE's several studies of comets, indicating that their origin must be different from that of planets, influenced CHARLIER who considerably extended some of them. The same applies to the sections of "*Théorie analytique*" dealing with the central limit theorem. This book is directly quoted by CHARLIER in his work on asymptotic expansions of probability densities and by HARALD CRAMÉR. In a sense, the particular sections may be considered as precursors of the entirely novel subdiscipline on "probabilities of large deviations". The element that attracts us most in the "*Théorie analytique*" is the close tie of the theory with the problems of science: it is the latter that originate non-trivial theoretical problems, the solutions of which generate further theoretical developments.

In general, even though contemporary probabilists and statisticians have gone far beyond LAPLACE in many directions, so that particular subjects treated in "*Théorie analytique*" and now are occasionally difficult to identify, we believe that the book is very much worth reading.

A substantial part of the work is devoted to the theory of generating functions. LAPLACE claims to have inherited this from LAGRANGE and LEIBNITZ. However, he proceeds to use (and abuse) the method on various difference, differential and integral operators with an enthusiasm which reappears only much later in the works of Heaviside. One finds in the book a substantial use of the method of characteristic functions, also called Laplace transforms or Fourier transforms. This method, also used by LAGRANGE and CAUCHY, presumably independently, was finally revived in the early 20th century by PAUL LÉVY, with great success.

The part of the *"Théory analytique"* relating to "fonctions de très grands nombres" gave birth to the method of steepest descent and to some understanding of asymptotic expansions. LAPLACE's proof of the central limit theorem cannot be considered rigorous, but it is almost rigorizable, as was finally shown by LIAPOUNOFF in 1901. A somewhat related result of LAPLACE concerns the behavior of the *a posteriori* distribution of a parameter given a large number of observations. Although occasionally LAPLACE used an argument of "fiducial" nature, already introduced by J. BERNOULLI and declared objectionable by LEIBNITZ, LAPLACE's treatment of the *a posteriori* distribution seems basically sound. He anticipated by about a century the proofs of S. BERNSTEIN and VON MISES to the effect that, under certain conditions, the *a posteriori* distribution tends to a normal limit. A pleasant detail here is a careful distinction made by LAPLACE between expected errors computed under the assumption that the observations are random variables and expected errors computed *a posteriori* assuming the observations fixed.

"Essai philosophique" ends with a historical note covering the period from PASCAL and FERMAT. Here LAPLACE points out the achievements of his several predecessors, including JACOB BERNOULLI (weak law of large numbers), DE MOIVRE (central limit theorem) and BAYES. Also, the note mentions the then recent developments regarding the method of least squares. The same subject is again discussed in another historical note in the *"Théorie analytique"*. It is with reference to least squares that LAPLACE conceived the fruitful ideas which, after being forgotten for a number of years, now serve as foundations of modern statistical theory: the idea that every statistical procedure is a game of chance played with Nature, the idea of a loss function and of risk, and the idea that risk may be used as the basis for defining optimality of the statistical method concerned. LAPLACE's thinking was directed towards the problem of estimation and the loss function he adopted is the absolute value of the error of the estimate. GAUSS was quick in recognizing the fruitfulness of these ideas in general, but adopted a more convenient loss function, namely the square of the error.

The details of the discussion conducted a century and a half ago, as well as the then prevailing styles of recognition of priority, are interesting and we feel compelled to introduce more quotations, from both LAPLACE and GAUSS, as follows. .

"In order to avoid this groping in the dark, Mr. LEGENDRE conceived the simple idea of considering the sum of squares of observational errors and of minimizing it. This provides directly the same number of final equations as there are unknowns. This learned mathematician is the first to publish the method. However, in fairness to Mr. Gauss, it must be observed that, several years before this publication, he had the same idea, that he himself used it customarily and that he communicated it to several astronomers. . . . Undoubtedly, the search for the most advantageous procedure (i. e. the procedure minimizing risk) for deriving the final equations is one of the most useful problems in the theory of probability. Its importance for physics and astronomy brought me to study it." ("*Théorie analytique*", 1820, p. 353).

"The estimation of a magnitude using an observation subject to a larger or to a smaller error can be compared, not inappropriately, to a game of chance in which one can only lose (and never win), and in which each possible error corresponds to a loss. The risk involved in such a game is measured by the probable loss, that is, by the sum of products of particular possible losses by their probabilities. However, what specific loss should one ascribe to any given error is by no means clear by itself. In fact, the determination of this loss depends, at least in part, on our evaluation. . . . Among the infinite variety of such functions, the one that is the simplest seems to have the advantage and this, unquestionably, is the square. Thus follows the principle just formulated.

LAPLACE treated the problem in a similar fashion, but he choose the absolute value of the error as the measure of loss. However, unless we are mistaken, this choice is surely not less arbitrary than ours." (CARL FRIEDRICH GAUSS, "*Abhandlungen zur Methode der kleinsten Quadrate*", Berlin, 1887, p. 6).

The end paragraph of the "*Essai philosophique*" begins with the statement: "Essentially, the theory of probability is nothing but good common sense reduced to mathematics. It provides an exact appreciation of what sound minds feel with a kind of instinct, frequently without being able to account for it." The history of least squares, as it emerges from the above quotations, is an illustration of this statement. First came the manipulative procedure of the method that two "esprits justes", first GAUSS and then LEGENDRE, advanced on purely intuitive grounds. Next came the efforts at a probabilistic justification of the procedure. Here the priority regarding the basic ideas seems to belong to LAPLACE who, however, was unlucky in the choice of his loss function. The last steps towards the present day foundations of the least squares method, beginning with the square error as the loss function, and culminating with the proof of the theorem about the minimum variance property of least squares estimates among all linear unbiased estimates, are due to GAUSS.

The difference between the GAUSS and the LAPLACE treatments of optimality of the least square solutions is that, in the article quoted, GAUSS considers the estimated parameters as unknown constants and minimizes the expected loss with regard to the random variation of the observations. On the contrary, in LAPLACE's treatment it is the parameters that are random variables with some *a priori* distribution. The method of proof of optimality used here was revived only very recently. Currently it is standard in the asymptotic decision theory.

Before concluding, we wish to express our hearty thanks to all the colleagues who consented to take part in the 1963 International Research Seminar, and to the University of California for providing the necessary facilities. Also we reiterate our expression of gratefulness to the National Science Foundation for the necessary financial help. Finally, cordial thanks are due to Dr. HEINZ GÖTZE of Springer-Verlag for his interest in the Seminar and to Springer-Verlag itself for its customary excellent publication of these Proceedings.

LUCIEN LE CAM   JERZY NEYMAN

Contents

*Unless otherwise indicated, all the papers in the present volume
were prepared with the partial support of the
U. S. National Science Foundation, Grant GP-10*

Contribution to the Theory of Epidemics*

By **R. Bartoszyński, J. Łoś** and **M. Wycech-Łoś**

University of California, Berkeley,
and
Mathematical Institute of the Polish Academy of Sciences, Warsaw

0. In the present paper we shall discuss the following model of epidemics proposed and studied by Neyman and Scott [1]. We assume that an individual, infected at a given moment, becomes infectious after a certain fixed period of incubation and that the interval of time during which he may infect others is of length 0. (These assumptions about the incubation period and period of infectiousness are not essential for our model, since we shall be interested only in sizes of "generations" of epidemics, which can be defined without using time coordinates.) Between the time the individuals get infected and become infectious they may travel over the habitat, denoted in this paper by $\mathcal{X}$. Thus, our model will depend upon

1. The family of probability measures $\mu_u(\cdot)$, $u \in \mathcal{X}$, governing the movements of individuals between the time they get infected and the time they become infectious. [We assume that the set $\mathcal{X}$ is a measure space and that all $\mu_u(\cdot)$ are defined on the same Borel σ-field of subsets of $\mathcal{X}$.] We interpret $\mu_u(X)$ as the probability that an individual infected at u shall become infectious at some point in the set X.

2. The family of probability distributions $p(k \mid x)$, $x \in \mathcal{X}$, $k = 0, 1, 2, \ldots$, where $p(k \mid x)$ is interpreted as the probability that an infectious at x will infect exactly k individuals. We assume that the functions $p(k \mid x)$ are measurable for every k. We also assume that all individuals travel and infect independently of each other.

We shall use the notation $N_n(u)$ for the size of the nth generation of an epidemic originated by a single individual who became infected at the point u. To achieve greater simplicity in stating theorems we shall often drop the phrase "originated by an individual who became infected at u," but the index u shall appear consistently in the formulas.

Intuitively speaking, one would expect that under some reasonable

---

* This investigation originated from a seminar on problems of health held at the Statistical Laboratory, University of California, Berkeley, and was supported (in part) by a research grant (No. GM-10525) from the National Institutes of Health, Public Health Service.

assumptions every epidemic should either become extinct or grow indefinitely (since no restrictions are introduced on the size of the population). The "physical" significance of this "explosion" property of epidemics makes it worthwhile to study in some detail the assumptions implying it. More precisely, in this paper we shall study conditions under which the epidemics have the property

$$P\left\{ \lim_{n\to\infty} N_n\,(u) = 0 \text{ or } \lim_{n\to\infty} N_n\,(u) = \infty\right\} = 1 . \qquad (*)$$

To avoid repetitions of $(*)$ we shall say that the epidemics have the property (E). (E stands for both "extinction" and "explosion.") We shall show that except for rather pathological cases, the property (E) holds under various plausible sets of assumptions.

1. We shall now present a formal description of our process. Suppose that we have a fixed method of numbering individuals of each generation, the method being independent of the space location of these individuals. If the nth generation of the epidemic originated at u consists of l individuals, then their location $\underset{\sim}{\mathbf{x}}_l = (x_1, x_2, \ldots, x_l)$ (where x_i is the location of the point where the ith individual of the nth generation becomes infectious) is a (random) point in the space

$$\mathscr{X}^l = \underbrace{\mathscr{X} \times \cdots \times \mathscr{X}}_{l \text{ times}} .$$

Thus the space of states of our process consists of all couples $z = (l, \underset{\sim}{\mathbf{x}}_l)$, where l is a positive integer and $\underset{\sim}{\mathbf{x}}_l \in \mathscr{X}^l$, with the additional point z_0 corresponding to $l = 0$ (epidemic becomes extinct). We assume that the initial state is $(1, u)$ and we denote by ${}^{(l)}\mu_u^{(n)}\,(B)$ the conditional probability, given l, that the members of the nth generation of the epidemic, the generation known to be composed of exactly l individuals, will be located at a point $\mathbf{x}_l \in B$, where B is a measurable subset of $\mathscr{X}^l$. Obviously, it is enough to study the values of measures ${}^{(l)}\mu_u^{(n)}\,(\cdot)$ only for sets B of the particular form $B_1 \times \cdots \times B_l$, where B_i, $(i = 1, \ldots, l)$ is a measurable subset of $\mathscr{X}$. We shall not study the process in full detail, for we shall need only certain lemmas concerning the behavior of measures ${}^{(l)}\mu_u^{(n)}\,(\cdot)$.

Definition 1.1. *A subset X of $\mathscr{X}$ will be called μ-positive if there exists a number $\eta > 0$ such that for all $y \in \mathscr{X}$ we have*

$$\mu_y\,(X) \geqq \eta . \qquad (1.1)$$

We shall denote by η_X the greatest lower bound of all numbers η with the property (1.1).

Lemma 1.1. *If the sets $X_1, \ldots, X_l$ are μ-possitive, then for every n*

$${}^{(l)}\mu_u^{(n)}\,(X_1 \times \cdots \times X_l) \geqq \eta_{X_1} \ldots \eta_{X_l} .$$

Proof. Suppose that the $(n - 1)$st generation consists of k individuals.

Let f be a function mapping the set $(1, \ldots, l)$ into $(1, \ldots, k)$ and let $\mathscr{F}$ be the class of all such functions. We shall interpret the function f as follows: for any integer s between limits $1 \le s \le l$, the number $f(s)$ identifies the particular member of the $(n-1)$st generation who infected the sth member of the nth generation of the epidemic.

Let $\pi\,(f \mid \underset{\sim}{\mathbf{x}}_k)$ denote the probability of a given function f, conditional given the location $\underset{\sim}{\mathbf{x}}_k$ of the k members of the $(n-1)$st generation. Obviously, this probability depends on l and can be computed from the probabilities $p\,(m \mid x)$ defined above, that an infectious at x will infect exactly m individuals. For any $\underset{\sim}{\mathbf{x}}_k$ we have $\Sigma_{f \in \mathscr{F}}\, \pi\,(f \mid \underset{\sim}{\mathbf{x}}_k) = 1$.

To prove the lemma we assume that the sets $X_1, X_2, \ldots, X_l$ are μ-positive and denote by $p_{n-1}^{(u)}(k)$ the probability that the $(n-1)$st generation will consist of exactly k individuals. Then

$$^{(l)}\mu_u^{(n)}\,(X_1, X_2, \ldots, X_l) =$$

$$= \sum_{k=1}^{\infty} p_{n-1}^{(u)}(k) \int_{\mathscr{X}^k} \sum_{f \in \mathscr{F}} \pi\,(f \mid \underset{\sim}{\mathbf{x}}_k) \prod_{j=1}^{l} \mu_{x_f(j)}\,(X_j)\, d^{(k)} \mu_u^{(n-1)}\,(\underset{\sim}{\mathbf{x}}_k)\,.$$

The product $\Pi_{j=1}^{l}\, \mu_{x_f(j)}\,(X_j)$ is uniformly bounded from below by $\eta_{x_1} \cdots \eta_{x_l}$, hence the lemma follows.

We shall need also a slight generalization of this lemma. Let Z be any event concerning generations of the epidemic earlier than the nth. Then we have

Lemma 1.2. *If the sets* $X_1, \ldots, X_l$ *are* μ-*positive, then for every* n

$$^{(l)}\mu_u^{(n)}\,(X_1 \ldots X_l \mid Z) \ge \eta_{x_1} \cdots \eta_{x_l}\,.$$

The proof is the same as that of Lemma 1.1, since the conditioning will affect only the probabilities π and $p_{n-1}^{(u)}(k)$ and the measure $^{(k)}\mu_u^{(n-1)}$, but not the integrand.

2. According to the assumptions about our model of epidemics, if extinction occurs at the nth generation [i.e., $N_n\,(u) = 0$, while $N_{n-1}\,(u) \ne 0$], then we also have $N_k\,(u) = 0$ for all $k > n$. The probability of extinction in the nth generation depends, of course, on the size of the $(n-1)$st generation. We shall denote $P\,[N_n\,(u) = l] = p_n^{(u)}(l)$, and $P\,[N_{n+1}\,(u) = 0 \mid N_n\,(u) = l] = p_{n+1}^{(u)}(0 \mid l)$. We have

$$P\,[N_{n+1}\,(u) = 0 \text{ and } N_n\,(u) = l] = p_{n+1}^{(u)}(0 \mid l)\, p_n^{(u)}(l)\,.$$

Thus the probability of extinction in exactly the nth generation equals

$$\sum_{l=1}^{\infty} p_{n+1}^{(u)}(0 \mid l)\, p_n^{(u)}(l)\,,$$

and the probability of ultimate extinction is

$$P\,[\lim N_n\,(u) = 0] = \sum_{n=1}^{\infty} \sum_{l=1}^{\infty} p_{n+1}^{(u)}(0 \mid l)\, p_n^{(u)}(l)\,.$$

Notice that we may express $p_{n+1}^{(u)}(0 \mid l)$ as

$$p_{n+1}^{(u)}(0 \mid l) = \int_{\mathscr{X}^l} \prod_{i=1}^{l} p(0 \mid x_i)\, d^{(l)} \mu_u^{(n)}(x_1, \ldots, x_l).$$

Now we shall prove

Theorem 2.1. *If for every l there exists a number $\alpha_l > 0$ such that for all n*

$$p_n^{(u)}(0 \mid l) \geqq \alpha_l,$$

then the epidemic has the property (E). We recall that the epidemic has the property (E) if

$$P\Big\{ \lim_{n \to \infty} N_n(u) = 0 \text{ or } \lim_{n \to \infty} N_n(u) = \infty \Big\} = 1.$$

Proof. We have

$$+\infty > \sum_{n=1}^{\infty} \sum_{l=1}^{\infty} p_{n+1}^{(u)}(0 \mid l)\, p_n^{(u)}(l) \geqq \sum_{l=1}^{\infty} \alpha_l \sum_{n=1}^{\infty} p_n^{(u)}(l),$$

hence for every $l > 0$ the series $\sum_{n=1}^{\infty} p_n^{(u)}(l)$ converges. By the Borel-Cantelli lemma, with probability one only a finite number of events $\{N_n(u) = l\}$ will occur. Thus, with probability one, either $N_n(u) = 0$ starting from some n [in which case $\lim_{n \to \infty} N_n(u) = 0$], or $\lim_{n \to \infty} N_n(u) = \infty$, for otherwise for some l_0 we would have an infinite sequence n_k such that $N_{n_k}(u) = l_0$, which completes the proof.

It is easy to show that the assumptions of Theorem 2.1 are satisfied if for some $\varepsilon > 0$ we have $p(0 \mid x) \geqq \varepsilon$ for all $x \in \mathscr{X}$ (in other words, at each point x the probability of not infecting anybody is at least ε). In fact

$$p_{n+1}^{(u)}(0 \mid l) = \int_{\mathscr{X}^l} \prod_{j=1}^{l} p(0 \mid x_j)\, d^{(l)} \mu_u^{(n)}(x_1, \ldots, x_l)$$

$$\geqq \varepsilon^l \int_{\mathscr{X}^l} d^{(l)} \mu_u^{(n)}(x_1, \ldots, x_l) = \varepsilon^l > 0$$

since $p(0 \mid x) \geqq \varepsilon$ uniformly in $x \in \mathscr{X}$. The last condition is rather strong. We shall show that the same conclusion may be obtained under considerably weaker assumptions.

Theorem 2.2. *If there exists a number $\varepsilon > 0$ such that the set $X_\varepsilon = \{x; p(0 \mid x) \geqq \varepsilon\}$ is μ-positive, then the epidemic has the property (E).*

Proof. We have

$$p_{n+1}^{(u)}(0 \mid l) = \int_{\mathscr{X}^l} \prod_{j=1}^{l} p(0 \mid x_j)\, d^{(l)} \mu_u^{(n)}(x_1, \ldots, x_l)$$

$$\geqq \int_{X_\varepsilon^l} \prod_{j=1}^{l} p(0 \mid x_i)\, d^{(l)} \mu_u^{(n)}(x_1, \ldots, x_l)$$

$$\geqq \varepsilon^l \int_{X_\varepsilon^l} \mu_u^{(n)}(x_1, \ldots, x_l) \geqq (\varepsilon \eta_{X_\varepsilon})^l > 0,$$

for on the set X_ε^l the integrand is bounded from below by ε^l, and by Lemma 1.1 we have

$$\int\limits_{X_\varepsilon^l} d^{(l)}\,\mu_u^{(n)}\,(x_1, \ldots, x_l) = {}^{(l)}\mu_u^{(n)}\,(X_\varepsilon, \ldots, X_\varepsilon) \geqq \eta_{X_\varepsilon}^l\,.$$

Thus, the assumptions of Theorem 2.1 are satisfied. This completes the proof.

3. Now we shall assume that the sets $X_\varepsilon = \{x;\, p\,(0 \mid x) \geqq \varepsilon\}$ are not μ-positive. We assume even something stronger, namely that for every $u \in \mathscr{X}$ we have $\mu_u\{x;\, p\,(0 \mid x) = 0\} = 1$, i.e., that the set where the probability of not infecting anyone is zero is of measure one for each μ_u.

In this case it is obvious that the epidemic may expire only in the first generation, and the probability of this event is $P\,(\lim_{n\to\infty} N_n\,(u) = 0) = p\,(0 \mid u)$. It follows also that the size of every generation is at least as large as the size of the preceding one.

In the above case the behavior of the epidemic depends on the functions $p\,(1 \mid x),\, x \in \mathscr{X}$. If we assume that for every $u \in \mathscr{X}$ we have $\mu_u\{x;\, p\,(1 \mid x) = 1\} = 1$, i.e., the set where every infectious individual is sure to infect exactly one person (this can be interpreted as if he himself remained infectious) is of measure one for every μ_u, then for every u we have $P\,\{N_n\,(u) = N_1\,(u)\} = 1$. Thus, we have to study the behavior of epidemics in the case when $\mu_u\,(x;\, p\,(1 \mid x) = 1) \neq 1$ for a "sufficiently large" set of points u.

Let u be a point such that $p\,(0 \mid u) = 0$, i.e., with probability one the epidemic will not expire in the first generation. Let Z_i denote the event $N_i\,(u) = N_{i-1}\,(u),\, i = 1, 2, \ldots.$ We shall prove the following lemma.

Lemma 3.1. *If for some $\varepsilon > 0$ the set $X_\varepsilon = \{x;\, p\,(1 \mid x) \leqq 1 - \varepsilon\}$ is μ-positive, then there exists a number $\alpha > 0$ such that for every system of indices $k_1 < k_2 < \cdots < k_n$ we have*

$$P\{Z_{k_n} \cap Z_{k_{n-1}} \cap \cdots \cap Z_{k_1}\} \leqq (1 - \alpha)^n\,.$$

Proof. We have

$$P\,(Z_{k_n} \cap Z_{k_{n-1}} \cap \cdots \cap Z_{k_1}) = P\,(Z_{k_n} \mid Z_{k_{n-1}} \cap \cdots \cap Z_{k_1})$$
$$\cdot P\,(Z_{k_{n-1}} \mid Z_{k_{n-2}} \cap \cdots \cap Z_{k_1}) \cdot \ldots \cdot P\,(Z_{k_1})\,.$$

Let us consider the general term of this product: $P\,(Z_{k_i} \mid Z_{k_{i-1}} \cap \cdots \cap Z_{k_1})$, and, for simplicity, denote $Z_{k_{i-1}} \cap \cdots \cap Z_{k_1} = \mathscr{Z}$. We have $P\,(Z_{k_i} \mid \mathscr{Z}) = 1 - P\,(\overline{Z}_{k_i} \mid \mathscr{Z})$. Now, the event $\overline{Z}_{k_i}$ occurs if and only if $N_{k_i}\,(u) > N_{k_{i-1}}\,(u)$. We can now write

$$P\,(\overline{Z}_{k_i} \mid \mathscr{Z}) = \sum_{l=1}^{\infty} P\,[\overline{Z}_{k_i} \mid \mathscr{Z},\, N_{k_i-1}\,(u) = l]\, p_{k_i-1}^{(u)}\,(l)\,.$$

Let us notice that for each l, the event $\overline{Z}_{k_i} = \{N_{k_i}\,(u) > N_{k_i-1}\,(u)\}$

under the condition $\mathscr{L} \cap N_{k_i-1}(u) = l$ could have happened if and only if not all of the l individuals of the $(k_i - 1)$st generation infected exactly one person. Hence we have

$$P\left(\overline{Z}_{k_i} \mid \mathscr{L} \cap N_{k_i-1}(u) = l\right) =$$

$$= 1 - \int_{\mathscr{X}^l} \prod_{j=1}^{l} p\left(1 \mid x_j\right) d^{(l)} \mu_u^{(k_i-1)}\left(x_1, \ldots, x_l\right).$$

Since $p\left(1 \mid x_j\right) \leq 1$ for every j, we have

$$\prod_{j=1}^{l} p\left(1 \mid x_j\right) \leq p\left(1 \mid x_1\right);$$

thus

$$P\left[\overline{Z}_{k_i} \mid \mathscr{L} \cap N_{k_i-1}(u) = l\right] \geq$$

$$\geq 1 - \int_{\mathscr{X}^l} p\left(1 \mid x_1\right) d^{(l)} \mu_u^{(k_i-1)}\left(x_1, \ldots, x_l\right)$$

$$= \int_{\mathscr{X}^l} \left[1 - p\left(1 \mid x_1\right)\right] d^{(l)} \mu_u^{(k_i-1)}\left(x_1, \ldots, x_l\right).$$

By the assumption, it follows that there exists an $\varepsilon > 0$ such that the set $X_\varepsilon = \left\{x; 1 - p\left(1 \mid x\right) \geq \varepsilon\right\}$ is μ-positive. On the other hand, by Lemma 1.2 we have

$$^{(l)}\mu_u^{(k_i-1)}\left(X_\varepsilon \times \mathscr{X} \times \cdots \times \mathscr{X}\right) \geq \eta_{X_\varepsilon} > 0,$$

hence

$$P\left(\overline{Z}_{k_i} \mid \mathscr{L} \cap N_{k_i-1}(u) = l\right) \geq \eta_{X_\varepsilon}\, \varepsilon = \alpha > 0,$$

and $P\left(\overline{Z}_{k_i} \mid \mathscr{L}\right) \geq \alpha > 0$. Thus, $P\left(Z_{k_i} \mid \mathscr{L}\right) \leq 1 - \alpha$, and the lemma follows, since the last bound is independent of the chosen system of indices $k_1, \ldots, k_n$. Now we can easily prove

Theorem 3.1. *If for every $u \in \mathscr{X}$ we have $\mu_u\left\{x; p\left(0 \mid x\right) = 0\right\} = 1$ and for some $\varepsilon > 0$ the set $\left\{x; p\left(1 \mid x\right) \leq 1 - \varepsilon\right\}$ is μ-positive, then the epidemic has the property* (E).

Proof. It follows from the lemma that under the assumption of the theorem we have $P\left(Z_{k_n} \cap \cdots \cap Z_{k_1}\right) \leq (1 - \alpha)^n$ for any n and any subsequence $k_1 < \cdots < k_n < \cdots$ of random events $Z_1, \ldots, Z_n, \ldots$. Thus we have $P\left(\cap_{i=1}^{\infty} Z_{k_i}\right) = 0$, which shows that in this subsequence at least one of the events $\overline{Z}_{k_i} = \left\{N_{k_i}(u) > N_{k_i-1}(u)\right\}$ must occur with probability one. Since we can select an infinite number of different subsequences from the sequence of natural numbers, we obtain the result that with probability one we shall have an infinite number of events consisting of an increase of the size of the generation, provided the epidemic did not expire in the first generation, which completes the proof.

4. The assumptions of both theorems proved in the preceding sections are satisfied in the following general case proposed by NEYMAN and SCOTT. Suppose that the measures μ_u satisfy the condition

C. For every $X \subset \mathscr{X}$ and every $u \in \mathscr{X}$, if $\mu_u(X) > 0$ then X is μ-positive.

In the particular case where $\mathscr{X}$ is the Euclidean space and measures μ_u are given by their densities $f(x \mid u)$, i.e., $\mu_u(X) = \int_X f(x \mid u)\, dx$, and if there exists a function $\varphi(x) > 0$ such that $f(x \mid u) \geq \varphi(x) > 0$ for all $u \in \mathscr{X}$ and all $x \in \mathscr{X}$, then condition (C) is satisfied. Indeed, if for some u we have $\mu_u(X) = \int_X f(x \mid u)\, dx > 0$, then the set X has a positive Lebesgue measure. On the other hand, $\mu_{u'}(X) \geq \int_X \varphi(x)\, dx = \eta$ for every $u' \in \mathscr{X}$, and since $\varphi(x) > 0$ and X is of positive Lebesgue measure, $\eta > 0$ as asserted. Hence X is μ-positive.

Thus, if we assume (C), we have the following situation in our epidemic. Either the set $\{x; p(0 \mid x) \geq \varepsilon\}$ is μ-positive for some $\varepsilon > 0$, or the set $\{x; p(0 \mid x) = 0\}$ is of measure 1 with respect to each measure μ_u. In the first case the assumptions of Theorem 2.1 are satisfied. In the second case we have another alternative: either for some $\varepsilon > 0$ the set $\{x; p(1 \mid x) \leq 1 - \varepsilon\}$ is μ-positive, or the set $\{x; p(1 \mid x) = 1\}$ is of measure 1 for each measure μ_u. In the first case the assumptions of Theorem 3.1 hold, in the second case we have the trivial case of stabilization on the level $N_1(u)$.

We shall show that the assertion of Theorems 2.1 and 3.1 do not hold under somewhat less restrictive assumptions concerning measures μ_x and distributions $p(\cdot \mid x)$, $x \in \mathscr{X}$. Namely, we shall construct our epidemics on the Euclidean plane $\mathscr{X}$ such that

(i) $\mu_x(X) > 0$ for every x in $\mathscr{X}$ and every set X of positive Lebesgue measure,

(ii) $p(0 \mid x) > 0$ for every x in $\mathscr{X}$,

(iii) $P[\lim_{n \to \infty} N_n(u) = 0 \lor \lim_{n \to \infty} N_n(u) = 1] = 1$ for every u in $\mathscr{X}$.

Let $X_1, X_2, \ldots$ be a sequence of disjoint sets of finite positive Lebesgue measure: $|X_i| < \infty$, $i = 1, 2, \ldots$, and let $a_1, a_2, \ldots$ be a sequence of numbers such that $0 \leq a_i \leq 1$, $i = 1, 2, \ldots$; $\sum_{i=1}^{\infty} a_i < \infty$. We define the measures μ_x by defining their densities $f(y \mid x)$ as follows

$$\text{for } x \text{ in } X_n, \quad f(y \mid x) = \begin{cases} \dfrac{1 - a_n}{|X_{n+1}|} & \text{for } y \text{ in } X_{n+1}, \\[2ex] \varphi_n(y) & \text{for } y \text{ not in } X_{n+1}, \end{cases}$$

where $\varphi_n(y)$ is a continuous positive function such that

$$\int_{\mathscr{X} - X_{n+1}} \varphi_n(y)\, dy = a_n.$$

For x not in $\bigcup_{n=1}^{\infty} X_n$ we define $f(y \mid x)$ in an arbitrary way, with the only restriction that (i) should be satisfied.

Let us define

$$p\,(1 \mid x) = \begin{cases} 1 - a_n & \text{for } x \text{ in } X_n\,, \\ \tfrac{1}{2} & \text{for } x \text{ not in } \bigcup_{n=1}^{\infty} X_n\,, \end{cases}$$

and $p\,(0 \mid x) = 1 - p\,(1 \mid x)$ for x in $\mathscr{X}$. [It follows that $p\,(k \mid x) = 0$ for all $k > 1$ and x in $\mathscr{X}$.]

Consider now an epidemic started at a certain point u. The probability that all generations will consist of one individual and their consecutive locations will be in the sets $X_1, X_2, \ldots$ is

$$P\,(1 \mid u) \int_{X_1} f\,(x \mid u)\, dx \prod_{k=1}^{\infty} (1 - a_k)^2 = \alpha > 0\,,$$

thus we have that $P\,(N_n\,(u) = 1 \text{ for all } n) \geq \alpha > 0$, for all u in $\mathscr{X}$.

But from the above assumptions, we have either $N_n\,(u) = 0$ or $N_n\,(u) = 1$, so it follows that

$$P\,[\lim_{n\to\infty} N_n\,(u) = 0 \vee \lim_{n\to\infty} N_n\,(u) = 1] = 1\,.$$

An easy modification will provide us with an example of epidemics such that

$$P\,[\lim_{n\to\infty} N_n\,(u) = k] > 0$$

for $k = 0, 1, 2, \ldots, \infty$.

Reference

[1] Neymann, J., and E. L. Scott: "A stochastic model of epidemics," published in the volume Stochastic Models in Medicine and Biology, pp. 45—83, edited by J. Gurland. Wisconsin: University of Wisconsin Press 1964.

Study of Some Statistical Models Introduced by Problems of Physics

By A. Blanc-Lapierre, P. Dumontet and B. Picinbono*

We are going to deal with two models motivated by problems of physics.

The first one is a *statistical model derived from a Poisson process.*

The second one concerns *the Gaussian tendency given by selective filtering.* This tendency is well known in the case of the time random functions $X(t) = \Sigma_j R(t - t_j)$ where the t_j have a uniform Poisson distribution. Here we study the case of more general random functions.

1. A particular statistical model derived from a Poisson process. Application to half-life of intermediary radioactive daughter product, or fluctuation of time delays in electronics, or time of flight fluctuations

Let us consider the random events ε_j occurring at times t_j (see Fig. 1, axis 1). We assume the t_j to be Poisson distributed with uniform density

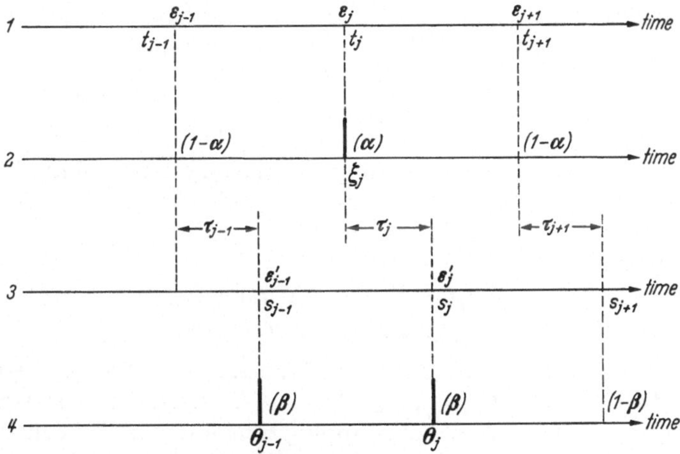

Fig. 1. General Model

ϱ_0. Each ε_j gives rise to a delayed event ε'_j at $s_j = t_j + \tau_j$, the τ_j being random variables independent of each other and of the t_j, obeying the same probability law (see Fig. 1, axis 3). We wish to deduce the common

* Faculty of Sciences of Algiers and Faculty of Sciences of Paris (Orsay)·

law of these τ_j from the study of correlations between random functions of time derived from the $\{\,t_j\,\}$ and the $\{\,s_j\,\}$.

We assume that the random variable τ possesses a probability density $g\,(\tau)$. We suppose the existence of counting losses in accordance with the imperfections of our counters. An event ε_j is detected with probability α and lost with probability $1 - \alpha$; likewise, an event ε_j' is detected with probability β. The events consisting of detection of any particular ε_j (or ε_j') are assumed to be independent of each other and of the random variables already introduced.

Finally, we introduce the time sequences S_1 and S_2: $S_1 = \{\xi_j\}$ where the ξ_j are the t_j detected (see Fig. 1, axis 2). $S_2 = \{\theta_j\}$ where the θ_j are the s_j detected (see Fig. 1, axis 4).

Then we can introduce the following random functions

$$X\,(t) = \sum_j a_j\,Q\,(t - t_j) = \sum_j R\,(t - t_j)\,, \tag{1.1}$$

$$Y\,(t) = \sum_j b_j\,Q\,(t - t_j - \tau_j) = \sum_j P\,(t - t_j)\,. \tag{1.2}$$

$a_j = 1$ if ε_j is detected and $a_j = 0$ if not; $b_j = 1$ if ε'_j is detected and $b_j = 0$ if not. Q is a sure real function standing for the response of the counting devices. R and P are random functions since they depend on the random variables a_j, b_j and τ_j.

Such a statistical model occurs in several physical problems: half-life of intermediary radioactive daughter product; fluctuation of time delays in electronics; time of flight in nuclear techniques.

Depending on the particular physical conditions, it is possible that the experiment gives us: either S_1 and S_2 *separately* [in this case we obtain separately $X\,(t)$ and $Y\,(t)$]; or S_1 and S_2 *together* [in this case we obtain only $Z\,(t) = X\,(t) + Y\,(t)$].

We discuss here only the first case, but the method is very similar for the second. Let us consider the two arbitrary sets of finite numbers of times:

$$[t_1, \ldots, t_k] \text{ and } [t'_1, \ldots, t'_{k'}] \tag{1.3}$$

and let us put

$$X\,(t_1) = X_1; \ldots; X\,(t_k) = X_k\,, \tag{1.4}$$

$$Y\,(t'_1) = Y_1, \ldots, Y\,(t'_{k'}) = Y_{k'}\,. \tag{1.5}$$

Let $\Phi\,(u_1, \ldots, u_k; t_1, \ldots, t_k; v_1, \ldots, v_{k'}; t'_1, \ldots, t'_{k'})$ be the characteristic function of $(X_1, \ldots, X_k; Y_1, \ldots, Y_{k'})$. Due to the statistical independence of the t_j in nonoverlapping time intervals, we can compute separately the contributions $\Delta\Psi\mu$ to $\Psi = \log\Phi$ from the events ε_j belonging to the different intervals, μ, $\mu + d\mu$. We find

$$\Delta\Psi\mu = \varrho_0\Big\{[\alpha e^{i\,[u_1\,Q\,(t_1-\mu)\,+\,\cdots]} + 1 - \alpha]\,\times$$

$$\times\,[\beta \int\limits_{-\infty}^{+\infty} e^{i\,[v_1\,Q\,(t'_1-\mu-\tau)\,+\,\cdots]} g\,(\tau)\,d\tau + 1 - \beta] - 1\Big\}\,d\mu \tag{1.6}$$

and

$$\Psi = \int \Delta \Psi \mu \, . \tag{1.7}$$

Let us compute $E\left[X\left(t\right) Y\left(t + \lambda\right)\right]$. Expanding $\Phi\left(u_1, v_1\right)$ in a series of u_1 and v_1, we obtain

$$E\left[X\left(t\right) Y\left(t + \lambda\right)\right] = \varrho_0 \alpha\beta \left[\varrho_0 a^2 + \int_{-\infty}^{+\infty} g\left(\tau\right) C\left(\lambda - \tau\right) d\tau\right] \tag{1.8}$$

where

$$a = \int_{-\infty}^{+\infty} Q\left(t\right) dt \text{ and } C\left(\lambda\right) = \int_{-\infty}^{+\infty} Q\left(\mu\right) Q\left(\mu + \lambda\right) d\mu \, . \tag{1.9}$$

Generally, the function $Q\left(t\right)$ is a short pulse, that is to say, Q vanishes for $t \neq 0$. Then, $C\left(\lambda\right)$ has the same property and (1.8) becomes

$$E\left[X\left(t\right) Y\left(t + \lambda\right)\right] = \varrho_0 \alpha\beta \left[\varrho_0 a^2 + g\left(\lambda\right) \int_{-\infty}^{+\infty} C\left(\tau\right) d\tau\right] \, . \tag{1.10}$$

If we can determine experimentally the correlation function $E\left[X\left(t\right) Y\left(t + \lambda\right)\right]$, we can, using equation (1.10), obtain the desired probability density g.

In fact, strictly speaking, we cannot measure the mathematical expectation $E\left[X\left(t\right) Y\left(t + \lambda\right)\right]$; we can measure only the time average

$$\mathscr{M}\left[T, \lambda\right] = \frac{1}{T} \int_0^T X\left(t\right) Y\left(t + \lambda\right) dt \, , \tag{1.11}$$

where T is equal to the duration of the measurement. So doing, we introduce in (1.10) the error

$$\mathscr{M}\left[T, \lambda\right] - E\left[X\left(t\right) Y\left(t + \lambda\right)\right].$$

To estimate the consequences of this error, we must compute its variance. This can be done by using the expression for $\log \Phi$ given in equations (1.6) and (1.7). The result was given by A. BLANC-LAPIERRE and P. DUMONTET [1]. By using their result, it is possible to discuss, in practical cases, the precision of the method.

It is interesting to mention here the work of G. LANDAUD and C. MABBOUX [2]. They study the random function $V\left(t\right)$ such that (i) $\left| V\left(t\right) \right| = 1$, (ii) the sign of V changes at each time ξ_j or θ_j. From the correlation function of V, it is also possible to get $g\left(\tau\right)$.

2. Gaussian tendency by selective filtering

2.1. First, let us consider an example. Let $X\left(t\right)$ be a random function

$$X\left(t\right) = \sum_j R\left(t - t_j\right) \tag{2.1}$$

derived from the Poisson distribution $\{t_j\}$ of common density ϱ_0 (R is assumed to be a sure real function). We can say that $X\left(t\right)$ results from the filtering of the sequence of Poisson pulses $\Sigma_j \delta\left(t - t_j\right)$ (δ stands for the Dirac function), by a *linear filter* (see [3], p. 342) whose impulse response is R. Let G be the gain of this filter (that is, the considered linear filter transforms $e^{2\pi i \nu t}$ into $G(\nu) e^{2\pi i \nu t}$ and G and R are Fourier transforms of each other). Let us assume that $|G|$ has the form represented in Fig. 2 and that $G(\pm \nu_0) = 1$. Let us put $X'(t) = X(t) - E[X(t)]$.

If the selectivity of the filter increases, that is, if its band width $\Delta \nu$ approaches zero, it is easily seen that $E|X'^2| \to 0$. But, if, instead of $X'(t)$ we consider

$$Y(t) = \frac{X'(t)}{\sqrt{\Delta \nu}}, \tag{2.2}$$

then $E[Y^2]$ remains finite if $\Delta \nu \to 0$. On the other hand, the narrower the band width $\Delta \nu$, the more extended in time $R(t)$ becomes. Then the

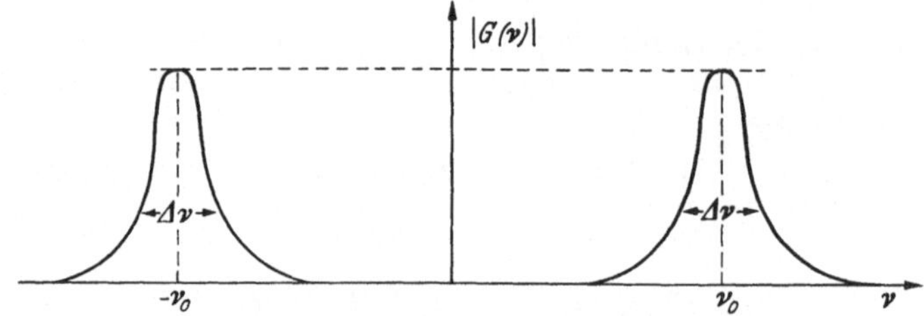

Fig. 2. Gain of the linear filter

pulses $R(t - t_j)$ overlap more and more. It is well known that under these conditions, $Y(t)$ approaches a Gaussian form. This result can be extended to a large class of random functions. *Here we want to find conditions on the spectral properties of a random function which insure its Gaussian tendency by selective filtering.*

2.2. *A useful property of the Gaussian random functions.* Let $X(t)$ be a real Gaussian function; the characteristic function $\varphi(u_1, \ldots, u_n)$ of $\{X(t_1) \ldots X(t_n)\}$ is equal to

$$\varphi(u_1, \ldots, u_n) = \exp\left[-\frac{1}{2}\sum_{ij} r_{ij} u_i u_j\right] \tag{2.3}$$

with

$$r_{ij} = E[X(t_i) X(t_j)]. \tag{2.4}$$

It is easy to prove that

$$\text{a) } E[X(t_1) X(t_2) \ldots X(t_{2k-1})] \equiv 0, \tag{2.5}$$

$$\text{b) } E[X(t_1) X(t_2) \ldots X(t_{2k})] = \Sigma \prod_\alpha E[X(t_{i_\alpha}) X(t_{j_\alpha})], \tag{2.6}$$

where Σ is the sum of the $(2k-1)!! = 1.3.5.\ldots(2k-1)$ terms of type $\prod_\alpha r_{i_\alpha j_\alpha}$ resulting from the grouping of the $2k$ i's into k groups of two, two terms not being considered distinct if differing only by the order of the i's in the groups or by the order of the groups in the grouping.

(Examples: a) $2k - 1 = 3$, $E[X(t_1) X(t_2) X(t_3)] = 0$,

b) $2k = 4$, $[(2k-1)!! = 3]$,

$$E[X(t_1) X(t_2) X(t_3) X(t_4)] = E[X(t_1) X(t_2)] E[X(t_3) X(t_4)] +$$
$$+ E[X(t_1) X(t_3)] E[X(t_2) X(t_4)] + E[X(t_1) X(t_4)] E[X(t_2) X(t_3)].$$

Let us write

$$F_X(t_1, \ldots, t_k) = E[X(t_1) \ldots X(t_k)] \qquad (2.7)$$

and let us first assume that F_X has a Fourier transform $f_X(\nu_1, \ldots, \nu_k)$; then we can replace equations (2.5) and (2.6) by the equivalent equations

$$f_X(\nu_1, \ldots, \nu_{2k-1}) \equiv 0, \qquad (2.8)$$

$$f_X(\nu_1, \ldots, \nu_{2k}) = \Sigma \prod_\alpha f_X(\nu_{i_\alpha}, \nu_{j_\alpha}). \qquad (2.9)$$

2.3. Let us now assume that $X(t)$ is a second order stationary function and has a spectral density $A_X(\nu)$. Then

$$f_X(\nu_i, \nu_j) = A_X(\nu_j) \, \delta(\nu_i + \nu_j) \qquad (2.10)$$

where δ, as before, stands for the Dirac function. *If $X(t)$ is not only a second order stationary function but if it is an all order stationary one,* $F_X(t_1, t_2, \ldots, t_k)$ must be, for every τ, equal to $F_X(t_1 + \tau, t_2 + \tau, \ldots, t_k + \tau)$. Hence $f_X(\nu_1, \ldots, \nu_k)$ vanishes except for $\nu_1 + \nu_2 + \ldots + \nu_k = 0$ and, finally, we have

$$f_X(\nu_1, \ldots, \nu_k) = B_X(\nu_1, \ldots, \nu_{k-1}) \, \delta(\nu_1 + \nu_2 + \ldots + \nu_k) \qquad (2.11)$$

where B_X is a function of $(K-1)$ of the ν. In the following we shall call "*stationary multiplicity of E_k,*" the multiplicity S_k defined by $\nu_1 + \ldots + \nu_k = 0$.

If, furthermore, $X(t)$ is Gaussian, from (2.8), (2.9) and (2.10), we have

$$f_X(\nu_1, \ldots, \nu_{2k'+1}) \equiv 0 \qquad (2.12)$$

$$f_X(\nu_1, \ldots, \nu_{2k'}) = \sum_\alpha A_X(\nu_{i_\alpha}) \, \delta(\nu_{i_\alpha} + \nu_{j_\alpha}). \qquad (2.13)$$

Then, f_X can differ from zero only on certain multiplicities $\mathcal{M}_{2k'}$ of the $E_{2k'}$ spaces. We shall call the $\mathcal{M}_{2k'}$ the "*Gaussian multiplicities*". For E_k with $k = 2k'$, there are $(2k'-1)!!$ Gaussian multiplicities defined by equations of the kind:

$$\nu_{i_1} + \nu_{j_1} = 0, \ldots, \nu_{i_{k'}} + \nu_{j_{k'}} = 0. \qquad (2.14)$$

Indeed, the Gaussian multiplicities $\mathcal{M}_{2k'}$ are included in the stationary multiplicity $S_{2k'}$.

Furthermore, if we put

$$X(t) = \int_{-\infty}^{+\infty} e^{2\pi i \nu t} \, dx(\nu) \qquad (2.15)$$

and if we suppose $X(t)$ to be a Gaussian stationary random function, then the distribution of the "complex weights" $E\,[dx\,(\nu_1)\ldots dx\,(\nu_{2k'})]$ on the Gaussian multiplicities of $E_{2k'}$ is not arbitrary. This distribution admits the density $A_X\,(\nu_{i_1})\,A_X\,(\nu_{i_2})\ldots A_X\,(\nu_{i_{k'}})$ which we shall call the "*Gaussian density*" on $\mathscr{M}_{2k'}$.

2.4. *Hypothesis on the filtering (hypothesis* A). We shall consider a class $\mathscr{S}_\lambda$ of linear filters depending on a parameter λ. In addition to very general convergence and regularity conditions, we assume that

a) for all λ,

$$G_\lambda\,(\nu) = 1 \text{ for } |\,\nu\,| = \nu_0, \tag{2.16}$$

b)

$$\Delta\nu\,(\lambda) = \int_{-\infty}^{+\infty} |G_\lambda\,(\nu)\,|\,^2 d\nu, \qquad \Delta f\,(\lambda) = \int_{-\infty}^{+\infty} |\,G_\lambda\,(\nu)\,|\,d\nu \tag{2.17}$$

exist and tend to zero if $\lambda \to \lambda_0$.

c) There exists a positive number "a" such that, for all values of λ, we have

$$\Delta f\,(\lambda) \leqq a\Delta\nu\,(\lambda). \tag{2.18}$$

2.5. *A theorem on the Gaussian tendency by selective filtering.* Let us put

$$Z_\lambda(t) = \frac{1}{[\Delta\nu\,(\lambda)]^{\frac{1}{2}}}\,Y_\lambda\,(t) \text{ where } Y_\lambda\,(t) = \mathscr{S}_\lambda\,[X\,(t)] \tag{2.19}$$

$Z_\lambda\,(t)$ does not tend to a limit in quadratic mean as $\lambda \to \lambda_0$; however, $E\,[\,|\,Z_\lambda^2\,(t)\,|\,]$ tends to a finite limit. Likewise, the covariance $E\,[Z_\lambda\,(t_i)\,Z_\lambda\,(t_j)]$ tends to $A_X\,(\nu_0)\,\cos\,2\,\pi\nu_0\,(t_i - t_j)$, which we shall call r_{ij}.

When speaking of Gaussian tendency, we have in mind the tendency of all moments of $X\,(t)$ to Gaussian moments. We have:

$$E\,[Z_\lambda\,(t_1)\ldots Z_\lambda\,(t_n)] = \frac{1}{[\Delta\,(\lambda)]^{n/2}} \int \ldots \int G_\lambda\,(\nu_1)\ldots G_\lambda\,(\nu_n)$$
$$f_X\,(\nu_1,\,\ldots,\,\nu_n)\,e^{2\pi i\,[\nu_1 t_1 + \,\cdots]}\,d\nu_1\ldots d\nu_n\,. \tag{2.20}$$

Picinbono [4] stated a theorem giving conditions for the tendency of all moments to Gaussian moments derived from the r_{ij}.

On $X\,(t)$, he introduces the following hypotheses (hypothesis B):

B-1. Obviously $X\,(t)$ is assumed to be an all order stationary real random function with $E\,[X] \equiv 0$.

B-2. For all n, the integral

$$\int \ldots \int |\,E\,[dx\,(\nu_1)\ldots dx\,(\nu_n)]\,| \tag{2.21}$$

is finite.

B-3. Hypothesis B-3 concerns the distribution, in the spaces E_n, of the "complex weights" $E\,[dx\,(\nu_1)\ldots dx\,(\nu_n)]$; for a particular E_n, these weights are in the corresponding stationary multiplicity S_n. Picinbono supposes that a) the weights contained in the Gaussian multiplicities $\mathscr{M}_n$

for $n = 2k$ even have a Gaussian density. b) Outside the Gaussian multiplicities, the weights are assumed to be distributed with finite densities either on the S_n themselves or on submultiplicities of the S_n defined by systems of equations $\Sigma_\beta \, \nu_{i\beta} = 0$ where the β are subsets of at least two distinct ν_i each ν_i being included in one and only one such subset.

Then PICINBONO proves:

Theorem. If Hypotheses A, B-1, B-2 and B-3 are satisfied, then

$$\lim_{\lambda \to \lambda_0} E\left[Z_\lambda\,(t_1)\ldots Z_\lambda\,(t_{2k+1})\right] = 0 , \tag{2.22}$$

$$\lim_{\lambda \to \lambda_0} E\left[Z_\lambda\,(t_1)\ldots Z_\lambda\,(t_{2k})\right] = \Sigma \prod_\alpha r_{i_\alpha j_\alpha} . \tag{2.23}$$

In other words, at the limit, the moments of $Z_\lambda\,(t)$ tend to be Gaussian moments.

The derivation of this result is fundamentally related to the following point: the contribution to the right member of equation (2.20) of the points outside the Gaussian multiplicities tends to zero as the filtering is more and more selective.

If n is odd, there is no Gaussian multiplicity in E_n and the odd moments vanish; if n is even, there is Gaussian tendency if and only if the densities on Gaussian multiplicities are themselves Gaussian.

Accordingly we must point out that the Gaussian tendency can exist only if the function $X\,(t)$ has, before filtering, a certain Gaussian character, namely that $X\,(t)$ has Gaussian densities on Gaussian multiplicities.

It is interesting to examine some cases for which the Gaussian tendency or the non-Gaussian tendency is obvious.

Examples. A. *Stationary random functions fulfilling the sufficient condition*

a) (already quoted in Section 2.1),

$$X\,(t) = \Sigma\,R\,(t - t_j), \tag{2.24}$$

b) $$Z\,(t) = X\,(t)\,Y\,(t), \tag{2.25}$$

where $X\,(t)$ and $Y\,(t)$ are two independent stationary Gaussian random functions.

c) $$Z\,(t) = X^2\,(t), \tag{2.26}$$

where $X\,(t)$ is a random function defined in (2.24) or a stationary Gaussian random function.

B. *Example of stationary random functions for which the Gaussian tendency is not true.* Let $X\,(t)$ and $Y\,(t)$ be two independent stationary Gaussian random functions whose spectral densities $A_X\,(\nu)$ and $A_Y\,(\nu)$ are assumed to be different. [We assume that neither $X\,(t)$ nor $Y\,(t)$ are constant in time.] Let $Z\,(t)$ be the random function equal to $X\,(t)$ or to

16   A. Blanc-Lapierre, P. Dumontet and B. Picinbono: Statistical Models

$Y(t)$ with probability 1/2, this choice being independent of $X(t)$ and of $Y(t)$. We have

$$A_Z(v) = \frac{1}{2}\left[A_X(v) + A_Y(v)\right]$$

and also

$$f_Z(v_1, \ldots, v_{2k}) = \frac{1}{2}\left[f_X(v_1, \ldots, v_{2k}) + f_Y(v_1, \ldots, v_{2k})\right].$$

Then, X and Y being Gaussian,

$$f_Z(v_1, \ldots, v_{2k}) = \frac{1}{2}\Sigma\left[\prod_\alpha A_X(v_{i_\alpha})\,\delta(v_{i_\alpha} + v_{j_\alpha}) + \prod_\beta A_Y(v_{i_\beta})\,\delta(v_{i_\beta} + v_{j_\beta})\right].$$

From this it is easy to see that f_Z is different from zero only on the Gaussian multiplicities; but on them it does not have a Gaussian density because

$$f_Z \neq \Sigma\prod_\alpha A_Z(v_{i_\alpha})\,\delta(v_{i_\alpha} + v_{j_\alpha})$$

so $Z(t)$ does not tend to the Gaussian form as may be seen directly.

References

[1] Blanc-Lapierre, A., et P. Dumontet: Etude d'un modèle statistique introduit par les techniques de temps de vol ou par l'étude des fluctuations de temps de transit, C. R. Acad. Sci. (Paris) **250**, 1216 (1960).
[2] Landaud, G., et C. Mabboux: Utilisation de la fonction aléatoire X (1) = ± 1 à l'étude des lois de désintégration de radio-éléments à filiations. C. R. Acad. Sci. (Paris), **250**, 3310 (1960).
[3] Blanc-Lapierre, A., et R. Fortet: Theorie des fonctions aléatoires. Paris: Masson et Cie 1953.
[4] Picinbono, B.: «Tendance vers le caractère gaussien par filtrage sélectif,» and «Remarques sur certaines fonctions aléatoires dérivées d'un processus de Poisson.» C. R. Acad. Sci. (Paris) **248**, 2280 (1959); **250**, 1174 (1960).

Stationary and Isotropic Random Functions

By A. BLANC-LAPIERRE and P. FAURE*

1. Definitions

We are going to deal with random functions $X(\underset{\sim}{M}) = X(x_1, \ldots, x_n)$ of a point $\underset{\sim}{M} \in E_n$ (E_n: n-dimensional Euclidean space). We assume:

(1) $E[X(\underset{\sim}{M})] \equiv 0$. $\hfill (1.1)$

(2) $X(\underset{\sim}{M})$ *to be a second order stationary random function*

(a) $E[X(\underset{\sim}{M}) X^*(\underset{\sim}{M} - \underset{\sim}{N})] = C[\underset{\sim}{N}]$. $\hfill (1.2a)$

(b) $X(\underset{\sim}{M})$ is continuous in quadratic mean . $\hfill (1.2b)$

In the study of many physical phenomena, stationary random functions are introduced which are also statistically isotropic. We call them *stationary and isotropic random functions* (abbreviated S.I.R.F.). Considering only the second order properties, they are characterized by: properties (1.1), (1.2a), (1.2b), the fact that the correlation function $C(\underset{\sim}{N})$ depends only on the length $r = |\underset{\sim}{N}|$.

In Section 2 we summarize some fairly well-known properties of the S.I.R.F. whose mathematical formulation depends on n, but not fundamentally.

In Section 3 we deal with properties less generally known and much more strongly connected with the number of dimensions of E_n. Particularly, we shall study carefully the relation between the n-dimensional power spectrum of $X(\underset{\sim}{M})$, $[\underset{\sim}{M} \in E_n]$ and the one-dimensional spectrum of the restriction of X on a straight line [for instance, $X(x_1, 0, 0, \ldots)$]. We call "restriction" of $X(\underset{\sim}{M}) = X(x_1, x_2, \ldots, x_n)$, on the x_1-axis, the function $X(x_1, 0, 0, \ldots)$. This problem is very relevant to the experimental analysis of second order properties of a physical stationary and isotropic phenomenon. Obviously, from the experimental point of view, it is much easier to analyze X on a straight line than in all the space.

2. General Properties [7]

Here we do not mention properties derived from the fact that $X(\underset{\sim}{M})$ is stationary but only properties derived from the fact that $X(\underset{\sim}{M})$ is *also* isotropic.

---

* Faculty of Sciences of Algiers and Faculty of Sciences of Paris (Orsay).

18   A. Blanc-Lapierre and P. Faure

a) *The correlation function $C(r)$ of a S.I.R.F. $X(\underset{\sim}{M})$ is a real function of r.* [This is an obvious consequence of $C(\underset{\sim}{M}) = C^*(-\underset{\sim}{M})$.]

b) *Spectrum and correlation function.* With obvious notations, we can write

$$C(\underset{\sim}{M}) = \int_{[\underset{\sim}{v}]} e^{2\pi i \underset{\sim}{v}\cdot \underset{\sim}{M}}\, dF(\underset{\sim}{v}) \;. \tag{2.1}$$

The real and nonnegative elements $dF(\underset{\sim}{v})$ represent the "spectral weights" distributed in the n-dimensional $\underset{\sim}{v}$-space.

From the fact that $C(\underset{\sim}{M})$ is invariant in an arbitrary rotation of $\underset{\sim}{M}$ around its origin, we obtain the invariance of the "weight distribution" $dF(\underset{\sim}{v})$ in an arbitrary rotation around the $\underset{\sim}{v}'$ origin ω. Using this invariance, it is possible to write (2.1) in a useful form, obtained by integrating in $\underset{\sim}{v}$-space on concentric spheres.

For simplicity, let us assume that $dF(\underset{\sim}{v})$ admits a density function $\Phi(\varrho)$; then

$$dF(\underset{\sim}{v}) = \Phi(\varrho)\, dv_1 \cdot dv_2 \cdots \cdot dv_n \tag{2.2}$$

where $\varrho = |\underset{\sim}{v}|$.

Then (2.1) becomes

$$C(r) = \int_0^\infty \left\{ \int_{S_{n,\,\varrho}} e^{2\pi i \underset{\sim}{M}\underset{\sim}{v}}\, d\sigma \right\} \Phi(\varrho)\, d\varrho \;, \tag{2.3}$$

where $S_{n,\varrho}$ is the n-dimensional sphere of radius ϱ in the $\underset{\sim}{v}$-space. It is easy to prove that

$$\int_{S_{n,\,\varrho}} e^{2\pi i \underset{\sim}{M}\underset{\sim}{v}}\, d\sigma = 2\pi \frac{\varrho^{n/2}}{r^{(n/2)-1}}\, J_{(n/2)-1}(2\pi r\varrho) \;, \tag{2.4}$$

where J stands for a Bessel function.

Introducing this expression in (2.3), we obtain

$$C(r)\, r^{(n/2)-1} = 2\pi \int_0^\infty \varrho \cdot \Phi(\varrho)\, \varrho^{(n/2)-1}\, J_{(n/2)-1}(2\pi r\varrho)\, d\varrho \;. \tag{2.5}$$

This proves that the functions $C(r)\, r^{(n/2)-1}$ and $\Phi(\varrho)\, \varrho^{(n/2)-1}$ are Hankel reciprocal transforms of order $(n/2)-1$. Besides, relation (2.5) may be inverted and it is possible to write

$$\Phi(\varrho)\, \varrho^{(n/2)-1} = 2\pi \int_0^\infty rC(r)\, r^{(n/2)-1}\, J_{(n/2)-1}(2\pi r\varrho)\, dr \;. \tag{2.6}$$

3. Properties strongly connected with n

1. *We shall say that $C(r)$ belongs to the class $\mathscr{C}_n$ if $C(r)$ is a correlation function of a S.I.R.F. $X(\underset{\sim}{M})$ defined on an n-dimensional space E_n* [1].

It is obvious that the function $C(r) = C(\sqrt{x_1^2 + x_2^2 + \cdots})$ need only be positive definite on E_n, or, in other words, to have a nonnegative Fourier transform on E_n.

But it is obvious that for a given $C(r)$, the fulfillment of these two equivalent conditions depends on the value of n. For any positive integer k, the class $\mathscr{C}_{n+k}$ is certainly included in the class $\mathscr{C}_n$. However, the converse is not true, which must be kept in mind when choosing $C(r)$ functions for particular models.

Example A.

$$C_A(r) = \frac{1-r^2}{(1+r^2)^2}. \tag{3.1}$$

It may be proved that $C_A \in \mathscr{C}_1 \{\Phi_A = 2\,\pi^2\,\varrho e^{-2\pi\varrho}\}$, but $C_A \notin \mathscr{C}_{1+p}$ for $(p > 0)$.

Example B. Let us consider

$$C_B = e^{-\pi r^2}, \qquad C_D = e^{-2\pi r}. \tag{3.2}$$

It may be shown that for all n

$$C_B \in \mathscr{C}_n \left\{\Phi_B = e^{-\pi\varrho^2}\right\}$$
$$C_D \in \mathscr{C}_n \left\{\Phi_D = \frac{\Gamma\left[(n+1)/2\right]}{2\,\pi^{(n+1)/2}\,(\varrho^2+1)^{(n+1)/2}}\right\}.$$

2. Relation between the n-dimensional spectrum of $X(\underset{\sim}{M})$ and the one-dimensional spectrum of the restriction of $X(\underset{\sim}{M})$ on a straight line. Physical

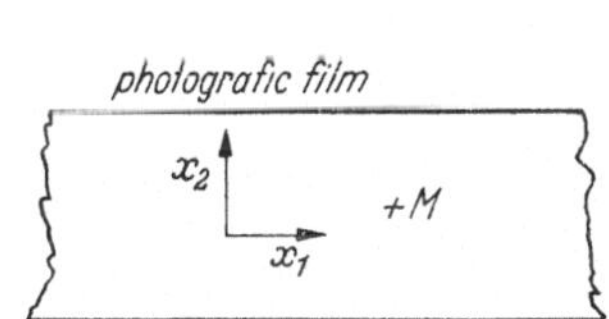

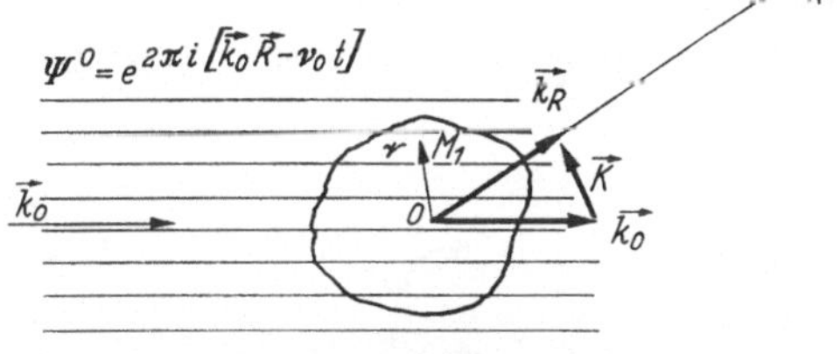

Fig. 1. Granularity of photografic films

Fig. 2. Scattering by a stationary and Isotropic Random medium

examples. Granularity of photographic films [2], [3]. Let $T(\underset{\sim}{M})$ be the transparency for the light *intensity* at a point $\underset{\sim}{M}$ of the film (Fig. 1); $T(\underset{\sim}{M})$ can be considered as a random function of $\underset{\sim}{M}$. Let us put

$$\mathscr{T}(\underset{\sim}{M}) = T(\underset{\sim}{M}) - E\left[T(\underset{\sim}{M})\right]. \tag{3.3}$$

With a good approximation, $\mathscr{T}(\underset{\sim}{M})$ can be assumed stationary and isotropic. The study of the spectral properties of $\mathscr{T}(\underset{\sim}{M})$ must give the two-dimensional spectral density $\Phi_2(\varrho)$. But the direct experimental determination of $\Phi_2(\varrho)$ is not easy; it is much easier, for instance, by using a photoelectric method, to obtain the one-dimensional spectral density $\varphi_1(u_1)$ of the restriction $\mathscr{T}(x_1, 0)$ of $\mathscr{T}$ on the x_1-axis. But if we do this, we must then be able to compute $\Phi_2(\varrho)$ from $\varphi_1(u_1)$.

Scattering of a plane wave by a volume $\mathscr{V}$ of a stationary and isotropic random medium [4], (Fig. 2). As an example, we consider an acoustical wave [5]. Let $\Psi^0(\underset{\sim}{R}, t)$ be the incident plane wave

$$\Psi^0(\underset{\sim}{R}, t) = e^{2\pi i\,(\underset{\sim}{k_0}\underset{\sim}{R} - \nu_0 t)}. \tag{3.4}$$

2*

20   A. Blanc-Lapierre and P. Faure

We consider the scattering at a large distance produced by a bounded volume $\mathscr{V}$ of a medium whose refractive index $n\,(x_1,\,x_2,\,x_3) = 1 + \mu\,(x_1,\,x_2,\,x_3)$, with $|\,\mu\,| << 1$, is stationary and isotropic. Then, in the following, $\mu\,(x_1,\,x_2,\,x_3)$ will be considered as a S.I.R.F.

At a distance R which is considerably greater than the dimensions of $\mathscr{V}$ and by using the first Born approximation, the scattered wave can be written

$$\Psi^1\,(\underset{\sim}{R})\,e^{-2\,\pi i\nu_0 t}$$

with

$$\Psi^1(\underset{\sim}{R}) = \frac{-1}{4\,\pi}\cdot\frac{e^{2\pi i k_0 R}}{R}\int_{\mathscr{V}} e^{2\pi i\,(\underset{\sim}{k}_0 - \underset{\sim}{k}_R)\underset{\sim}{M}_1}\,f\,(\underset{\sim}{M}_1)\,d\,[\underset{\sim}{M}_1]\,, \tag{3.5}$$

where $d\,[\underset{\sim}{M}_1]$ is the elementary volume, $R = |\,\underset{\sim}{R}\,|$, and $\underset{\sim}{k}_R$ is the vector of length $\underset{\sim}{k}_0$ in the direction of $\underset{\sim}{R}$. Also $\underset{\sim}{K} = \underset{\sim}{k}_R - \underset{\sim}{k}_0$ is such that $|\,\underset{\sim}{K}\,| = 2\,k_0\sin\theta/2$ (Fig. 2)

$$f\,(\underset{\sim}{M}_1) = -\,8\,\pi^2 k_0^2\mu\,(\underset{\sim}{M}_1)\,. \tag{3.6}$$

Assuming that $\mu\,(\underset{\sim}{M}_1)$ is a S.I.R.F., let $C_\mu\,(r)$ be its correlation function. There is a "correlation distance a" such that $C_\mu\,(r)$ vanishes except when $r \lesssim a$.

Let us consider two extreme cases.

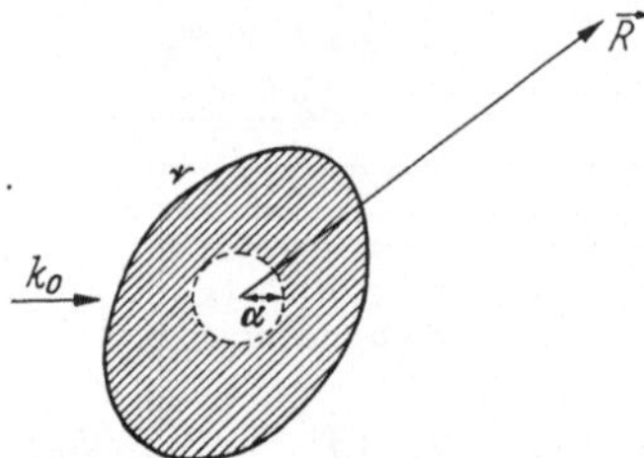

Fig. 3. Volume considerably larger than the "correlation distance a"

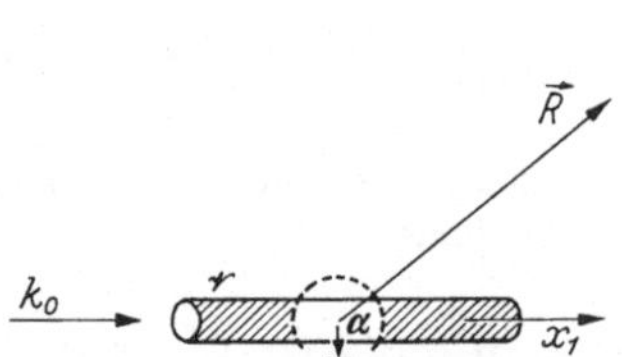

Fig. 4. Cylinder alined along the direction of the propagation

1. *The volume $\mathscr{V}$ is, in all directions, considerably larger than the "correlation distance a"* (Fig. 3). Then, it is easy to prove that

$$E\big\{\,|\,\Psi^1\,(\underset{\sim}{R})\,|^2\,\big\} = \frac{4\,\mathscr{V}\pi^2\,k_0^4}{R^2}\,s_\mu\,(K)\,, \qquad \text{with } K = |\,\underset{\sim}{K}\,|\,, \tag{3.7}$$

where $s_\mu\,(K)$ is the three-dimensional spectral density of the S.I.R.F. $\mu\,(\underset{\sim}{M}) = \mu\,(x_1,\,x_2,\,x_3)$. *Then, by studying experimentally the scattering at a large distance, we can obtain the three-dimensional spectral density $s_\mu\,(K)$, under the obvious condition that K may be taken experimentally in a sufficiently large range.*

2. Let us now assume that the volume $\mathscr{V}$ is a cylinder alined along the direction of the incident wave propagation and that its longitudinal

size is much larger than "a", ist transversal size being much smaller than "a" (Fig. 4). Then it can be proved that

$$E\left\{\,|\,\Psi^1\,(\underset{\sim}{R})\,|\,^2\right\}=\frac{4\,\pi^2\,k_0^4\,\mathscr{V}\Sigma}{R^2}\,s_\mu^1\,(K_1)\,,\qquad(3.8)$$

where $s_\mu^1\,(K_1)$ is the one-dimensional spectral density of the restriction $\mu\,(x_1,\,0,\,0)$ of $\mu\,(x_1,\,x_2,\,x_3)$ and Σ the surface of the transversal section of $\mathscr{V}$; K_1 is the projection of $\underset{\sim}{K}$ on the frequency axis corresponding to x_1.

In the present case, by studying experimentally the scattering at a large distance, we can obtain the one-dimensional spectral density $s_\mu^1\,(K_1)$, the condition being that K_1 may be taken experimentally in a sufficiently large range (and this can be experimentally difficult because, for large values of K, most of the scattered energy is concentrated inside the angle $\theta = 1/k_0 a$).

Here again, if we want to compare the two situations, 1 and 2, we have to obtain a relation between the three-dimensional spectral density $s_\mu\,(K)$ and the one-dimensional spectral density $s_\mu^1\,(K_1)$.

Relation between $\Phi\,(\varrho)$ (spectral density in F_n) and $\varphi\,(u_1)$ [spectral density of the restriction of $X\,(t)$ on the x_1-axis] [1]. We can write

$$C(\,x_1) = \int\limits_{-\infty}^{+\infty} e^{2\pi i u_1 x_1}\,\varphi\,(u_1)\,du_1 \qquad(3.9)$$

and

$$C\,(r) = \int\limits_{-\infty}^{+\infty} \ldots \int\limits_{-\infty}^{+\infty} e^{2\pi i\,\underset{\sim}{R}\,\underset{\sim}{\nu}}\,\Phi\,(\varrho)\,du_1 \ldots du_n \qquad(3.10)$$

$$(n \text{ times})$$

with $\varrho = \sqrt{u_1^2 + u_2^2 + \cdots + u_n^2}$. Putting $x_2 = x_3 = \cdots = x_n = 0$ in (3.10), we get

$$C\,(x_1) = \int\limits_{-\infty}^{+\infty} \ldots \int\limits_{-\infty}^{+\infty} e^{2\pi i u_1 x_1}\,\Phi\,(\varrho)\,du_1 \ldots du_n\,. \qquad(3.11)$$

$$\langle n \text{ times}\rangle$$

From (3.9) and (3.11), we can write

$$\varphi\,(u_1) = \int\limits_{-\infty}^{+\infty} \ldots \int\limits_{-\infty}^{+\infty} \Phi\,(\varrho)\,du_2 \ldots du_n\,. \qquad(3.12)$$

$$\langle(n-1) \text{ times}\rangle$$

Equation (3.12) can be interpreted in the following way: The mean power corresponding to the $(u_1,\,u_1 + du_1)$ interval of the u_1-axis in the one-dimensional spectrum is equal to the weight belonging to the domain limited by the two planes u_1 and $u_1 + du_1$ in the $\underset{\sim}{\nu}$-space. In other words, $\varphi\,(u_1)$ is the projection of $\Phi\,(\varrho)$ on the u_1-axis.

Let us write: $\varrho_1^2 = \varrho^2 - u_1^2 = u_2^2 + u_3^2 + \ldots + u_n^2$. By integrating

$n - 2$ times and by using the expresion of the extension of the $(n - 1)$-dimensional sphere, we obtain

$$\varphi(u_1) = \frac{2\pi^{(n-1)/2}}{\Gamma\left(\frac{n-1}{2}\right)} \int_0^\infty \Phi\left(\sqrt{u_1^2 + \varrho_1^2}\right) \varrho_1^{n-2} \, d\varrho_1 . \tag{3.13}$$

Then, introducing again the variable ϱ

$$\varphi(u_1) = \frac{2\pi^{(n-1)/2}}{\Gamma\left(\frac{n-1}{2}\right)} \int_{u_1}^\infty \varrho\Phi(\varrho) \, (\varrho^2 - u_1^2)^{(n-3)/2} \, d\varrho . \tag{3.14}$$

To obtain $\Phi(\varrho)$ from $\varphi(u_1)$, equation (3.14) must be inverted.

Let $\mathscr{T}_n$ be the transformation $[\Phi \to \varphi]$ defined in E_n by

$$\varphi(u_1) = \mathscr{T}_n\left[\Phi(\varrho)\right] . \tag{3.15}$$

According to the properties of the n-dimensional spectral density Φ, it is easy to prove that we can differentiate under the integral in (3.14); so we obtain

$$\frac{d\varphi(u_1)}{du_1} = -\frac{2\pi^{(n-1)/2}}{\Gamma\left(\frac{n-1}{2}\right)} (n-3) \, u_1 \int_{u_1}^\infty \varrho\Phi(\varrho) \, (\varrho^2 - u_1^2)^{(n-5)/2} \, d\varrho . \tag{3.16}$$

Then we obtain

$$\varphi'(u_1) = -2\pi u_1 \mathscr{T}_{n-2}\left[\Phi(\varrho)\right] \tag{3.17}$$

and, finally

$$\mathscr{T}_{n-2p}\left[\Phi(\varrho)\right] = A^p\left[\varphi(u_1)\right] \tag{3.18}$$

with

$$A = -\frac{1}{2\pi u_1} \frac{d}{du_1} .$$

Now we must distinguish two cases according to the parity of n.

Case 1. n *even*: $n = 2m$. Then, according to (3.18) and putting $p = m - 1$, we obtain

$$\mathscr{T}_2\left[\Phi(\varrho)\right] = A^{m-1}\left[\varphi(u_1)\right] . \tag{3.19}$$

But, for $\mathscr{T}_2$, we have as a consequence of (3.14):

$$\mathscr{T}_2\left[\Phi(\varrho)\right] = 2 \int_{u_1}^\infty \frac{\varrho\Phi(\varrho) \, d\varrho}{\sqrt{\varrho^2 - u_1^2}} . \tag{3.20}$$

This is an Abel integral equation and, by using known results concerning these equations, we obtain the solution [6]

$$\Phi(\varrho) = -\frac{1}{\pi} \int_\varrho^\infty \frac{d}{du_1} A^{m-1}\left[\varphi(u_1)\right] \frac{du_1}{\sqrt{u_1^2 - \varrho^2}} . \tag{3.21}$$

Case 2. n *odd*: $n = 2m + 1$. Then we have:

$$\mathscr{T}_1\left[\Phi(\varrho)\right] = A^m\left[\varphi(u_1)\right] . \tag{3.22}$$

But $\mathscr{T}_1\left[\Phi(\varrho)\right] = \Phi(u_1)_{u_1=\varrho}$; hence

$$\Phi\left(\varrho\right) = \left\{A^{m}\left[\varphi\left(u_{1}\right)\right]\right\}_{u_{1}=\varrho}. \tag{3.23}$$

Equations (3.21) and (3.23) solve our problem for an arbitrary value of n. Let us now consider some particular cases.

Particular cases. a) $n = 2$.

$$\Phi\left(\varrho\right) = -\frac{1}{\pi}\int_{\varrho}^{\infty}\frac{d}{du_{1}}\left[\varphi\left(u_{1}\right)\right]\frac{du_{1}}{\sqrt{u_{1}^{2}-\varrho^{2}}} \tag{3.24}$$

This formula was used by B. PICINBONO [2] and M. SAVELLI [3] and applied to the study of the granularity of photographic films.

b) $n = 3$.

$$\Phi\left(\varrho\right) = \left\{-\frac{1}{2\,\pi u_{1}}\frac{d}{du_{1}}\left[\varphi\left(u_{1}\right)\right]\right\}_{u_{1}=\varrho}. \tag{3.25}$$

Remarks. 1. $\Phi\left(\varrho\right)$ *is surely nonnegative. So, for* $n = 3$, *as a consequence of* (3.25), $\varphi\left(u_{1}\right)$ *is certainly a nonincreasing function of* $\mid u_{1}\mid$. This result is strongly connected with the fact that $n = 3$. This can easily be seen. Due to isotropy, we can consider the spectral distribution of the weights $dF\left(\underset{\sim}{v}\right)$ to be constituted by concentric spheres, each of uniform density. As a consequence of the spherical zone surface, the contribution of the b-radius sphere to $s_{\mu}^{1}\left(u_{1}\right)$ is constant if $\mid u_{1}\mid < b$ and null if $\mid u_{1}\mid > b$. If we note that $s_{\mu}^{1}\left(u_{1}\right)$ can depend only on spheres with radius $b \geq \mid u_{1}\mid$, it is obvious that $s_{\mu}^{1}\left(u_{1}\right)$ is a nonincreasing function of $\mid u_{1}\mid$.

2. The result pointed out in the first remark can be generalized in the following way: The necessary and sufficient condition for a correlation function $C\left(r\right)$ of an n-dimensional S.I.R.F. to be a correlation function of an $n+2$-dimensional S.I.R.F is that its n-dimensional spectrum be a nonincreasing function of ϱ.

References

[1] FAURE, P.: Sur quelques résultats relatifs aux fonctions aléatoires stationnaires isotropes introduites dans l'étude expérimentale de certains phénomènes de fluctuations. C. R. Acad. Sci. (Paris) **244**, 842 (1957). Thèse de Doctorat de 3ième cycle, Algiers, 1957.)
— Déduction de certaines propriétés statistiques d'une fonction aléatoire stationnaire isotrope définie dans un espace à plusieurs dimensions de l'étude de sa trace sur une courbe de cet espace. C. R. Acad. Sci. (Paris) **244**, 998 (1957). Erratum **244**, 1843. (These publications contain the mathematical results on the stationary and isotropic random functions.)

[2] PICINBONO, B.: Modèle statistique sugéré par la distribution de grains d'argent dans les films photographiques. C. R. Acad. Sci. (Paris) **240**, 2206 (1955).

[3] SAVELLI, M.: Doctoral Thesis 1958.

[4] BLANC-LAPIERRE, A.: Remarques sur la diffusion et sur la propagation des ondes dans les milieux aléatoires. XIV Assembly of the Union Radioscientifique Internationale, Tokyo 1963.

[5] CHERNOV, LEV. A.: Wave Propagation in a Random Medium (translated from Russian by R. A. Silverman). New York: McGraw-Hill 1960.

[6] RIESZ, F., and B. S. NAGY: Leçons d'Analyse Fonctionnelle. Budapest 1963.

On the Estimation of a Multivariate Location Parameter with Squared Error Loss

By **Robert Cogburn**

University of California, Berkeley

1. Introduction

The purpose of this paper is to discuss the mean squared error properties of a certain family of estimators of a multivariate location parameter. The estimators are an obvious generalization of those introduced by Stein [3] for the multivariate normal problem.

In Section 2 bounds on the mean squared error are derived, making only minimal assumptions on the underlying distribution. The model discussed is not parametric, but some assumptions on moments of the distribution are made. At the end of Section 2 there is a brief discussion of errors introduced by a departure from the model.

Section 3 discusses applications of the results of Section 2 to Bayesian problems in the situation that only partial information about the nature of the *a priori* distribution is available. In particular, the "empirical Bayes" problem of Robbins [2] is of this type. My starting point was this work of Robbins, however the approach here is different in two respects. First, parametric assumptions are avoided. There is, consequently, no hope of estimating a Bayes solution. Second, to fill in this gap, another method of making use of partial information about the *a priori* distribution is introduced.

Finally, it should be pointed out that squared error does not play any special role, other than its inherent mathematical simplicity, in the key results of this paper. Similar results should hold for suitably modified estimators and any nondecreasing continuous function of the distance between estimate and parameter for loss.

2. The estimator and its mean squared error

Let R be r dimensional Euclidean space, S be a linear subspace of dimension s, $r = s + t$, and T be the t dimensional orthogonal compliment to S. For $x \in R$, let x_S, x_T denote the projections of x on S, T, respectively. To simplify notation, we write xy for the inner product of x and y and x^2 for $|x|^2$.

Let $\xi \in R$, Y be a random variable in R and $X = Y + \xi$. We assume

$$(H) \qquad EY = 0, \qquad E\,Y_T^2 = \tau^2 \; known,$$

and construct the estimator δ of ξ

$$\delta(X) = X_S + \Lambda X_T, \; where \; \Lambda = \frac{(X_T^2 - \tau^2)^+}{X_T^2}\,.$$

Bounds on the mean squared error of δ ($M.S.E.\ \delta$) are given in the following theorem. First we introduce the notation

$$\alpha^2 = \sup_{|\xi|=1} \frac{E\,(Y_T\,\xi)^2}{\tau^2}\,, \qquad\qquad \beta^2 = \sup_{|\xi|=1} \frac{E\,(Y_T\,\xi)^4}{\tau^4}$$

$$\gamma = \frac{E\,|\,Y_T^2 - \tau^2\,|}{\tau^2}\,, \qquad\qquad \eta = \frac{E\,(Y_T^2 - \tau^2)^2}{\tau^4}$$

$$\lambda = \frac{\xi_T^2}{\tau^2 + \xi_T^2} \; and \; \sigma^2 = E\,Y_S^2\,.$$

Theorem. *Let* (H) *hold. Then finite constants* A, B *exist such that*

1. M.S.E. $\delta \leq \sigma^2 + \tau^2\{\lambda + A\,\sqrt{1-\lambda}\,\sqrt{\gamma + \alpha}\}$,
2. M.S.E. $\delta \leq \sigma^2 + \tau^2\{\lambda + B\,(1-\lambda)\,\sqrt{\eta + \alpha^2 + \beta^2}\}$.

Corollary 1. *Let $\mathscr{J}$ be a family of uniformly square integrable distributions and let (H) hold. If the components, Y_i, of Y are independent with distributions drawn from $\mathscr{J}$, $\min_i EY_i^2 \geq \varepsilon$ for some fixed $\varepsilon > 0$ and, as $t \to \infty$,*

$$\max_i EY_i^2 = o\,(t),$$

then

$$M.S.E.\ \delta \leq \sigma^2 + \tau^2\,(\lambda + o\,(1)).$$

Corollary 2. *Let (H) hold. If the components, Y_i, of Y are independent and $\min_i EY_i^2 \geq \varepsilon$ for some fixed $\varepsilon > 0$ and $\max_i EY_i^4 = k$, and, as $t \to \infty$, $k/t = O\,(1)$, then*

$$M.S.E.\ \delta \leq \sigma^2 + \tau^2\left(\lambda + O\left(\sqrt{\frac{k}{t}}\right)\right).$$

Proof of Corollary 1. Since $\tau^2 \geq t\,\varepsilon$, $\max EY_i^2/\tau^2 \to 0$. The uniform square integrability hypothesis then implies the hypothesis of the Lindeberg-Feller theorem, hence $\mathscr{L}\,((\Sigma Y_i)\,_T/\tau) \to N\,(0,1)$, as $t \to \infty$. By Raikov's theorem, $Y_T^2/\tau^2 \to 1$ in probability, and since $E\,(Y_T^2/\tau^2) = 1$, it follows [7, p. 528] that the Y_T^2/τ^2 are uniformly integrable. But then $\gamma \to 0$ as $t \to \infty$. On the other hand,

$$\alpha^2 \leq \frac{\max EY_i^2}{\tau^2}\,.$$

The corollary follows from the first inequality in the theorem.

Proof of Corollary 2. The hypotheses imply that

$$\varepsilon \leq \min EY_i^2 \leq \max EY_i^2 \leq \sqrt{k} \leq \frac{k}{\varepsilon}\,,$$

and it follows easily that

$$\alpha = 0\left(\frac{k}{t}\right), \qquad \beta = 0\left(\frac{k^2}{t^2}\right), \qquad \eta = 0\left(\frac{k}{t}\right),$$

and the corollary follows from the second inequality in the theorem.

Proof of theorem. From the definition of δ, $\Lambda X_T \in T$, hence

$$(\delta\,(X) - \xi)^2 = (X_S - \xi_S)^2 + (\Lambda X_T - \xi_T)^2\,. \tag{1}$$

A direct computation establishes that

$$E\,(\lambda X_T - \xi_T)^2 = \tau^2 \lambda\,.$$

Let $W = (\Lambda - \lambda)^2\, X_T^2/\tau^2$. From the expansion $\Lambda X_T - \xi_T = (\Lambda - \lambda)\, X_T +$ $+ (\lambda X_T - \xi_T)$ and Schwartz's inequality, it follows that

$$E\,(\Lambda X_T - \xi_T)^2 \leqq \tau^2\,(\lambda + 2\,\sqrt{\lambda\,EW} + EW)\,. \tag{2}$$

Combining (1) and (2),

$$\text{M.S.E. } \delta \leq \sigma^2 + \tau^2\{\lambda + \sqrt{EW}\,(2 + \sqrt{EW})\}\,. \tag{3}$$

The inequalities in the theorem follow from (3) and estimates of EW obtained below [Formulas (9) and (12)].

Let $Z = X_T^2/\tau^2$ and note that $EZ = (\xi_T^2 + \tau^2)/\tau^2 = (1 - \lambda)^{-1}$. Expanding and simplifying the expression for W, we obtain

$$W = \begin{cases} \dfrac{(Z - EZ)^2}{Z\,(EZ)^2}\,, & \text{for } Z \geqq 1\,, \\[2ex] \lambda^2 Z\,, & \text{for } Z \leqq 1\,. \end{cases}$$

Then elementary inequalities yield

$$W \leqq \frac{Z^2}{(EZ)^2} - \frac{2}{EZ} + \min\left\{\frac{1}{Z}, \max\left(\frac{2}{EZ}, 1\right)\right\}\,. \tag{4}$$

Now

$$\frac{E\,|\,Z - EZ\,|}{EZ} = \frac{E\,|\,Y_T^2 - \tau^2 + 2\,\gamma\,Y_T\,\xi_T\,|}{\tau^2 + \xi_T^2}$$

$$\leqq \gamma + \frac{2\,|\,\xi_T\,|}{\tau^2 + \xi_T^2}\left[E\left(\frac{Y_T\,\xi_T}{|\,\xi_T\,|}\right)^2\right]^{\frac{1}{2}} \leqq \gamma + \alpha\,, \tag{5}$$

and for $EZ > 2$, by Chebychev's inequality,

$$P\,[2\,Z < EZ] = P\,[2\,Y_T^2 + 4\,Y_T\xi_T < \tau^2 - \xi_T^2]$$

$$\leqq P\,[4\,Y_T\xi_T < -\tau^2\,(EZ - 2)]$$

$$\leqq \frac{16\,\xi_T^2\,\alpha^2}{\tau^2\,(EZ - 2)^2} = (1 - \lambda)\,\alpha^2\,\frac{16\,EZ\,(EZ - 1)}{(EZ - 2)^2}\,. \tag{6}$$

It follows from (4), (5) and the first upper bound in the lemma following the theorem that for $1 \leqq EZ \leqq 2$,

$$EW \leq (1 - \lambda)\,(\gamma + \alpha)\,, \tag{7}$$

and from (4), (5), (6) and the second upper bound in the lemma that for $EZ \geq 2$,

$$EW \leq (1-\lambda) \min \left\{ (EZ-1)\,(\gamma+\alpha),\, \gamma+\alpha+\alpha^2 \frac{EZ\,(EZ-1)}{(EZ-2)^2} \right\}. \tag{8}$$

Since $\alpha \leq 1, \gamma \leq 2$, (7) and (8) imply that there is a constant, a, such that

$$EW \leq a^2\,(1-\lambda)\,(\gamma+\alpha) \leq 3\,a^2. \tag{9}$$

To obtain the second inequality, observe that by the c_r inequality [1, p. 155]

$$\frac{V(Z)}{EZ} \leq 2\,\frac{E\,(Y_T^2-\tau^2)^2+4\,E\,(Y_T\,\xi_T)^2}{\tau^2\,(\tau^2+\xi_T^2)} \leq 2\,(\eta+4\,\alpha^2) \tag{10}$$

and for $EZ > c/(c-1),\ c>1$,

$$P\,[cZ < EZ] \leq P\,[2\,cY_T\xi_T < -\tau^2\,((c-1)\,EZ-c)] \tag{11}$$
$$\leq \frac{16\,c^4\,|\,\xi_T\,|^4\,\beta^2}{\tau^4\,[(c-1)\,EZ-c]^4} = (1-\lambda)^2\,\beta^2 \frac{16\,(EZ)^2\,(EZ-1)^2}{[(c-1)\,EZ-c]^4}.$$

Combining (10), (11) and the third and fourth upper bounds of the lemma, it follows that there exists a constant b such that

$$EW \leq b^2\,(1-\lambda)^2\,(\eta+\alpha^2+\beta^2). \tag{12}$$

Lemma. *Let Z be any nonnegative random variable with finite expectation. Then for $h \geq 2$,*

$$\frac{1}{EZ} \leq E \min \left\{ \frac{1}{Z},\, \frac{h}{EZ} \right\} \leq \frac{1}{EZ} + (h-1)\,\frac{E\,|\,Z-EZ\,|}{(EZ)^2}$$

and for $EZ \geq 2$,

$$\frac{1}{EZ} \leq E \min \left\{ \frac{1}{Z},\, 1 \right\} \leq \frac{1}{EZ}$$
$$+ \min \left\{ (EZ-1)\,\frac{E\,|\,Z-EZ\,|}{(EZ)^2},\, \frac{E\,|\,Z-EZ\,|}{(EZ)^2} + P\,[2\,Z < EZ] \right\}.$$

If, moreover, the variance of Z, $V(Z)$, is finite, then for $h \geq 2$,

$$E \min \left\{ \frac{1}{Z},\, \frac{h}{EZ} \right\} \leq \frac{1}{EZ} + h\,\frac{V(Z)}{(EZ)^3},$$

and for $EZ \geq 2$ and any $c \geq 1$,

$$E \min \left\{ \frac{1}{Z},\, 1 \right\} \leq \frac{1}{EZ} + c\,\frac{V(Z)}{(EZ)^3} + P\,[cZ < EZ].$$

Proof. Let $f_h(z) = \min\{1/z,\, h/EZ\}$ and $l(z) = (2/EZ) - (z/(EZ)^2)$, $z \geq 0$. Then for $h \geq 2$, $l \leq f_h$, and the lower bounds in the lemma follow since $El(Z) = 1/EZ$. The first and third upper bounds follow from the elementary inequality

$$f_h(z) - l(z) \leq \min \left\{ (h-1)\,\frac{|\,z-EZ\,|}{(EZ)^2},\, h\,\frac{(z-EZ)^2}{(EZ)^3} \right\}.$$

The second follows from the first, taking $h = EZ$, and from

$$f_{EZ}(z) - l(z) \leq \begin{cases} \dfrac{|z - EZ|}{(EZ)^2}, & \text{for } 2z \geq EZ, \\[2mm] 1, & \text{for } z < EZ. \end{cases}$$

The fourth follows from the above inequality and

$$f_{EZ}(z) - l(z) \leq \frac{c(z - EZ)^2}{(EZ)^3}$$

for $cz \geq EZ$ and $c \geq 1$. The lemma is proved.

To examine the result of a departure from the assumptions under (H), first note that if δ^*, δ are estimators, then, letting $D = \delta^* - \delta$,

$$|\,\text{M.S.E. } \delta^* - \text{M.S.E. } \delta\,| \leq E(D^2) + 2\sqrt{E(D^2)(\text{M.S.E. } \delta^*)}\ .$$

Now suppose that δ is the estimator introduced in the first part of this section, but that in fact $EY_T^2 = \tau^2(1 + h)$. Let

$$\delta^*(X) = X_S + \frac{[X_T^2 - \tau^2(1+h)]^+}{X_T^2} X_T.$$

An easy calculation shows that $|\,D\,| \leq |\,h\,|\,\tau$. Hence M.S.E. $\delta \leq$ M.S.E. $\delta^* + \tau^2 h^2 + 2\tau\,|\,h\,|\,\sqrt{\text{M.S.E. } \delta^*}$. The theorem gives a bound of order $\sigma^2 + \tau^2$ on the M.S.E. δ^*, hence the relative increase in the bound on M.S.E. δ over the bound on M.S.E. δ^* is at most $0(h)$ for small h.

The other possibility is that $EY = \zeta \neq 0$. This situation can be reduced to the one treated within the model by replacing. Y by $Y^* = Y - \zeta$ and ξ by $\xi^* = \xi + \zeta$. Then $Y^* + \xi^* = Y + \xi = X$, and a bound on $E(\delta(X) - \xi^*)^2 = \varepsilon$, say, is provided by the theorem (and the above discussion if $EY_T^2 \neq \tau^2$). Letting $\delta^* = \delta - \zeta$, $\varepsilon = \text{M.S.E. } \delta^*$ at ξ, hence

$$\text{M.S.E. } \delta \leq \varepsilon + 2\,|\,\zeta\,|\,\sqrt{\varepsilon} + \zeta^2\ .$$

3. Application to Bayesian problems

In this section ξ and Y are both random variables in r dimensional Euclidean space, $X = Y + \xi$, and the problem still is to estimate ξ, based on the observation X, with squared error loss. We treat only the simplest of various hypotheses allowing the application of the previous section. In particular, we assume that

(H') *Y is independent of ξ, the components, Y_i, of Y are independent and the distribution of each Y_i is drawn from a given family $\mathscr{F}$ of distributions; and each distribution in $\mathscr{F}$ has mean 0 and a variance in the interval $[m, M]$ for some fixed $0 < m \leq M < \infty$. Moreover, we assume $\tau^2 = EY_T^2$ known.*

The subspaces S, T and the estimate δ are as in Section 2, $\sigma^2 = EY_S^2$, and we let

$$\varrho = E\lambda = E\,\frac{\xi_T^2}{\tau^2 + \xi_T^2}\,.$$

Applying the corollaries of Section 2, we obtain

1. *If (H') holds and $\mathscr{J}$ is a family of uniformly square integrable distributions, then*

$$\text{M.S.E. } \delta \leq \sigma^2 + \tau^2\,(\varrho + o\,(1))$$

as $t \to \infty$.

2. *If $\max_i EY_i^4 \leq k$, then*

$$\text{M.S.E. } \delta \leq \sigma^2 + \tau^2\left[\varrho + 0\left(\sqrt{\frac{k}{t}}\right)\right]$$

as $t \to \infty$.

These results show that a substantial reduction in M.S.E. δ over the value $\sigma^2 + \tau^2$ for the invariant estimate $\iota\,(X) = X$ may be effected if information about the distribution of ξ allows a choice of S such that s is small compared to r and such that ϱ is small.

In particular, in the "empirical Bayes" formulation it is assumed, in addition to (H'), that the components of ξ are independent and identically distributed and that the components of Y are both independent and identically distributed with common variance one, say. Let Ψ^2 be the variance of the components of ξ, if it exists. Let S be the equiangular line. Then $t = r - 1$ and $E\,\xi_T^2 = (r - 1)\,\Psi^2$. From the concavity of λ as a function of ξ_T^2 it follows that $\varrho \leq \Psi^2/(1 + \Psi^2)$. As $r \to \infty$, $\xi_T^2/(r - 1) \to \Psi^2$ a.s., hence $\varrho \to \Psi^2/(1 + \Psi^2)$. Thus, as $r \to \infty$,

$$\frac{\text{M.S.E. } \delta}{r} = \frac{\Psi^2}{1 + \Psi^2} + o\,(1)$$

while, if the components of Y have a finite fourth moment,

$$\frac{\text{M.S.E. } \delta}{r} = \frac{\Psi^2}{1 + \Psi^2} + 0\left(\frac{1}{\sqrt{r}}\right).$$

In the case that Y and ξ are both normally distributed, $r\Psi^2/(1 + \Psi^2)$ is precisely the Bayes risk.

References

[1] LoÉVE, M.: Probability Theory, 3rd edition. Princeton: Van Nostrand 1963.

[2] ROBBINS, H.: The empirical Bayes approach to statistical decision problems. Ann. Math. Statist. 35, No. 1 (1964).

[3] STEIN, C.: Inadmissibility of the usual estimator for the mean of a multivariate normal distribution. Proceedings of the Third Berkeley Symposium on Mathematical Statistics and Probability. 1, 197 (1956).

Some Notes on Laplace*

By F. N. DAVID

University College, London and University of California, Berkeley

I should begin, I suppose, by confessing that my talk of today was concocted under something of a misapprehension. When I heard that we were to commemorate JAMES BERNOULLI, THOMAS BAYES and PIERRE-SIMON DE LAPLACE, I envisaged an exercise similar to that carried out by my own college each year in memory of JEREMY BENTHAM, and I prepared myself accordingly. I am sorry if my talk does not fit in with the general scheme but nevertheless and in spite of the modern attitude summarised so aptly by the phrase "History is bunk," I propose to speak the traditional words of commemoration and to say "Let us now praise famous men and in particular PIERRE-SIMON DE LAPLACE, Pair de France, Grand officier de la Légion d'honneur, l'un des quarante de l'Académie française, de l'Académie des Sciences, membre du Bureau des Longitudes de France; des Sociétés royales de Londres et de Göttingen, des Académies des Sciences de Russie, de Danemark; de Suède, de Prusse, des Pays-Bas, d'Italie; mathematician, astronomer, physicist, chemist and politician. He was honored in his generation and a glory in his days. His memory shall not be blotted out."

In a famous essay (Urn Burial), Sir THOMAS BROWNE (1605—1682) remarks "What song the Syrens sang, or what name Achilles assumed when he hid himself among women, though puzzling questions, are not beyond all conjecture." In considering the life and works of any famous man we find ourselves in a similar situation; we are confronted by puzzling questions the answers to which are not beyond all conjecture. But it is important to remember that we *do* have to conjecture and that a variety of opinions is possible. Of no person perhaps is this so easily said as of LAPLACE, the conjectures about whose works and conduct vary from the indifferently good to the downright derogatory, and for whom the aphorism of "damning with faint praise" might have been coined. For few men have been so disliked either by their contemporaries or by their biographers, and it is difficult to arrive at a balanced viewpoint in the light of this almost universal condemnation. Moreover we may note that much of the personal and scientific material, including all his correspond-

* This investigation was supported (in part) by a research grant (No. GM-10525) from the National Institutes of Health, Public Health Service.

ence with English scientists, was annihilated at Caen in the Battle of Normandy in 1944, and such letters as we have are those which have been preserved in the collections of letters of the other scientists with whom he corresponded, and the ones presented by his son to the Academy. I have taken as my main sources the funeral orations spoken by SIMEON-DENIS POISSON (1781—1890) and JEAN-BAPTISTE BIOT (1774—1882) both pupils of LAPLACE, the éloge given in the Chamber of Peers by the Marquis de PASTORET, the éloge historique given in the Academy of Sciences by the Baron FOURIER, and the correspondence between LAPLACE's great-great grandson, M. le Comte de COLBERT-LAPLACE and KARL PEARSON. It is due to KARL PEARSON's historical instinct that the British bombardment of Caen was not a complete disaster since a certain amount of the material there was recorded in *Biometrika* in 1929.

Now as an amateur historian who works in the history of probability and statistics not from the necessity of earning bread and butter, but merely as a hobby I have been interested more in motives than in facts. I would dislike to be misunderstood in this statement. I think it is of interest to find out, if one can, who first formalised a new mathematical concept, but to me it is even more interesting to speculate why the mathematician concerned felt the necessity to do so. It would be no help if one could project oneself backwards through time and actually ask the person concerned. What we think depends not only on our genetical makeup, but also on the outside influences at work at the particular point in the space-time continuum at which we find ourselves. Thus when, centuries from now, someone rises to commemorate the discoverer of confidence intervals, it would be stimulating to have on record why he pursued that particular problem at that particular time-stimulating but no more, because he is as much a prisoner of his time and circumstance as LAPLACE was of his. LAPLACE is particularly interesting from this kind of angle. He lived through the Revolution, the Napoleonic interlude, the Bourbon restoration. He was one of a coterie of mathematicians and natural philosophers several of whom were guillotined and he himself must have gone in fear of his life for a time. If one thinks of life as a random walk through time, the violence and horror of the Revolution must have affected the direction and depth of his thoughts in powerful fashion. One is reminded of Dr. JOHNSON's dictum "Depend on it, Sir, when a man knows he is to be hanged in two weeks it concentrates his mind wonderfully."

PIERRE-SIMON DE LAPLACE was born in Beaumont-en-Auge, in Basse-Normandie on March 23rd, 1749. His great-great-grandson says it was a family joke that the house on which there is a plaque is not his birth-place; this was outside the village but still within the commune. His father, neither a peasant, as reported by some biographers, nor a farmer

as reported by others, was concerned in some way with the manufacture
of cider. The family were, however, of very modest circumstances, since
their house and small holding was rented, being purchased by Laplace
at a later date. Laplace was destined for the church and was sent to a
school founded and conducted by the order of the Benedictines, at the
age of six, say about 1755. (The military academy at Beaumont to which
tradition and his biographers have always sent him was not founded
until 1776, after he had gone to Paris.) His uncle, a priest and a mathematician, taught there, but he died before the boy was ten years old.
Writing nearly one hundred years later one of his biographers, Puiseux,
says he had a precocious intellect and, above all, a prodigious memory.
In 1765 at sixteen years of age he left school and entered the College of
Arts, refounded and conducted by the Jesuits, at Caen, with the idea of
continuing study in the humanities and taking the clerical robe. The
Society of Jesus played however its usual beneficial rôle and it was here
that the young Laplace found his life-work, and was directed to the
path which he must follow.

There were two teachers of mathematics at Caen, Christopher
Gadbled and Pierre Le Canu. Laplace went to both their courses;
they became "more friends than masters" and he said to have made
rapid progress. The usual almost certainly apocryphal story is told about
him. — One day he found a book on the higher mathematics. He read it
with avidity. From that day his vocation was fixed. *Achille a trouvé ses
armes.* — The only new thing in the story is the substitution of "the
higher mathematics" for "Euclid's propositions." Leaving Caen at about
19 years old, Laplace was, for a short time, a teaching assistant at
Beaumont. He left for Paris in a few weeks (1768) with a letter of introduction to D'Alembert from Pierre Le Canu who had a small acquaintance with the savant.

Laplace is now about to make his appearance in the mathematical
world so we may pause a moment perhaps and review it briefly. D'Alembert (1717—1783) then a Perpetual Secretary of the French Academy
and Member of the Academy of Sciences, was one of the most powerful
men on the scientific scene. Euler (1707—1783), unhappy in Berlin,
had just left for St. Petersburg (1766), and was beginning to go blind.
Legendre (1752—1833) and about whom we know little, must still have
been at school in the *Collége Mazarin*, while the greatest pure mathematician since Fermat and Newton had just succeeded Euler at Berlin.
Joseph-Louis Lagrange (1736—1813) was of French extraction — one
of his ancestors was Descartes — but his family had settled in Turin,
he received his education there, and thought of himself as Italian. When
speaking of France to French mathematicians he always referred to
"your country." He became professor of mathematics at the Artillery

School at Turin and at the age of 22 (1758) founded, with two other scientists, the Academy of Sciences at Turin. He won the Grand Prize of the French Academy of Sciences in 1764 for the solution of an astronomical problem (the libration of the moon — why does the moon always present the same face to the earth?) and he won similar prizes in 1766, 1772, 1774 and 1778. EULER and D'ALEMBERT, impressed by the young man's work and inventive genius in the Calculus of Variations, wrote to him as an equal. Actually he was by far their superior. It was on D'ALEMBERT's recommendation that LAGRANGE succeeded EULER in Berlin.

When LAPLACE arrived in Paris (1768) he went to D'ALEMBERT to present his letter of recommendation. LAPLACE's great-great-grandson was told the story when taken carriage rides by his grandmother who was brought up in LAPLACE's household from birth. In the family circle, according to M. COLBERT-LAPLACE, PIERRE-SIMON would tell of his reception by D'ALEMBERT. D'ALEMBERT who, because of his position and authority, had many such letters of recommendation, received the young man coldly, put a large tome in his hands, perhaps his latest work, and told him to come back when he had read it. LAPLACE returned several days later; and was received even more coldly by D'ALEMBERT, who didn't bother to hide that he thought LAPLACE wasn't telling the truth. The matter was resolved by the young man going away and writing an essay on the principles of mechanics which finally convinced D'ALEMBERT; he got him a post teaching mathematics in the Military Academy. The reorganization of the Ecole Militaire in 1776 left LAPLACE without a job and for a time he was in straitened circumstances. The situation was resolved when he was appointed an examiner in 1783 for the passing out examination of the military cadets. (According to AUGUST FOURNIER in his *Life of Napoleon I*, NAPOLEON transferred to the Military Academy of Paris from that at Brienne in 1784 and was examined for his commission by LAPLACE in September 1785.) LAPLACE an ambitious young man of 19 with his way to make, began to write research papers on his arrival in Paris and there must at this time have been some contact by letter with LAGRANGE. For LAPLACE's first published work appears in the *Miscellaneous Papers of the Royal Society of Turin* dated 1766—1769 although LAPLACE says the paper was written in the month of March 1771. It appears to be one of the numerous rejected memoirs with which he bombarded the Academy of Sciences in Paris during his beginning years and it was to a very large extent reprinted in his first Academy paper of 1774 — *Memoir on rucurso-recurrent series and their uses in the theory of chance*. How many papers he presented and had rejected one doesn't know. He wrote to CONDORCET (1743—1794) in December 1771, ... "Reviewing the different papers that I have already presented to the Academy of Sciences ———" and the preface to the Academy volume of 1774 says "These two memoirs of M. DE LA PLACE have been chosen from a very large number which he has presented to the Academy over a period of three years ...". He actually got a foothold in the Academy as an "adjoint" in April 1773 after having been passed over once in 1772. Because he felt he had been slighted by being passed over he got D'ALEMBERT to write to LAGRANGE about the possibility of a position with the Academy of Sciences in Berlin but LAGRANGE, although seemingly favorable to the idea, did not pursue it. That LAPLACE had not yet learnt to conduct himself with the judicious decorum of his later years is clearly seen in the answer by LAGRANGE to a letter by CONDORCET in July 1774: "I am a little surprised by what you have said to me about M. DE LA PLACE: this is just, it seems to me, the fault of a young man who always exaggerates his first success; but the presumptuousness diminishes as knowledge grows."

These first papers of LAPLACE, the one I have already mentioned and the one I will shortly touch on; are on the theory of chance. The first one deals with *Duration of Play* and the *Differences of Zero Problem*, both exhaustively dealt with by DE MOIVRE. Why did he write on chance at this stage of his career given as is very likely, that he had learnt nothing of it from the Jesuits? Well D'ALEMBERT was obsessed with the contradictions which he thought he had found in elementary probability theory — the same fallacies which had engaged PASCAL and ROBERVAL a century earlier — and there is no doubt but that he would have talked to the young man and that the young man would have listened. Why did he not start on a new problem instead of beating over the old ground? Well D'ALEMBERT was not a clear thinker and could probably give him no idea of where to go next, and men's thoughts at that time were concerned not with gambling in any of its many guises but with the coming struggle for social justice. So the young man solves problems already solved by DE MOIVRE. Why did he not follow the fashion of the times and write about degree of belief? I can only suggest here that to someone coming new to the theory the problems posed by dicing, gambling, lotteries and the like were more clear-cut. That LAPLACE must have made the acquaintance of CONDORCET shortly after he arrived in Paris is interesting and lends credence to the accusation that LAPLACE was a cultivator of social position. CONDORCET was a marquis of the old nobility; more important perhaps he was a friend and disciple of the elderly VOLTAIRE then busy commuting across the border between Ferny and Geneva. CONDORCET's work (and probably his conversation also), is typical of his time. Men were becoming obsessed with social justice, with the credibility of witnesses, with the probability that a tribunal will reach a correct decision and so on, all discussed by the mathematicians from the point of view that a mathematical probability can be translated as a degree of belief.

Many of the mathematicians in France, but not D'ALEMBERT, were attempting to apply probability theory in this way and I would suggest that mathematicians, like their lay brothers, are influenced more than they know by outside events.

Now JAMES BERNOULLI (1654—1705) had promised that he would apply the theory of chance to civil, economic and moral affairs and had posed the problem of inverse probability. His nephew NICHOLAS had written (1709) on applications of probability in jurisprudence so that the idea of inverse probability used in this way was familiar and it was probably natural that LAPLACE should turn to their writings in an attempt to formulate something new about the then current problems, although he did not in fact achieve this in the moral sphere. He gives us, in the second paper of 1774, a "Memoir on the probability of the causes of events, (An event may be produced by any one of a number of different causes C_n each having a probability *a priori* π_n. If the probability of the event given the cause C_n is p_n, the probability of the cause given the event can be deduced provided the *a priori* probabilities are known.) This is just BAYES theorem and was applied by LAPLACE to the problem: An urn contains an infinite number of black and white tickets in unknown ratio. p and q tickets are drawn of which p are white and q are black. What is the probability of drawing m white and n black in some pre-assigned order in the next $m + n$ drawings? Under the hypotheses of the equal distribution of ignorance (all compositions equally likely), the probability of one further white given p and q is $(p + 1)/(p + q + 2)$. He has some trouble with the beta-function using an approximation he got from EULER. From this he later *(Théorie Analytique des Probabilitiés)* goes on to deduce the odds are $(p + 1) : 1$ that the sun will rise tomorrow given that it has already risen p days previously without missing. LAPLACE did recognize there that what he had done was not, in his own eyes, altogether admissible for he says "but this conclusion is made incomparably strong when, knowing all the phenomena, the principle regulating days and seasons, one sees that nothing, at the actual moment, is able to stop its course."

In contrast to the work going on in France, which was, at that moment, sterile in its results, what, up to now, had been the work of LAGRANGE, who at this time was living in Berlin and who had managed to achieve a popularity denied to the pompous and somewhat humorless EULER? From the time he started teaching at the age of nineteen, LAGRANGE's chief interest was, and remained, the analytical development of mechanics as opposed to the geometrical methods in practice everywhere. But possibly due to prompting from D'ALEMBERT, possibly due to queries by LAPLACE, he appears after reaching Berlin in 1766 to have turned his attention to the calculus of hazard, as it was then called. Not however for him the spur which urged the French mathematicians, he was inspired by an English mathematician. Thomas Simpson (1710—1761), the tinker, had made what little reputation he had by a shameless and unacknowledged plagiary of de Moivre's *Doctrine of Chances*. In 1757 when he was Professor of Mathematics at the Royal Military Academy at Woolwich, he published a book with the title *"Miscellaneous Tracts on some curious and very interesting subjects in Mechanics, Physical-Astronomy and Speculative Mathematics."* Lagrange's interest in applying calculus to mechanics would have led him to read this book, since the development of such studies came largely from Newton, then dead only thirty years (1727). One of these Miscellaneous Tracts was entitled *"An Attempt to show the Advantage arising by Taking the Mean of a Number of Observations in Practical Astronomy,"* and in it we have the first error distribution, *discrete and triangular*, but still it is there; and as far as I can tell it was Simpson's, his only original contribution to the probability calculus. In the *Miscellanea Taurinensia* of 1770—1773 Lagrange reproduces the whole of Simpson's work, without acknowledgement, using the same problems as did Simpson and leaving one in no doubt that he had just "lifted" it. He *did* introduce the terminology "curve of errors," and he did extend Simpson's work by proceeding to the limit and showing that the probability that the sample mean lies between assigned limits, which is a sum for the discrete error distribution, can be replaced by an integral. This last piece of work was in line with his work in mechanics. Quite why Lagrange plagiarised Simpson is difficult to conjecture. He was, in ability and achievement, the leading mathematician of Europe and he had no need to do this. Perhaps he felt the need of quick publication of something profound to impress Frederick the Great.

Laplace was obviously inspired by this idea of error distributions although he writes in his paper previously referred to *("Memoir on the probability of the causes of events")* that he has not seen Lagrange's paper, and tries to solve the problem for the mean of three observations and a negative exponential error distribution. He doesn't get very far. It seems unquestionable to me that Laplace needed, very strongly, the stimulus of other men's creative ideas. He, himself, was not an originator like Fermat or Newton, but a synthesizer and improver, a mathematician unrivalled in beating out and extending the paths pointed out by other men's intuition. Once he got the idea, he was magnificent in generalization and application, but he seems to have needed to be started off. And in his early days at any rate he was not altogether scrupulous as to how he did this. He read his first paper on astronomy to the Academy of Sciences in 1773 — it was published in 1776 — and demonstrated that apart from secular periodic variations, the mean distances of the planets from the Sun are invariable. This was a good original piece of work for a young man of 24; he succeeded where NEWTON had been unsure and although his method of attack had been foreshadowed by LAGRANGE his pertinacity brought him reputation and started his successful career. The warmth with which the paper was received also perhaps inspired him to cheat a little. LAGRANGE communicated a memoir on the same subject (Secular inequalities of planetary movement) in 1774, to the Academy of Sciences in Paris. This memoir was sent to LAPLACE, because of his previous lecture on the subject, and he sat himself to work right away to extend LAGRANGE's results. Legitimate, so far, but he then wrote up his own work and hurriedly arranged for it to be published well before that of LAGRANGE. He himself writes "I had thought for a long time of doing this research but the small usefulness in Astronomy of the calculation, added to the difficulty of it, made me abandon the idea and I own that I would not have taken it up again without reading the excellent memoir which M. DE LAGRANGE has just sent to the Academy and which will appear in one of the following volumes." He adds "I should naturally have waited so that the researches of M. DE LAGRANGE could be published before mine but having just written on the subject, I want to communicate this as a supplement to it, which is all that was wanting to make it complete, besides rendering to the memoir of M. DE LAGRANGE all the justice it deserves." LAGRANGE took all this very well, and indeed, apart from his lapse over THOMAS SIMPSON, he seems to have been an admirable person in every way. He wrote with the usual florid politeness of the day to LAPLACE, and bowed himself out of that particular line of research, but he remained friendly with him all his life.

Up to this date LAPLACE had contributed little to the calculus of probabilities which would win him recognition from his contemporaries whereas his astronomy researches had brought a certain reputation, so it is not surprising in fact that he pursued the latter for a few years and only worked on probability when it tied in with his other work or when a small problem arose which interested him. Various memoirs appeared until 1786. He gives us a discussion on generating functions; improves on methods suggested by both LEGENDRE and LAGRANGE for approximating to integrals by series expansions, and does a little vital statistics, taking up the problem of the inequality of male and female birth ratios which intrigued DE MOIVRE and DANIEL BERNOULLI, and followed a certain Baron MOHEAU in trying to determine the population of a country given the annual number of births. The review paper of 1781 was possibly the forerunner of the *Theory of Probability* volumes.

During these first twenty years that he was in Paris he probably felt himself insecure. He cultivated the society of nobility, and perhaps learnt from them the formal polite manners of the period but he did not known how to conduct himself among his fellow mathematicians and appears to

vary between arrogance of achievement and obsequiousness when called
to account for it. Thus we find him writing in 1777 to D'ALEMBERT
"I have always pursued mathematics from taste rather than the desire
for a vain reputation, which I don't want anyway. My greatest amuse-
ment is to study the work of others, to see their invention at work in
overcoming obstacles. I put myself in their place and I ask myself how
I would overcome these same obstacles. ... If I am fortunate enough to
add something to what they have done, I attribute it to their first
attempts. ..." If he had really said this in print each time that he
improved on other people's work probably his contemporaries would not
have complained as they did, none louder than D'ALEMBERT. D'ALEM-
BERT's wrath overflowed when LAPLACE presented a paper on the tides
to the Academy in 1777 without giving him what he considered enough
acknowledgement and both he and LAPLACE wrote to LAGRANGE com-
plaining each of the other's conduct. LEGENDRE was also unhappy about
LAPLACE's ambience. LEGENDRE read a paper to the Academy in 1784
(and printed in 1787) introducing his polynomials, in the course of solving
the problem of the shape of the surface formed by the rotation of a
homogeneous fluid. LAPLACE had a memoir printed in the volume
published in 1785 (work supposedly done in 1782) on the same subject.
This made LEGENDRE write at the top of his memoir, and remembering
the formal politeness of the time, he was obviously a very angry man,
"The proposition which is the object of this memoir being demonstrated
in a much more general way in a memoir that M. DE LA PLACE has
already published in the volume for 1782, I should observe that the date
of my memoir is previous to his and that the proposition which appears
here has inspired M. DE LA PLACE to go deeper into the matter." LAPLACE
had added much to LEGENDRE's results but he himself gives no inkling
of *how* much. Again in the memoirs of the Academy for 1789, LEGENDRE
writes "There are found in a memoir of M. DE LA PLACE printed at the
beginning of this volume, researches analogous to mine. So I give notice
that my memoir was submitted in 1790 and that the date of that of
M. DE LA PLACE is later." Obviously the young man of ability had found
the trick of getting on in the world. The only puzzle is how he, LAPLACE,
managed to get this trick of prior publication worked on so many
occasions. Because there must have been others which have not been
brought to one's attention.

LAPLACE's great-great-grandson wrote that his grandmother told him
that LAPLACE and his contemporaries thought in a similar way to each
other during this prerevolutionary epoque and that they found the ideas
of 1789 corresponded with their own ideals of social justice. This may
have been so. We may however note that the Marquis DE CONDORCET,
one of the aristocrats who, at first, led the revolution, had obviously very

little time for him. We may also note that one of his close friends during this period was the nobleman LAVOISIER, the hated tax-gatherer, who had considerable patronage at his disposal. LAVOISIER and LAPLACE worked together on specific heat (1782—1784) and on the generation of electricity 1781. There is no report of LAPLACE having any political interests at this time and indeed it would have been very foolish for a young man with his way to make to indulge himself in this way. It was possibly for this reason that he confined himself to the factual applications of probability theory. But *by a man's friends ye shall know him* and one notes his close friendship with LAVOISIER at this time. Also about this time LAPLACE worked on the velocity of sound and on capillary action. He became an associate of the Academy in 1783 and finally pensionary in 1785. Having achieved this step he was settled for life and would possibly, had events permitted it, have contented himself with a research career. He married in 1788 with MARIE-CHARLOTTE DE COURTY DE ROMANGES who was of the minor nobility, and two children, first a girl and then a boy were born in quick succession to one another. Finally LAGRANGE, not liking Berlin after the death of FREDERICK in 1786, accepted the offer of LOUIS XVI to come to Paris in 1787, so that both LAPLACE's public and private life might appear harmoniously determined. He had already written two long review papers, one a *Memoir on Probability* printed in 1781 in the Histoire de l'Academie de Paris and one, the Celestial Mechanics in embryo, the publication of which was paid for by SAVOR, president of the Paris Parlement. So that he was beginning to recognize for himself that his metier lay in the bringing together of other men's ideas, newly applying the analytical tools supplied by LAGRANGE and LEGENDRE and with his own incomparable assimilative powers, deepening and extending these notions wherever he turned his attention. But events outside his control intervened, although they should have been compatible with his ideas of social justice.

The Revolution started quietly enough in 1789, but its development saw the dispersal of the old Academies. The members met together and divided the Archives between them thus making certain that a goodly portion of them would be lost. LOUIS XVI and MARIE-ANTOINETTE were guillotined in 1793, and after this it was hardly safe to walk the streets of Paris. The arrest of CONDORCET hiding from arrest was typical. "Who are you?" "A carpenter." "Show me your hands. You're no carpenter." Whereupon he was taken off to meet death by poison in prison. Savor who paid to have LAPLACE's work published, and BAILLY were guillotined in 1794. BORDA, LAVOISIER, COULOMB, BRISSON and DELAMBRE were arrested as "men not sufficiently trustworthy either in their republican sympathies or in their hatred of royalty" and suffered the same fate. It is said, I haven't been able to confirm it, that LAPLACE was asked by the remnants of the Academy to make a list of the contents of LAVOISIER's laboratory for the benefit of posterity but that he was afraid and refused. If this story is true then the story that he had great personal charm must also be true since Madame LAVOISIER, then the Countess RUMFORD, received him in her salon after the Restoration. LAPLACE presumably retired with his young family to Melun and it was while hiding near his house that BAILLY was caught. Remembering the story of the pension to BAILLY's widow one wonders a little.

In any period of revolution it is enough to be able to say simply "I survived" and had LAPLACE done anything but lie low he would almost certainly have suffered the fate of so many of his contemporaries. One may recall that ROBESPIERRE, the sea-green incorruptible, thundered that the Revolution had no use for scientists. LAGRANGE, friend and protégé of MARIE-ANTOINETTE, was more free since he was an Italian; he said what he thought, but was tolerated by the revolutionaries and appointed to several paid committees and to the ephemeral École Normale. When the École Polytechnique was founded in 1797 he planned the mathematics courses and became the first professor. Of LAPLACE, in obscurity, we know little but he is supposed to have written at this time his popular version of the Celestial Mechanics (Exposition du Systeme du Monde). In December 1794 the Terror had abated somewhat and LAPLACE became assistant to LAGRANGE at the Ecole Normale. When the revolutionaries started to try to rebuild, LAGRANGE and LAPLACE were both appointed to the newly created Bureau of Longitudes (1795), LAPLACE making himself useful in the calculation of artillery tables. They were also the first two resident mathematicians of the National Institute for Science and Arts formed in the same year.

Taking things all round LAPLACE came out of the Revolutionary period remarkably well. He started work again and the first two parts of the *Celestial Mechanics* appeared in 1799. With remarkable prescience LAPLACE sent copies to the Citoyen BONAPARTE. There is no record of any meeting between them since BONAPARTE was examined by LAPLACE in 1785, but since LAPLACE had allayed revolutionary suspicion at the Bureau of Longitudes by his fervent calculation of gun-ranges and since BONAPARTE was an artillery officer it is possible that there was some contact. BONAPARTE wrote (October 19, 1799) "I have welcomely received, citizen, the example of your beautiful work which you have just sent me. The first six months that I have free will be used in reading it. If you have nothing better to do please come and dine at my house tomorrow." What went on at the dinner we do not know, but shortly afterwards BONAPARTE seized power as a member of the consulate and LAPLACE was named as Minister of the Interior on November 12th. On the evening of the first day of his appointment he obtained a pension for the widow of BAILLY who was destitute, and Madame LAPLACE herself took the first installment of the pension to Madame BAILLY. Since this was the only thing of its kind that LAPLACE ever did one wonders a little about the circumstances of BAILLY's arrest.

During his short tenure (six weeks) as Minister I can find only two directives issued to him by BONAPARTE. The first charges him to send troups of comedians and dancing girls to the army in Egypt. The second asks him to write a memorandum on suitable dates for national holidays. A pompous minute from LAPLACE in reply exists. (One of the holidays he recommended was July 14th.) The next day LUCIEN BONAPARTE was nominated Minister but he did not last longer than eighteen months. LAPLACE was removed to the Senate, of which he was successively vice-president and president, becoming Chancellor in 1803. In this last capacity in 1803 he writes enthusiastically "I have just proclaimed as Emperor of France, to the acclamation of the people, the hero for whom twenty years ago I had the advantage of opening the career that he has pursued with so much glory and good fortune for France." NAPOLEON obviously liked him and loaded him with honours — Grand Officer of the Legion of Honor at its foundation in 1802, count of the Empire in 1808, Dignitary of the Order of the Reunion at its inception. Madame LAPLACE became Lady-in-Waiting to Elisa-Bonaparte Grand Duchess of Tuscany in 1804. In common with the majority of Frenchmen LAPLACE's passion for social justice had undergone a radical change. The climate of the outside world did not however change the direction of LAPLACE's researches — it was for CAUCHY to exemplify the change of emphasis in mathematical thought, which leaned to an examination and rigorisation of the series expansions used in the pre-revolutionary period and finally became quite abstract, almost as though it was deliberately turned from contact with the outside world. LAPLACE, little burdened by cares of office, which don't seem to have been onerous, started the stream of astronomical papers again flowing and continued publication of the Celestial Mechanics with Book 3 in 1802, Book 4 in 1805, with the fifth and last in 1825. Books 3 and 4 were sent to NAPOLEON who returned gracious thanks. NAPOLEON also asked for reports to be written on the monetary system and on weights and measures. Presumably LAPLACE did this.

LAPLACE's work on *Probability Theory* first appeared in 1812 with a dedication to NAPOLEON. I have noted that a long review paper on probability theory appeared in 1781 and from a letter to LAGRANGE it appears that LAPLACE was meditating on possibly turning it into a book. The second edition appeared in 1814 together with the Philosophical Essay previously printed separately and written originally for the pupils of the École Normale. The edition which we know and to which we usually refer was printed in 1820. As usual LAPLACE sent a copy to

Napoleon, then on his way to Moscow (August 1st, 1812) and received a gracious reply, one of 80 letters which Napoleon wrote during the few days he was at Vitebsk. (With Napoleon as his artillery A.D.C. was Laplace's only son.) There is a story about Laplace and weather prediction for this campaign which I cannot verify but which may bear repeating. Napoleon left Moscow in flames on September 20th and moved westward. He *said* he was looking for the Russian army, a manoeuvre he executed with complete success until the night of November 6th, 1812. On November 7th he wrote from Mikhailovka saying hundreds of horses had died during the night from the cold which was 16° or 18° below freezing. On November 14th he wrote "we are without cavalry, without artillery, without transport." Later when writing *Considerations on the Art of War* he described, among many others, the Russian campaign. He said "If the cold had not started two weeks earlier than usual, the army would have reached Smolensk without loss," and goes on "We knew that it would be cold in December and January, but I was assured, from the examination of the temperatures of the twenty preceding years, that the thermometer never descended below six degrees of freezing in November." Rumour will have it that it was Laplace who examined the temperature records and gave the advice. I think rumor is probably untrue in that if Napoleon had been able to blame someone notable for his Russian failure he would have done so. With a person like Laplace so disliked and envied by his fellow scientists, any story is suspect.

It fell to Laplace, as Chancellor of the Senate, to sign the instrument deposing Napoleon in 1814. Given the favor which Napoleon had shown him a greater man would perhaps have resigned rather than do this, particularly since by now he had no monetary worries, but this was not his way. What he did when he heard that Napoleon had landed at Golfe-Juan on March 1st, 1815, history does not relate but he does not appear to have been very active, scientifically or politically, during that summer. He did not return to the Tuileries from Arceuil, from his country house, during the One Hundred Days. With the Bourbon Restoration he pops up again, created first a peer of France; than a marquis with the Grand Cross of the Legion of Honor all at the hands of Louis XVIII. This would imply, I think, that there was some underground contact between Laplace and Louis XVIII prior to 1814, possibly when many Frenchmen began to turn against Napoleon which was just after the Russian campaign. The events of the 100 days, which so nearly saw Napoleon successful must have given him furiously to think. At the time of the Restoration he was 67 and his life work was to all intents and purposes finished. He busied himself at the Bureau of Longitudes, he held evening parties where he collected young scientists together, he enjoyed what had always been his, a happy family life. His health remained good; although his eyesight troubled him a bit, but his powerful memory remained unimpaired and it was not until the last two years of his life that he became at all ill. He died in 1827 on March 5th. His official last words were reported to be "What we do know is small; what we are ignorant of is immense." Someone probably made this up in the same way as NEWTON is reported to have said that he had been like a boy playing with pebbles on the sea-shore while the ocean of truth lay undiscovered in front of him. DE MORGAN states his very last words were "Man follows only phantoms." One is reminded of Sir WALTER RALEIGH the night before he met the king's executioner.

> *"Even such is Time, that takes in trust*
> *Our youth, our joys, our all we have,*
> *And pays us but with earth and dust;*
> *Who in the dark and silent grave,*
> *When we have wandered all our ways,*
> *Shuts up the story of our days."*

Did the dying man remember friends betrayed, unfair advantages stolen, disloyalties committed, or was he just thinking of the days when he was young and eager and wore the black robe as he ran through the green fields of Beaumont? We shall never know.

Now it is easy to read LAPLACE and to point out that he borrowed wholeheartedly from his contemporaries and his predecessors — in probability particularly DE MOIVRE and JAMES BERNOULLI — and in so doing to forget that although he was not a creative genius, he was without peer in any age or country as a coordinator and extender of other men's ideas. KARL PEARSON, who was the most fervent admirer of LAPLACE, once said in a lecture "It is unlikely that he could have failed to read either THOMAS WRIGHT, in Durham's *Original Theory of the Universe* 1726 or KANT's *Theory of the Heavens* 1755: Yet I remind my audience that the *Mécanique Céleste* did not profess to be an original memoir, but a gigantic treatise on the mechanics of the universe, in which the author not only included all that was already known in his day, but an immense amount of new matter. Is it more reasonable to blame LAPLACE than to blame EUCLID, many of whose propositions must have been known long before his day? ... He certainly did not set out to steal but ... his treatment of his own countryman LEGENDRE and of our English THOMAS YOUNG shows that he was far from careful not to wound the susceptibilities of men who made very real contributions to knowledge. The remarks which the *Mécanique Céleste* calls forth apply as strongly, if not more so to the *Théorie Analytique des Probabilitiés*. LAPLACE put together all that was known of the subject in his day, and added to and developed his material. But only those intimately acquainted with what Montmort, De Moivre, the Bernoullis, Condorcet and Lagrange had achieved, can fully grasp how much he owed to them not only for fundamental principles but for suggestions for further research."

This I think sums up the situation fairly, and here I will leave M. le Marquis de la Place to the judgment of history. In my own opinion, in spite of himself, we owe him much.

Extension of the Kolmogorov-Smirnov Test
to Regression Alternatives

By Jaroslav Hájek

Mathematical Institute of the Czechoslovak Academy of Science
and
Statistical Laboratory University of California, Berkeley

0. Summary

The Kolmogorov-Smirnov test for regression alternatives is defined in Section 3, where an artificial numerical example is also given. The limiting distribution is derived in Section 4 by means of techniques developed in Section 1 (conditions for convergence in distribution in $C\,[0,\,1]$) and Section 2 (extension of a Kolmogorov inequality to the case of sampling without replacement from a finite population). The auxiliary results just mentioned also offer some interest in their own right.

1. Convergence in distribution in $C\,[0,\,1]$ of stochastic processes

Let us have a sequence of stochastic processes $\{X_\nu\,(t,\,\omega_\nu),\,0 \le t \le 1\}$, $1 \le \nu < \infty$, defined on some measurable spaces $(\Omega_\nu,\,\mathscr{F}_\nu)$ and assume that all sample paths $X_\nu\,(\cdot,\,\omega_\nu),\,1 \le \nu < \infty,\,\omega_\nu \in \Omega_\nu$, are continuous functions on $[0,\,1]$. Moreover, we shall assume that the subvectors $[X_\nu\,(t_1),\,\ldots,$ $X_\nu\,(t_n)],\,1 \le n < \infty,\,0 \le t_1,\,\ldots,\,t_n \le 1$, converge in distribution to the corresponding subvectors $[X\,(t_1),\,\ldots,\,X\,(t_n)]$ of some process $\{X\,(t),$ $0 \le t \le 1\}$

$$\mathscr{L}\,[X_\nu\,(t_1),\,\ldots,\,X_\nu\,(t_n)] \to \mathscr{L}\,[X\,(t_1),\,\ldots,\,X\,(t_n)]\,, \qquad (1.1)$$

$$(1 \le n < \infty,\,0 \le t_1,\,\ldots,\,t_n \le 1,\,\nu \to \infty)\,.$$

We suppress here the dependence of random variables on elementary events ω_ν or ω because our interest is centered on respective n-dimensional distributions.

The implications of relation (1.1) are rather poor. We cannot even infer that there is a version of the process $X\,(t)$, say $X\,(t,\,\omega_0)$ such that every path $X\,(\cdot,\,\omega_0)$ again is a continuous function on $[0,\,1]$ for every $\omega_0 \in \Omega_0$. To see this it suffices to take $X_\nu\,(t,\,\omega_\nu) = f_\nu\,(t),\,\omega_\nu \in \Omega_\nu$, where $\{f_\nu\,(t)\}$ is a sequence of continuous functions converging pointwise to a discontinuous function.

The existence of a version of $X\,(t)$ with continuous paths is not very consequential either, but it is a prerequisite for the formulation of a stronger kind of convergence of processes $X_\nu(t)$ to the process $X(t)$. Let h be a continuous functional defined on the usual Banach space $C[0, 1]$ of continuous function $X(t)$ on $[0, 1]$. This means that for every $\varepsilon > 0$ and $x(\cdot)$ there is a $\delta > 0$ such that

$$[\max_{0 \leq t \leq 1} |x(t) - y(t)| < \delta] \Rightarrow \{|h[x(\cdot)] - h[y(\cdot)]| < \varepsilon\} \tag{1.2}$$

for any function $y(t)$ from $C[0, 1]$. We give some examples of continuous functionals:

$$h_M[x(\cdot)] = \max_{0 \leq t \leq 1} x(t), \tag{1.3}$$

$$h_\delta[x(\cdot)] = \max_{|t-s| < \delta} |x(t) - x(s)|, \qquad \delta > 0, \tag{1.4}$$

$$h_G[x(\cdot)] = \int_0^1 x(t)\, dG(t), \tag{1.5}$$

where $G(t)$ is any function of finite variation. Obviously, h_G is a linear functional, while h_M and h_δ are nonlinear.

Definition 1.1. A sequence of processes $\{X_\nu(t), 0 \leq t \leq 1\}, 1 \leq \nu < \infty$, with continuous paths everywhere, will be said to converge in distribution in $C[0, 1]$ to a process $\{X(t), 0 \leq t \leq 1\}$, if there is a version of $X(t)$, say $X(t, \omega_0)$, such that the paths $X(\cdot, \omega_0)$ are continuous everywhere, and if for every continuous functional $h = h[x(\cdot)]$ on $C[0, 1]$, $h[X_\nu(\cdot)]$ converges in distribution to $h[X(\cdot)]$:

$$\mathscr{L}\{h[X_\nu(\cdot)]\} \to \mathscr{L}\{h[X(\cdot)]\}, \qquad \nu \to \infty. \tag{1.6}$$

Remark 1.1. If the paths $X(\cdot)$ are continuous everywhere and the functional h is continuous, then $h[X(\cdot)]$ is a random variable (measurable function).

The question of convergence in distribution of stochastic processes in connection with statistical problems was first raised by Doob (1949). His conjecture was then proved by Donsker (1952). First general results in terms of weak convergence of measures were obtained by Prohorov (1952) who utilized earlier results by Alexandrov (1941—43). The matter was further developed by Le Cam (1957), Veradarajan (1958), Driml (1959), Prohorov (1961), Bartoszynski (1962) and others.

In this paper a somewhat simpler formulation of necessary and sufficient conditions for convergence in distribution in $C[0, 1]$ are given, and the proof is deliberately self-sufficient (and therefore rather long), because I believe that some reader with prevalently statistical background might find it convenient.

Theorem 1.1. Assume that (1.1) holds true. Then $\{X_\nu(t), 0 \leq t \leq 1\}$ converge to $\{X(t), 0 \leq t \leq 1\}$ in distribution in $C[0, 1]$ if and only if

$$\lim_{\delta \to 0} \limsup_{\nu \to \infty} P(\max_{|t-s| < \delta} |X_\nu(t) - X_\nu(s)| > \varepsilon) = 0 \qquad \text{for every } \varepsilon > 0. \tag{1.7}$$

Condition (1.7) is satisfied if

$$\lim_{\delta\to 0}\frac{1}{\delta}\limsup_{\nu\to\infty}\ \max_{0\le s\le 1-\delta}\ P\ (\max_{s\le t\le s+\delta}\ |X_\nu(t)-X_\nu(s)|\ge\varepsilon)=0,\qquad (\varepsilon>0),\qquad (1.8)$$

where the max within the parentheses refers to t only, while s is assumed fixed.

If the paths of the processes $X_\nu(t)$ are linear on intervals $[k/N_\nu, (k+1)/N_\nu]$, $k=1,\ldots,N_\nu$, $N_\nu\to\infty$ for $\nu\to\infty$, then (1.7) is implied by

$$\lim_{\delta\to 0}\frac{1}{\delta}\limsup_{\nu\to\infty}\ \max_{1\le k\le N_\nu-m_{\nu\delta}+1}\ P\ (\max_{k\le j<k+m_{\nu\delta}}\ |X_\nu\!\left(\frac{k-1}{N_\nu}\right)-X_\nu\!\left(\frac{j}{N_\nu}\right)|>\varepsilon]=0,$$
$$(\varepsilon>0),\qquad (1.9)$$

where the $m_{\nu\delta}$'s are integers such that

$$\lim_{\nu\to\infty}\frac{m_{\nu\delta}}{N_\nu}=\delta,\qquad (\delta>0).\qquad (1.10)$$

Proof. Necessity of (1.7). Relation (1.6) must hold for every functional h_δ defined by (1.4). Thus, for every $\varepsilon>0$

$$\limsup_{\nu\to\infty} P\{h_\delta[X_\nu(\cdot)]>\varepsilon\}\le P\{h_\delta[X(\cdot)]>\tfrac{1}{2}\varepsilon\}.\qquad (1.11)$$

On the other hand, as all continuous functions on $[0, 1]$ are equicontinuous, $h_\delta[X(\cdot)]\to 0$ for $\delta\to 0$ everywhere in $C[0,1]$, so that

$$\lim_{\delta\to\infty} P\{h_\delta[X(\cdot)]>\tfrac{1}{2}\varepsilon\}=0,\qquad (\varepsilon>0).\qquad (1.12)$$

Now (1.11) and (1.12), in view of (1.4), yield (1.7) immediately.

Sufficiency of (1.7). We shall first prove the existence of a continuous path version of the limiting process $X(t)$. In view of (1.1), we know the finite-dimensional distributions of $X(t)$. These are, obviously consistent, and hence we may construct the process in the space of all functions on $[0, 1]$, say $R^{[0,1]}$, by the Kolmogorov procedure. We obtain a probability distribution, say, $P(\cdot)$, on the Borel field $\mathscr{B}$ of subsets of $R^{[0,1]}$ generated by subsets $A_{tc}=\{x(\cdot):x(t)<c\}$, $t\in[0,1]$, $-\infty<c<\infty$, such that the distribution law of coordinate random variables $\{X(t_1),\ldots,X(t_n)\}$ will coincide with the law on the right side of (1.1). Now denote by C, $C\subset R^{[0,1]}$, the subset of continuous functions. If for every two subsets B_1 and B_2 from $\mathscr{B}$

$$[C\cap B_1=C\cap B_2]\Rightarrow[P(B_1)=P(B_2)],\qquad (1.13)$$

then the probability measure $P(\cdot)$ defined on $\mathscr{B}$ may be carried over to the Borel field $\mathscr{A}$, consisting of subsets $C\cap B$, $B\in\mathscr{B}$. Denoting the corresponding measure on $\mathscr{A}$ by $P_C(\cdot)$, we have

$$P_C(A)=P(B),\qquad (A\in\mathscr{A},\ B\in\mathscr{B},\ A=C\cap B),\qquad (1.14)$$

where B is any subset such that $A=C\cap B$. In view of (1.13) the value of $P_C(A)$ does not depend on the choice of B. Obviously the coordinate random variables $\{X(t_1), \ldots, X(t_n)\}$ would have the same distribution in $(C, \mathscr{A})$ under $P_C(\cdot)$ as in $(R^{[0,1]}, \mathscr{B})$ under $P(\cdot)$. So, if we prove (1.13) we can construct a version of the process $X(t)$ possessing continuous paths everywhere.

DOOB (1937) showed that (1.13) holds true if and only if (the proof is quite simple)

$$[B \supset C, B \in \mathscr{B}] \Rightarrow [P(B) = 1] . \tag{1.15}$$

(Recall that C does not belong to $\mathscr{B}$.) We know that to each event B, $B \in \mathscr{B}$, there corresponds a countable subset of $[0, 1]$, say I_B, $I_B \subset C$ $[0, 1]$, such that

$$[x(\cdot) \in B, x(t) = y(t), t \in I_B] \Rightarrow [y(\cdot) \in B] . \tag{1.16}$$

[See, for example, ŠPAČEK (1955).] Without loss of generality, we may assume that I_B is dense in $[0, 1]$. Now if B contains C, then it also contains all functions $x(\cdot)$ which are uniformly continuous on the set I_B, i.e., for each $\varepsilon > 0$ there is a $\delta > 0$ such that

$$[\,|t-s| < \delta, t \in I_B, s \in I_B] \Rightarrow [\,|x(t) - x(s)| < \varepsilon] . \tag{1.17}$$

Denote the latter set by $C(I_B)$. So

$$[B \supset C] \Rightarrow [B \supset C(I_B)] . \tag{1.18}$$

However, the set $C(I_B)$ is measurable, so that it suffices to show that

$$P[C(I_B)] = 1 . \tag{1.19}$$

Let us take an increasing sequence of finite subsets $I_1 \subset I_2 \subset \cdots$ of $[0, 1]$ such that

$$I_B = \bigcup_1^\infty I_n . \tag{1.20}$$

Now, in view of (1.1),

$$P[\max_{\substack{|t-s|<\delta \\ t, s \in I_n}} |X(t) - X(s)| > \varepsilon]$$

$$\leq \limsup_{\nu \to \infty} P[\max_{\substack{|t-s|<\delta \\ t, s \in I_n}} |X_\nu(t) - X_\nu(s)| > \tfrac{1}{2}\varepsilon] \tag{1.21}$$

$$\leq \limsup_{\nu \to \infty} P[\max_{|t-s|<\delta} |X_\nu(t) - X_\nu(s)| > \tfrac{1}{2}\varepsilon] .$$

Consequently,

$$P[\max_{\substack{|t-s|<\delta \\ t, s \in I_B}} |X(t) - X(s)| > \varepsilon]$$

$$= \lim_{n \to \infty} P[\max_{\substack{|t-s|<\delta \\ t, s \in I_n}} |X(t) - X(s)| > \varepsilon] \tag{1.22}$$

$$\leq \limsup_{\nu \to \infty} P[\max_{|t-s|<\delta} |X_\nu(t) - X_\nu(s)| > \tfrac{1}{2}\varepsilon] .$$

Now combining (1.22) and (1.7), we obtain

$$\lim_{\delta \to 0} P\,[\max_{\substack{|t-s|<\delta \\ t,s \in IB}} |\,X(t) - X(s)\,| > \varepsilon] = 0\,, \qquad (\varepsilon > 0)\,. \qquad (1.23)$$

This is, however, equivalent to (1.19). So we have proved the existence of a continuous-path version of the limiting process $X(t)$.

Now we have to show that (1.6) holds true for any continuous functional h, provided the paths of $X(t)$ are continuous. Let $X^{(k)}(t)$ be a process derived from the process $X(t)$ as follows:

$$X^{(k)}(t) = X\left(\frac{i}{2^k}\right) + 2^k\left(t - \frac{i}{2^k}\right)\left[X\left(\frac{i+1}{2^k}\right) - X\left(\frac{i}{2^k}\right)\right],$$
$$\text{for } \frac{i}{2^k} \leq t \leq \frac{i+1}{2^k}\,. \qquad (1.24)$$

Obviously, within the intervals $[i/2^k, (i+1)/2^k]$, we have that $X^{(k)}(t)$ is linear.

Let the processes $X_\nu^{(k)}(t)$ be derived from the processes $X_\nu(t)$ in the very same way. Now, from (1.24) and (1.7) it follows that

$$\lim_{k \to \infty} P\,[\max_{0 \leq t \leq 1} |\,X^{(k)}(t) - X(t)\,| > \varepsilon] = 0, \qquad (\varepsilon > 0)\,, \qquad (1.25)$$

$$\lim_{k \to \infty} \sup_\nu P\,[\max_{0 \leq t \leq 1} |\,X_\nu^{(k)}(t) - X_\nu(t)\,| > \varepsilon] = 0, \quad (\varepsilon > 0)\,. \qquad (1.26)$$

Now, since the functional h is continuous, (1.25) and (1.26) imply

$$\lim_{k \to \infty} P\,\big\{\,|\,h\,[X^{(k)}(\cdot)] - h\,[X(\cdot)]\,| > \varepsilon\big\} = 0\,, \qquad (\varepsilon > 0)\,, \qquad (1.27)$$

$$\lim_{k \to \infty} \sup_\nu P\,\big\{\,|\,h\,[X_\nu^{(k)}(\cdot)] - h\,[X_\nu(\cdot)]\,| > \varepsilon\big\} = 0\,, \qquad (\varepsilon > 0)\,. \qquad (1.28)$$

Now, note that h applied to $X^{(k)}(t)$ and $X_\nu^{(k)}(t)$ may be considered as a continuous functional of the vectors $\big\{X(i/2^k),\, i = 0, 1, \ldots, 2^k\big\}$ and $\big\{X_\nu(i/2^k),\, i = 0, 1, \ldots, 2^k\big\}$. Bearing this in mind, we conclude from (1.1) that

$$\mathscr{L}\big\{h\,[X_\nu^{(k)}(\cdot)]\big\} \to \mathscr{L}\big\{h\,[X^{(k)}(\cdot)]\big\}, \qquad \nu \to \infty\,. \qquad (1.29)$$

Suppose that y is a continuity point of the distribution of $h\,[X(\cdot)]$. Then for every $\varepsilon > 0$ there exists a $\delta > 0$ such that

$$P\,\big\{h\,[X(\cdot)] < y + \delta\big\} - \tfrac{1}{4}\varepsilon \leq P\,\big\{h\,[X(\cdot)] \leq y\big\} \leq$$
$$\leq P\,\big\{h\,[X(\cdot)] < y - \delta\big\} + \tfrac{1}{4}\varepsilon\,. \qquad (1.30)$$

Furthermore, in view of (1.27) and (1.28), there exists a k_0 such that

$$P\,\big\{h\,[X(\cdot)] < y + \delta\big\} \geq P\,\big\{h\,[X^{(k_0)}(\cdot)] < y + \tfrac{2}{3}\delta\big\} - \tfrac{1}{4}\varepsilon\,, \qquad (1.31)$$

$$P\,\big\{h\,[X(\cdot)] < y - \delta\big\} \leq P\,\big\{h\,[X^{(k_0)}(\cdot)] < y - \tfrac{2}{3}\delta\big\} + \tfrac{1}{4}\varepsilon\,, \qquad (1.32)$$

and, uniformly in ν,

50   Jaroslav Hájek

$$P\left\{h\left[X_\nu^{(k_0)}(\cdot)\right] < y + \tfrac{1}{3}\,\delta\right\} \geq P\left\{h\left[X_\nu(\cdot)\right] \leq y\right\} - \tfrac{1}{4}\,\varepsilon, \qquad (1.33)$$

$$P\left\{h\left[X_\nu^{(k_0)}(\cdot)\right] < y - \tfrac{1}{3}\,\delta\right\} \leq P\left\{h\left[X_\nu(\cdot)\right] \leq y\right\} + \tfrac{1}{4}\,\varepsilon. \qquad (1.34)$$

Now, (1.29) implies that there exists ν_0 such that

$$P\left\{h\left[X_\nu^{(k_0)}(\cdot)\right] < y + \tfrac{1}{3}\,\delta\right\} \leq P\left\{h\left[X^{(k_0)}(\cdot)\right] < y + \tfrac{2}{3}\,\delta\right\} + \tfrac{1}{4}\,\varepsilon, \qquad (1.35)$$

$$P\left\{h\left[X_\nu^{(k_0)}(\cdot)\right] \leq y - \tfrac{1}{3}\,\delta\right\} \geq P\left\{h\left[X^{(k_0)}(\cdot)\right] < y -\right.$$
$$\left. - \tfrac{2}{3}\,\delta\right\} - \tfrac{1}{4}\,\varepsilon, \qquad (\nu \geq \nu_0). \qquad (1.36)$$

Upon combining (1.30) through (1.36), we can see that

$$P\left\{h\left[X_\nu(\cdot)\right] \leq y\right\} - \varepsilon \leq P\left\{h\left[X(\cdot)\right] \leq y\right\}$$
$$\leq P\left\{h\left[X_\nu(\cdot)\right] \leq y + \varepsilon\right\}, \qquad (\nu \geq \nu_0). \qquad (1.37)$$

However, (1.37) is equivalent to

$$\lim_{\nu \to \infty} P\left\{h\left[X_\nu(\cdot)\right] \leq y\right\} = P\left\{h\left[X(\cdot)\right] \leq y\right\} \qquad (1.38)$$

at each continuity point y, which, in turn, is equivalent to (1.6).

Sufficiency of (1.8) **and** (1.9). For a given $\delta > 0$, divide the interval into n intervals $[s_i, s_{i+1}]$, $1/\delta \leq n < 1/\delta + 1$, $s_i = i\delta$, $i = 0, 1, \ldots, n-1$, $s_n = 1$. We can easily see that for any function $x(t)$

$$[\max_{|t-s|<\delta} |x(t) - x(s)| > \varepsilon] \qquad (1.39)$$

$$\Rightarrow [\max_{s_i \leq s \leq s_{i+1}} |x(s) - x(s_i)| > \tfrac{1}{3}\,\varepsilon \text{ for at least one } i = 0, \ldots, n-1].$$

Consequently,

$$P[\max_{|t-s|<\delta} |X_\nu(t) - X_\nu(s)| > \varepsilon]$$

$$\leq \sum_{i=0}^{n-1} P[\max_{s_i \leq s \leq s_{i+1}} |X_\nu(s) - X_\nu(s_i)| > \tfrac{1}{3}\,\varepsilon] \qquad (1.40)$$

$$\leq \left(\frac{1}{\delta} + 1\right) \max_{0 \leq s \leq 1-\delta} P[\max_{s \leq t \leq s+\delta} |X_\nu(t) - X_\nu(s)| > \tfrac{1}{3}\,\varepsilon].$$

Now it is obvious that (1.8) implies (1.7). The proof of sufficiency of (1.9) is then immediate.

The proof is terminated.

Remark 1.2. Consider the sequence of probability measures $\left\{P_\nu(\cdot)\right\}$ induced in $(C, \mathscr{A})$ by processes $\left\{X_\nu(t), 0 \leq t \leq 1\right\}$. It is easy to show that $\left\{P_\nu(\cdot)\right\}$ is relatively compact (every subsequence contains weakly convergent subsubsequences) under conditions (1.1) and (1.7). According to Prohorov (1956) it amounts to showing that for each $\varepsilon > 0$ there is a compact subset K_ε of $C[0, 1]$ such that

$$\inf P_\nu(K_\varepsilon) > 1 - \varepsilon. \qquad (1.41)$$

In fact, as $X_\nu(0)$ converge in distribution to $x(0)$, there is a constant M_ε such that

$$\inf_\nu P[\,|X_\nu(0)|<M_\varepsilon] > 1 - \tfrac{1}{2}\varepsilon. \tag{1.42}$$

Furthermore, (1.7) is equivalent to

$$\lim_{\delta\to 0}\sup_\nu P[\max_{|t-s|<\delta}|X_\nu(t)-X_\nu(s)|>\varepsilon]=0. \tag{1.43}$$

Thus, we may choose a sequence $\delta_n\to 0$ such that

$$\sup_\nu P\left[\max_{|t-s|<\delta_n}|X_\nu(t)-X_\nu(s)|>\frac{1}{n}\right]<(\tfrac{1}{2})^{n+1}\varepsilon. \tag{1.44}$$

Now, it suffices to put

$$K_\varepsilon=\left\{x(\cdot):|x(0)|<M_\varepsilon\right\}\cap\bigcap_{n=0}^{\infty}\left\{x(\cdot):\max_{|t-s|<\delta_n}|x(t)-x(s)|\le\frac{1}{n}\right\}. \tag{1.45}$$

In this way, utilizing the results of PROHOROV (1956), we could shorten considerably the "sufficiency" part of the proof.

Remark 1.3. BARTOSZYNSKI (1962) proved that a process $\{X(t),\ 0\le t\le 1\}$ may be realized in $(C,\mathscr{A})$ if and only if for every $\varepsilon>0$ there is a function $\Psi_\varepsilon(\delta)$ such that $\Psi_\varepsilon(\delta)\downarrow 0$ as $\delta\downarrow 0$ and for every finite subset $I=\{t_1,\ldots,t_n\}$ of $[0,1]$

$$P[\max_{\substack{|t_i-t_j|<\delta\\ t_i,\,t_j\in I}}|X(t_i)-X(t_j)|>c]\le\Psi_\varepsilon(\delta). \tag{1.46}$$

In our case we may simply put

$$\Psi_\varepsilon(\delta)=\lim_{\nu\to\infty}\sup P[\max_{|t-s|<\delta}|X_\nu(t)-X_\nu(s)|>\tfrac{1}{2}\varepsilon]. \tag{1.47}$$

Then $\Psi_\varepsilon(\delta)\downarrow 0$ as $\delta\downarrow 0$ is insured by (1.7), and by (1.1):

$$P[\max_{\substack{|t_i-t_i|<\delta\\ t_j,\,t_j\in I}}|X(t_i)-X(t_j)|>\varepsilon]$$

$$\le\lim_\nu\sup P[\max_{\substack{|t_i-t_j|<\delta\\ t_i,\,t_j\in I}}|X_\nu(t_i)-X_\nu(t_j)|>\tfrac{1}{2}\varepsilon] \tag{1.48}$$

$$\le\lim_\nu\sup P[\max_{|t-s|<\delta}|X_\nu(t)-X_\nu(s)|>\tfrac{1}{2}\varepsilon]$$

$$=\Psi_\varepsilon(\delta).$$

Thus we could shorten the proof of existence of $C[0,1]$ version of $X(t)$ this way also.

Remark 1.4. If we guess that the limiting process $x(t)$ is Gaussian with a certain mean value function $m(t)$ and covariance kernel $K(t,s)$, then (1.1) is implied by

$$\mathscr{L}\left[\sum_{i=1}^{n}\lambda_i x_\nu(t_i)\right]\to\mathscr{L}\left[\sum_{i=1}^{n}\lambda_i x(t_i)\right]$$

$$=N\left[\sum_{i=1}^{n}\lambda_i m(t_i),\ \sum_{i=1}^{n}\sum_{j=1}^{n}\lambda_i\lambda_j K(t_i,t_j)\right], \tag{1.49}$$

$$(1\le n<\infty,\ -\infty<\lambda_1,\ldots,\lambda_n<\infty)$$

where $N(a, \sigma^2)$ denotes the normal distribution with mean value a and variance σ^2.

Observe that $\sum_{i=1}^{n} \lambda_i x(t_i)$ represent a special kind of linear functionals in $C[0, 1]$. If the convergence in distribution for these linear functionals is supplemented by the convergence of the functional $h_\delta[x(\cdot)]$ [see (1.4)], which amounts to the condition (1.7), then the convergence for all other continuous functionals follows.

2. An extension of Kolmogorov's inequality

Let us have a finite population consisting of N values $\{c_1, c_2, \ldots, c_N\}$. Assume that

$$\sum_{i=1}^{N} c_i = 0, \qquad (n < N), \tag{2.1}$$

and draw a sequence of n, $(n < N)$ values $Y_1, \ldots, Y_n$ by simple random sampling without replacement. From (2.1) it follows that

$$EY_i = 0, \qquad (i = 1, \ldots, n). \tag{2.2}$$

Now, if random variables $(Y_1, \ldots, Y_k)$, $k < n$, are fixed, then $(Y_{k+1}, \ldots, Y_n)$ represents again a simple random sample of size $n - k$ from the population $\{c_1', \ldots, c_{N-k}'\}$ where c_i' are identical with values c_i excepting those selected as $Y_1, \ldots, Y_k$, $(Y_i = c_{r_i}, i = 1, \ldots, k)$. Consequently,

$$E(Y_{k+1} + \cdots + Y_n \mid Y_1, \ldots, Y_k) \tag{2.3}$$

$$= \frac{n-k}{N-k} \sum_{i=1}^{N-k} c_i' = \frac{n-k}{N-k}\left[\sum_{i=1}^{N} c_i - \sum_{j=1}^{k} Y_j\right] = -\frac{n-k}{N-k}(Y_1 + \cdots + Y_k),$$

in view of (2.1).

Lemma 2.1. *For any* $r \geq 1$ *and* $\varepsilon > 0$ *it follows that*

$$P\left(\max_{1 \leq k \leq n} |Y_1 + \cdots + Y_k| > \varepsilon\right) < \frac{E|Y_1 + \cdots + Y_n|^r}{\varepsilon^r \left(1 - \dfrac{n}{N}\right)^r}. \tag{2.4}$$

Proof. Introduce disjoint events $A_1, \ldots, A_n$ defined as follows:

$$A_k = \Big\{(Y_1, \ldots, Y_n): \max_{1 \leq j \leq k-1} |Y_1 + \cdots + Y_j| \leq \varepsilon,$$

$$|Y_1 + \cdots + X_k| > \varepsilon\Big\}. \tag{2.5}$$

Then

$$E|Y_1 + \cdots + Y_n|^r = \int |Y_1 + \cdots + Y_n|^r \, dP$$

$$\geq \sum_{k=1}^{n} \int_{A_k} |Y_1 + \cdots + Y_n|^r \, dP. \tag{2.6}$$

Now, since A_k depends only on $Y_1, \ldots, Y_k$,

$$\int_{A_k} |Y_1 + \cdots + Y_n|^r \, dP = \int_{A_k} E\{|Y_1 + \cdots + Y_n|^r \mid Y_1, \ldots, Y_k\} \, dP,$$

$$(1 \leq k \leq n). \tag{2.7}$$

As for $r \geq 1$ the function $|x|^r$ is convex in x, upon applying the Jensen inequality we get

$$E\{\,|\,Y_1 + \cdots + Y_n\,|^r\,|\,Y_1, \ldots, Y_k\}$$
$$\geq |\,E\{Y_1 + \cdots + Y_n\,|\,Y_1, \ldots, Y_k\}\,|^r$$
$$= |\,Y_1 + \cdots + Y_k + E\{Y_{k+1} + \cdots + Y_n\,|\,Y_1, \ldots, Y_k\}\,|^r.$$

Recalling (2.3), we conclude that

$$E\{\,|\,Y_1 + \cdots + Y_n\,|^r\,|\,Y_1, \ldots, Y_k\} \geq$$
$$\geq |\,Y_1 + \cdots + Y_k - \frac{n-k}{N-k}\,(Y_1 + \cdots + Y_k)\,|^r \tag{2.8}$$
$$\geq \left(1 - \frac{n}{N}\right)^r |\,Y_1 + \cdots + Y_k\,|^r, \qquad (1 \leq k \leq n)\,.$$

Now, combining (2.6), (2.7) and (2.8), we obtain

$$E\,|\,Y_1 + \cdots + Y_n\,|^r \geq \left(1 - \frac{n}{N}\right)^r \sum_{i=1}^{n} \int_{A_k} |\,Y_1 + \cdots + Y_k\,|^r \, dP\,. \tag{2.9}$$

However, according to (2.5), $|\,Y_1 + \cdots + Y_k\,| > \varepsilon$ on A_k, so that (2.9) implies

$$E\,|\,Y_1 + \cdots + Y_n\,|^r \geq \left(1 - \frac{n}{N}\right)^r \varepsilon^r \sum_{i=1}^{n} \int_{A_k} dP$$
$$= \left(1 - \frac{n}{N}\right)^r \varepsilon^4 \, P\left(\bigcup_{i=1}^{n} A_k\right) \tag{2.10}$$
$$= \left(1 - \frac{n}{N}\right)^r \varepsilon^4 \, P\,(\max_{1 \leq k \leq n} |\,Y_1 + \cdots + Y_n\,| > \varepsilon)\,.$$

Noting that (2.10) is equivalent to (2.4), our proof is finished.

Case $r = 2$. Since for simple random sampling without replacement

$$E\,(Y_1 + \cdots + Y_n)^2 = n\left(1 - \frac{n}{N}\right) \frac{1}{N-1} \sum_{i=1}^{N} c_i^2\,, \tag{2.11}$$

the inequality (2.4) for $r = 2$ yields

$$P\,(\max |\,Y_1 + \cdots + Y_k\,| > \varepsilon) < \frac{n \sum_{i=1}^{N} c_i^2}{\varepsilon^2\,(N-1)\left(1 - \dfrac{n}{N}\right)} \tag{2.12}$$

provided (2.1) holds true.

Case $r = 4$. We have [see ISERLIS (1931)]

$$E\,(Y_1 + \cdots + Y_n)^4$$
$$= \frac{n\,(N-n)}{N\,(N-1)\,(N-2)\,(N-3)} \left\{ [N\,(N+1) - 6n\,(N-n)] \sum_{i=1}^{N} c_i^4 + \right. \tag{2.13}$$
$$\left. + 3\,(N-n-1)\,(n-1) \left[\sum_{i=1}^{N} c_i^2\right]^2 \right\}\,.$$

Applying (2.13) to (2.4) we obtain

$$P\left(\max_{1\le k\le n}|Y_1+\cdots+Y_k|>\varepsilon\right.$$

$$\le\left\{\frac{n}{N}\left(1-\frac{n}{N}\right)\sum_{i=1}^{N}c_i^4+3\left[\frac{n}{N}\left(1-\frac{n}{N}\right)\sum_{i=1}^{N}c_i^2\right]^2\right\}\frac{1+o(1)}{\varepsilon^4\left(1-\dfrac{n}{N}\right)^4},\qquad(2.14)$$

$(o\,(1)$ refers to $n\to\infty$, $N-n\to\infty)$.

Formula (2.14) will be useful in the sequel.

3. Kolmogorov-Smirnov type test for regression alternatives

Let the observations $X_1,\ldots,X_N$ be decomposed as

$$X_i=\alpha+\beta c_i+E_i,\qquad(i=1,\ldots,N),\qquad(3.1)$$

where α and β are unknown parameters, $c_1,\ldots,c_N$ are some known (or approximately known) constants and $E_1,\ldots,E_N$ are independent random variables with common but unknown continuous distribution.

We suggest the following statistic for testing the hypothesis $\beta=0$ against the alternative $\beta>0$:

First, arrange the observations according to ascending magnitude,

$$X_{D_1}<X_{D_2}<\cdots<X_{D_N}.\qquad(3.2)$$

Then form the corresponding sequence of the c_i's,

$$c_{D_1},\,c_{D_2},\,\ldots,\,c_{D_N}\qquad(3.3)$$

and compute the statistic

$$K=\frac{\displaystyle\max_{1\le k\le N}(k\bar c-c_{D_1}-\cdots-c_{D_k})}{\left[\displaystyle\sum_{i=1}^{N}(c_i-c)^2\right]^{\frac{1}{2}}}\qquad(3.4)$$

where

$$\bar c=\frac{1}{N}\sum_{i=1}^{N}c_i.\qquad(3.5)$$

In Section 4 we shall show that the limiting distribution of the statistic K coincides with the limiting distribution of the well-known Kolmogorov-Smirnov statistic for the two-sample problem, i.e., that

$$P\,(K>\lambda)\to e^{-2\lambda^2}\qquad(3.6)$$

in large samples under conditions of Corollary 4.1.

Example. Let us have model

$$X_i=\alpha+\beta i+E_i,\qquad(i=1,\ldots,g),\qquad(3.7)$$

yielding the following observations:

i	1	2	3	4	5	6	7	8	9
x_i	124	131	134	127	128	140	136	149	137

Rearranging the observations, we get

x_{D_i}	124	127	128	131	134	136	137	140	149
c_{D_i}	1	4	5	2	3	7	9	6	8

Now $\bar{c} = 5$ and the successive sum of c_{D_i}'s are as follows:

k	$5\,k - c_{D_1} - \cdots - c_{D_k}$
1	4
2	5
3	5
4	8
5	10
6	8
7	4
8	3
9	0

On the other hand,

$$\sum_{i=1}^{9} (c_i - \bar{c})^2 = 2\,(1^2 + 2^2 + 3^2 + 4^2) = 60$$

so that

$$K = \frac{10}{\sqrt{60}} = 1.29 \; .$$

If we correct for discontinuity, then

$$K^* = \frac{9.5}{\sqrt{60}} = 1.23 \; .$$

The critical values for $\alpha = 0.05$ is $K_{0.05} = 1.22$, so that the result is significant on this level.

In conclusion let us show that for

$$c_i = 0 \text{ for } i = 1, \ldots, n, \tag{3.8}$$
$$= 1 \text{ for } i = n + 1, \ldots, N$$

the K-test coincides with the usual Kolmogorov-Smirnov tests for two samples of sizes n and $N - n$, respectively. Actually for the c_i's given by (3.8), we have

$$\bar{c} = \frac{N - n}{N} \tag{3.9}$$

and

$$\sum_{i=1}^{N} (c_i - \bar{c})^2 = \frac{n\,(N - n)}{N} \; . \tag{3.10}$$

Moreover,

$$k\bar{c} - c_{D_1} + \cdots + c_{D_k} = (N - n)\,[S_N\,(X_{D_k}) - S_{N-n}\,(X_{D_k})] \; ,$$

where $S_N\,(x)$ and $S_{N-n}\,(x)$ denote the empirical distribution functions corresponding to the total sample and the second sample (of size $N - n$), respectively. As, furthermore,

$$NS_N(x) = nS_n(x) + (N - n) S_{N-n}(x) ,$$

where $S_n(x)$ corresponds to the first sample, we also have

$$k\bar{c} - (c_{D_t} - \cdots - c_{D_k}) = \frac{n(N-n)}{N} [S_n(X_{D_k}) - S_{N-n}(X_{D_k})] .$$

Consequently, in view of (3.4),

$$K = \left| \frac{n(N-n)}{N} \right|^{\frac{1}{2}} \max_{1 \le k \le N} | S_n(X_{D_k}) - S_{N-n}(X_{D_k})|$$

$$= \left| \frac{n(N-n)}{N} \right|^{\frac{1}{2}} \max_{-\infty < x < \infty} | S_n(x) - S_{N-n}(x) |$$

because the maximum is attained at some of the points $X_{D_1}, \ldots, X_{D_N}$.

4. Limiting distribution

Let us consider a sequence of testing problems with N_ν observations and regression constants c_{ν_i}, $i = 1, \ldots, N_\nu + x$ $\nu = 1, 2, \ldots$. For simplicity assume that

$$\sum_{i=1}^{N_\nu} c_{\nu i} = 0 , \qquad (1 \le \nu < \infty) , \tag{4.1}$$

$$\sum_{i=1}^{N_\nu} c_{\nu i}^2 = 1 , \qquad (1 \le \nu < \infty) , \tag{4.2}$$

and, furthermore,

$$\lim_{\nu \to \infty} \max_{1 \le i \le N_\nu} c_{\nu i}^2 = 0 . \tag{4.3}$$

The last condition is usually called the Noether condition.

Now let us recall briefly some results of [5]. Denote

$$V_{\nu i} = X_{\nu D_{\nu i}} , \qquad (1 \le i \le N_\nu) , \tag{4.4}$$

and, inversely,

$$X_{\nu i} = V_{\nu R_{\nu i}} , \qquad (1 \le i \le N_\nu) . \tag{4.5}$$

The vector $(V_{\nu 1}, \ldots, V_{\nu N_\nu})$, $V_{\nu 1} < \cdots < V_{\nu N_\nu})$ is usually called the ordered sample, and the vector $(R_{\nu 1}, \ldots, R_{\nu N_\nu})$ is the vector of ranks. Given a square integrable function $\varphi(u)$, $0 \le u \le 1$, and an N_ν, let us define

$$\varphi_\nu^0 \left(\frac{i}{N_\nu + 1} \right) = N_\nu \int_{(i-1)/N_\nu}^{i/N_\nu} \varphi(u) \, du , \qquad (1 \le i \le N_\nu) , \tag{4.6}$$

and introduce statistics

$$S_\nu^0(\varphi) = \sum_{i=1}^{N_\nu} c_{\nu i} \varphi_\nu^0 \left(\frac{R_{\nu i}}{N_\nu + 1} \right) , \qquad (1 \le \nu < \infty) . \tag{4.7}$$

From Corollary 2.1 of [5] we have

Lemma 4.1. Let conditions (4.1), (4.2) and (4.3) be satisfied, and let $\varphi(u)$, $0 \le u \le 1$, be a square integrable function. Then the distribution of the statistic $S_\nu^0(\varphi)$ given by (4.7) with φ_ν^0 defined by (4.6) is asymptotically normal with zero mean and variance

$$\sigma^2(\varphi) = \int_0^1 \varphi^2(u)\, du - [\int_0^1 \varphi(u)\, du]^2. \tag{4.8}$$

In this paper we shall be interested in functions $\varphi_t(u)$ defined as follows.

$$\varphi_t(u) = 0 \text{ if } 0 \leq u \leq t, \tag{4.9}$$
$$= 1 \text{ if } t < u \leq 1.$$

Obviously,

$$\begin{aligned}
S_\nu^0(\varphi_t) &= \sum_{i=1}^{N_\nu} c_{\nu i}\, \varphi_{t\nu}^0\left(\frac{R_{\nu i}}{N_\nu + 1}\right) \\
&= \sum_{i=1}^{N_\nu} c_{\nu D_{\nu i}}\, \varphi_{t\nu}^0\left(\frac{i}{N_\nu + 1}\right) \\
&= \sum_{i=k+1}^{N_\nu} c_{\nu D_{\nu i}} + N_\nu\left(\frac{k}{N_\nu} - t\right) c_{\nu D_{\nu k}}, \qquad \left(\frac{k-1}{N_\nu} \leq t \leq \frac{k}{N_\nu}\right).
\end{aligned} \tag{4.10}$$

Or, equivalently, in view of (4.1),

$$S_\nu^0(\varphi_t) = -\sum_{i=1}^{k-1} c_{\nu D_{\nu i}} - N_\nu\left(t - \frac{k-1}{N_\nu}\right) c_{\nu D_{\nu k}}, \qquad \left(\frac{k-1}{N_\nu} \leq t \leq \frac{k}{N_\nu}\right). \tag{4.11}$$

Consider now a sequence of stochastic processes $\{X_\nu(t), 0 \leq t \leq 1\}$ defined as follows:

$$X_\nu(t) = S_\nu^0(\varphi_t), \qquad (0 \leq t \leq 1, 1 \leq \nu < \infty), \tag{4.12}$$

where $S_\nu^0(\varphi_t)$ is given by (4.11). Obviously the process $\{X_\nu(t), 0 \leq t \leq 1\}$ is a mapping of the vector of ranks $(R_{\nu 1}, \ldots, R_{\nu N_\nu})$, or, equivalently, of "antiranks" $(D_{\nu 1}, \ldots, D_{\nu N_\nu})$, into the space $C[0, 1]$ of continuous function, and the paths $X_\nu(\cdot)$ are linear within the intervals $[k/N_\nu, (k + 1)/N_\nu]$, $\nu = 0, \ldots, N_\nu - 1$.

Next we shall show that the processes $\{X_\nu(t), 0 \leq t \leq 1\}$ converge in distribution in $C[0, 1]$ to a process we shall call a Brownian bridge:

Definition 4.1. A stochastic process $\{X(t, \omega), 0 \leq t \leq 1\}$ will be called a Brownian bridge, if it is Gaussian with parameters

$$E[X(t)] = 0, \qquad\qquad (0 \leq t \leq 1), \tag{4.13}$$
$$E[X(t)\, X(s)] = t(1 - s), \qquad\qquad (0 \leq t \leq s \leq 1), \tag{4.14}$$

and if all paths $X(\cdot, \omega)$ are continuous functions on $[0, 1]$.

In terms of the function $\varphi_t(u)$ introduced by (4.9) the covariance (4.14) may be expressed as follows:

$$E[X(t)\, X(s)] = \int_0^1 \varphi_t(u)\, \varphi_s(u)\, du - [\int_0^1 \varphi_t(u)\, du][\int_0^1 \varphi_s(u)\, du]. \tag{4.15}$$

Theorem 4.1. Under conditions (4.1) through (4.3) the sequence of processes $\{X_\nu(t), 0 \leq t \leq 1\}$ defined by (4.12) and (4.11), converge in distribution in $C[0, 1]$ to a Brownian bridge defined above.

Proof. According to Theorem 1.1, we have first to prove that (1.1) holds true. In view of Remark 1.4, (1.1) will be proved if we show that for any n and $t_1, \ldots, t_n \in [0, 1]$, $-\infty < \lambda_1, \ldots, \lambda_n < \infty$,

$$\mathscr{L}\left[\sum_{i=1}^n \lambda_i X_\nu(t_i)\right] \to \mathscr{L}\left[\sum_{i=1}^n \lambda_i X(t_i)\right], \qquad \text{as } \nu \to \infty. \tag{4.16}$$

However, in view of (4.13) through (4.15), $\sum_{i=1}^n \lambda_i X(t_i)$ is normal with zero mean and variance

$$\sigma^2(\psi) = \int_0^1 \psi^2(u)\, du - [\int_0^1 \psi(u)\, du]^2 \tag{4.17}$$

where

$$\psi(u) = \sum_{i=1}^n \lambda_i \varphi_{t_i}(u), \qquad 0 \leq u \leq 1. \tag{4.18}$$

On the other hand, from (4.12) and (4.11) it follows that

$$\sum_{i=1}^n \lambda_i X_\nu(t_i) = S_\nu^0(\psi) \tag{4.19}$$

where ψ is defined by (4.18). Thus (4.16) follows simply by Lemma 4.1.

Now we need to show that condition (1.9) is satisfied. In view of (4.11)

$$X_\nu\left(\frac{k-1}{N_\nu}\right) - X_\nu\left(\frac{j}{N_\nu}\right) = \sum_{i=k}^j c_{\nu D_{\nu i}} = \sum_{i=1}^{j-k+1} Y_{\nu i} \tag{4.20}$$

where we have put, for simplicity,

$$Y_{\nu i} = c_{\nu D_{\nu k+i-1}}, \qquad (1 \leq i \leq j-k+1). \tag{4.21}$$

Since $c_{\nu D_{\nu 1}}, \ldots, c_{\nu D_{\nu N_\nu}}$ represents a random permutation of $\{c_{\nu 1}, \ldots, c_{\nu N_\nu}\}$, $Y_{\nu 1}, \ldots, Y_{\nu j-k+1}$ represents a simple random sample without replacement from the population $\{c_{\nu 1}, \ldots, c_{\nu N_\nu}\}$. Consequently, by Lemma 2.1 and (2.14), we have

$$P\left[\max_{k \leq j \leq k+m_{\nu\delta}} |X_\nu\left(\frac{k-1}{N_\nu}\right) - X_\nu\left(\frac{j}{N_\nu}\right)| > \varepsilon\right]$$

$$= P\left(\max_{1 \leq i < m_{\nu\delta}} |Y_{\nu 1} + \cdots + Y_{\nu i}| > \varepsilon\right) \tag{4.22}$$

$$\leq \left\{\frac{m_{\nu\delta}}{N_\nu}\left(1 - \frac{m_{\nu\delta}}{N_\nu}\right)\sum_{i=1}^{N_\nu} c_{\nu i}^4 + 3\left|\frac{m_{\nu\delta}}{N_\nu}\left(1 - \frac{m_{\nu\delta}}{N_\nu}\right)\sum_{i=1}^{N_\nu} c_{\nu i}^2\right|^2\right\} \cdot$$

$$\cdot \frac{1 + o(1)}{\varepsilon^4\left(1 - \dfrac{m_{\nu\delta}}{N_\nu}\right)^4}$$

where $o(1)$ refers to

$$m_{\nu\delta} \to \infty, \quad N_\nu - m_{\nu\delta} \to 0, \text{ as } \nu \to \infty, \tag{4.23}$$

which is satisfied for every $0 < \delta < 1$, in view of (1.10). Now, utilizing the conditions (4.2), we get

$$P\left[\max_{k\leq j<k+m_{\nu\delta}} \Big| X_\nu\left(\frac{k-1}{N_\nu}\right) - X_\nu\left(\frac{j}{N_\nu}\right)\Big| > \varepsilon\right]$$

$$< \frac{\delta \max_{1\leq i\leq N_\nu} c_{\nu i}^2 + 3\,\delta^2\,(1-\delta)}{\varepsilon^4\,(1-\delta)^3}\,[1 + o_\nu(1)] \tag{4.24}$$

where $o_\nu(1) \to 0$ if $\nu \to \infty$. The right side being independent of k, we have, in view of (4.3)

$$\frac{1}{\delta}\limsup_{\nu\to\infty}\ \max_{1\leq k\leq N_\nu - m_{\nu\delta}+1}\ P\left[\max_{k\leq j<k+m_{\nu\delta}}\Big| X_\nu\left(\frac{k-1}{N_\nu}\right) - X_\nu\left(\frac{j}{N_\nu}\right)\Big| > \varepsilon\right]$$

$$= \frac{\delta}{\varepsilon^4\,(1-\delta)^2}\,. \tag{4.25}$$

Consequently the condition (1.9) is satisfied and the proof is thereby terminated.

Corollary 4.1. Assume that

$$\lim_{\nu\to\infty}\frac{\max_{1\leq i\leq N_\nu}(c_{\nu i} - \bar c_\nu)^2}{\sum_{i=1}^{N_\nu}(c_{\nu i} - \bar c_\nu)^2} = 0 \tag{4.26}$$

and consider the statistics K_ν defined by (3.4). Then

$$\lim_{\nu\to\infty} P\,(K_\nu < \lambda) = 1 - e^{-2\lambda^2}\,, \qquad (\lambda > 0)\,, \tag{4.27}$$

and

$$\lim_{\nu\to\infty} P\,(\,|\,K_\nu\,| < \lambda) = \sum_{r=-\infty}^{\infty} (-1)^k\, e^{-2k^2\lambda^2}\,, \qquad (\lambda > 0)\,. \tag{4.28}$$

Proof. In view of (4.12) and (4.11),

$$K_\nu = \max_{0\leq t\leq 1} X_\nu(t) = h_M\,[X_\nu(\cdot)]\ \ [h_M \text{ is given by } (1.3)] \tag{4.29}$$

and

$$|\,K_\nu\,| = \max_{0\leq t\leq 1}|\,X_\nu(t)\,| \tag{4.30}$$

which are continuous functionals in $C\,[0, 1]$ applied to sample paths $X_\nu(\cdot)$. The rest follows from the definition of convergence in distribution in $C\,[0, 1]$ and the well-known formulas for the Brownian bridge $X(t)$ [see DOOB (1949)]:

$$P\,[\max_{0\leq t\leq 1} X(t) < \lambda] = 1 - e^{-2\lambda^2}\,, \tag{4.31}$$

and

$$P\,[\max_{0\leq t\leq 1}|\,X(t)\,| < \lambda] = \sum_{k=-\infty}^{\infty} (-1)^k\, e^{-2k^2\lambda^2}\,. \tag{4.32}$$

Remark 4.1. (4.16) is sufficient for (1.1) because it implies the convergence of corresponding characteristic functions:

$$\varphi_\nu \left(\lambda_1, \ldots, \lambda_n\right) = E \left\{\exp\left[i \sum_{i=1}^{n} \lambda_i\, X_\nu\left(t_i\right)\right]\right\}$$
$$\to \varphi\left(\lambda_1, \ldots, \lambda_n\right) = E \left\{\exp\left[i \sum_{i=1}^{n} \lambda_i\, X\left(t_i\right)\right]\right\}.$$

$$(4.33)$$

References

[1] BARTOSZYNSKI, L.: On a certain characterization of measures in C [0, 1]. Bull. Acad. pol. Sci. Cl. 3. **10**, No. 8, 445 (1962).

[2] DONSKER, M. D.: Justification and extension of Doob's heuristic approach to the Kolmogorov-Smirnov theorems. Ann. Math. Statist. **23**, 277 (1952).

[3] DOOB, J. L.: Heuristic approach to the Kolmogorov-Smirnov theorems. Ann. Math. Stat. **20**, 393 (1949).

[4] DRIML, M.: Convergence of compact measures on metric spaces. Transactions of the 2nd Prague Conf. Inf. Th. **1959**, 71.

[5] HÁJEK, J.: Asymptotically most powerful rank-order tests. Ann. Math. Statist. **33**, 1124 (1962).

[6] ISSERLIS, L.: On the moment distributions of moments in the case of samples drawn from a limited universe. Proc. roy. Soc. **132**, 586 (1931).

[7] LE CAM, L.: Convergence in distribution of stochastic processes. Univ. Calif. Publ. Statistics. **2**, No. 11, 207 (1957).

[8] PROHOROV, JU. V.: Convergence of stochastic processes and limiting theorems of probability theory. Probability Theory **1**, 177 (1956).

[9] — The method of characteristic functionals. Proc. Fourth Berkeley Symposium on Mathematical Statistics and Probability **1961**, 403.

[10] SMIRNOV, N. V.: Estimation of the difference of two empirical distribution functions from two independent samples (in Russian). Bull. Mosk, Univ. **2**, 3 (1939).

[11] ŠPAČEK, A.: Regularity properties of random transforms. Czech. Math. J. **5**, 143 (1955).

[12] VARADARAJAN, V. S.: Convergence of stochastic processes. Thesis. Indian Stat. Institute 1958.

First-Passage Percolation, Subadditive Processes, Stochastic Networks, and Generalized Renewal Theory

By J. M. HAMMERSLEY and D. J. A. WELSH

Oxford University

1. Miscellaneous problems involving stochastic networks

1.1. *Origins of first-passage percolation problems.* In 1957, BROADBENT and HAMMERSLEY gave a mathematical formulation of percolation theory. Since then much work has been done in this field and has now led to first-passage percolation problems. In the following two examples we contrast the early formulation with its more recent developments.

Example 1.1.1. Suppose the trees in a large orchard are planted at the vertices of a square lattice, that the distance between trees is such as to make it possible for a diseased tree to infect *only* its four nearest neighbors, and that, moreover, an infected tree has (independently for each neighbor), a probability p of infecting that neighbor. Percolation theory is concerned with the probability $P_N(p)$, that under these conditions disease from one given tree will spread to more than N other trees.

Example 1.1.2. Suppose more generally that a given tree once infected will not infect a given neighboring tree until time u (possibly infinite) has elapsed, where u is a nonnegative random variable independently but identically distributed for each pair of neighboring trees. First-passage percolation theory considers the time at which infection first spreads outside a given region.

Still more generally, first-passage percolation theory considers the following problem. Let g be a connected graph with a countable set of nodes $\{x_i\}_{i=0}^{\infty}$ and arcs $\{l_{ij}\}$ joining x_i to x_j. For given $i, j,$ the arc l_{ij} may or may not exist. To each arc l_{ij} which does exist we assign a nonnegative random variable u_{ij} drawn from a distribution U_{ij} and we may regard u_{ij} as the time taken for an abstract particle to travel from x_i to x_j (or in the opposite direction if the arc l_{ij} is not directed). What then is the shortest time of travel between two prescribed nodes (say i_0 and j_0) of the graph. This shortest time, $t(i_0, j_0)$ is called the *first-passage time* between these nodes. This first-passage time is a random variable, as will be shown rigorously in Section 2. The distribution of $t(i_0, j_0)$ will depend on which distributions U_{ij} are used. The bulk of the work below is devoted to determining properties of these first-passage times in certain simple idealized situations.

1.2. *The shortest route problem and PERT networks.* A well-known problem in graph theory is

Example 1.2.1. The shortest route problem. We are given a connected graph g with nodes $\{x_i\}_{i=0}^{\infty}$ and arcs $\{l_{ij}\}$, which may or may not exist, and we assign to each existing arc a number $c\,(l_{ij}) \geq 0$ which is called the *length* of l_{ij}. How can we find a path μ from a vertex x_0 to a vertex x_n such that the total length $\Sigma_{l \varepsilon \mu}\, c\,(l)$ should be as small as possible?

This problem has many applications: for example, let X be a set of localities and L a set of roads connecting these localities. We shall suppose that all road intersections have been included in X. For a given road l the number $c\,(l)$ may signify its length in miles, the cost of traveling over it, or the time taken to travel along it (with due allowance made for traffic conditions); and we have to look for the shortest, cheapest, or quickest means of traveling between two localities.

First-passage percolation problems on a graph are merely the extension that includes the case where $c\,(l)$ is a random variable. There are many algorithms for selecting the shortest route in the nonrandom version of this problem. See the review article by POLLACK and WIEBENSON (1960). It is an interesting and much more difficult problem to decide a) if there exists an optimal general strategy when the lengths are random variables and b) what is this strategy?

Alternatively one may search for the *kth* best route through a network. Solutions of this problem have important applications. If for any reason the shortest (best) route is unavailable, then alternate routes are desirable. Or, it may be acceptable in some problems to use any route whose "length" is within, say, 10 per cent of the shortest route. Such alternate routes are useful in road traffic studies or in determining alternate message routes in communication networks and in information retrieval. See KOCHEN et al. (1962) and POLLACK (1963).

1.3. *PERT networks and critical paths.* MALCOLM et al. (1959) introduced the concept of a PERT network as a technique for measuring and controlling development progress for the Polaris Fleet Ballistic Missile Programme. Since then similar techniques have been introduced by many organizations and a vast literature of a not very mathematical nature has sprung up. See BIGELOW (1962) for a bibliography.

A PERT network is a directed acyclic graph. In the PERT model such a network represents a partial ordering of the many individual *jobs* (arcs of the network) that together comprise some *project* (the complete network), the partial ordering coming from the requirement that all inward pointing jobs at a node must be completed before any outward pointing job at the node can be started. If jobs are assigned duration times (that is, arcs are assigned lengths), the length of the longest path from the origin to the terminal represents the duration time of the entire network. Such a path is called a *critical path*.

One of the fundamental problems in this work is the estimation of the expected critical path length. The usual practice in this situation is to replace the random variables by their expected values, thus reducing the problem to that of a "longest route problem" for which many algorithms exist *provided* the graph is acyclic. This, however, produces an estimate g of the critical length e which in most cases is highly optimistic $(g \leq e)$. FULKERSON (1962) has derived a method for obtaining a better approximation f satisfying $g \leq f \leq e$.

The relation between PERT theory and first-passage percolation is as follows. Most of the algorithms for solving the shortest route problem (Example 1.2.1) can be readily adopted to the solution of a longest route problem on a directed acyclic graph. Hence if a general strategy could be obtained for the randomized shortest route problem it is reasonable to expect that an adaptation of this strategy would solve the PERT network problem. Also, many of the techniques used in first-passage theory are applicable to the PERT network problem.

1.4. *The relation between ordinary and first-passage percolation.* Ordinary percolation theory was originally defined as follows. Let g be an arbitrary connected graph, let x_0 be a specified node (the origin), let each arc of g be closed or open with probability $1 - p$, p respectively, independently for each arc, where p is a fixed constant $0 \leq p \leq 1$. Fluid is supplied at the origin and can only flow along open arcs. The basic problem of percolation theory is to determine $P_N(p)$, the probability that fluid will spread from the origin to at least N nodes of the graph g. Obviously $P_N(p)$ is a nondecreasing function of p such that for all N,

$$0 = P_N(0) \leq P_N(p) \leq P_N(1) = 1 . \tag{1.4.1}$$

A closely linked problem to the above is to obtain the probability, $P_p(x_i)$, that under the same conditions fluid will spread from the origin to the node x_i of g. We may frame this problem in terms of first-passage percolation theory as follows.

Let the arcs of g have "length" 0 or 1 with probability p, and $1 - p$ respectively. Then the expected first-passage time between x_0 and x_i is the expected minimum number of arcs of g which have to be opened for fluid from x_0 to reach x_i. Or, we may say that $P_p(x_i)$ is the probability that the first-passage time between x_0 and x_i is zero.

For a general review of ordinary percolation theory and its many physical applications, see FRISCH and HAMMERSLEY (1963).

1.5. *First-passage percolation theory as a generalization of renewal theory.* In Example 1.1.2 instead of studying the first-passage time of infection between two trees of the orchard we could equally well consider the inverse problem of finding the maximum distance infection will have spread in time not exceeding t. This maximum distance problem on a graph is a generalization of ordinary renewal theory. For, consider the special case when the underlying graph g has nodes at the nonnegative integer points of the real line, while its arcs are merely directed lines from $(i, 0)$ to $(i + 1, 0)$. If these arcs have lengths drawn independently from a fixed distribution U, then the expected maximum distance traveled in time not exceeding t is exactly the renewal function associated with U. See W. L. Smith (1958).

This problem is treated in Section 5 and it will be seen that elementary renewal type theorems also hold for various regular graphs. A more difficult unsolved problem is to define a randomly irregular graph.

2. General formulation of first-passage percolation

2.1. *The probability space* (Ω, B, P). Consider a countably infinite connected graph g with arcs $\{l_i\}_{i=1}^{\infty}$ and nodes $\{P_{ij}\}$. The node P_{ij}, if it exists, is the intersection of l_i, l_j. To each arc l_i independently assign a nonnegative random variable u_i drawn from an associated distribution U_i. This u_i is called the *time coordinate* of l_i. The sequence $\{u_i\}_{i=1}^{\infty}$ defines a configuration of time coordinates on g which may be represented by an infinite vector $\omega \equiv (u_1, u_2, \ldots)$ called the *time state* of g. The set of all possible time states of g forms a set which we call the *phase space* and usually denote by Ω_g.

If $r \equiv l_{i_1}, l_{i_2}, \ldots, l_{i_n}$ is any connected path on g (such a path is *connected* if and only if $P_{i_j i_{j+1}}$ exists for all $j = 1, 2, \ldots, n - 1$), then the *time coordinate of r under ω* is defined by

$$t(r, \omega) = u_{i_1} + u_{i_2} + \ldots + u_{i_n}. \tag{2.1.1}$$

Hereafter, path will always mean a connected path.

Further, if R is any nonnull set of paths on g, the first-passage time of R under ω is defined by

$$t_R(\omega) = \inf_{r \in R} t(r, \omega). \tag{2.1.2}$$

If there exists an $r_0 \in R$ such that $t(r_0, \omega) = t_R(\omega)$ we say that the first-passage time $t_R(\omega)$ has a *route* and this route is r_0. To prove the existence of a route is in general a difficult unsolved problem. However, in the majority of practical cases, the route can be shown either to exist absolutely or to exist with probability 1. In the latter case we can replace Ω_g by the relevant subset of probability 1. Accordingly, throughout the remainder of this section we shall assume the existence of the route of all first-passage times considered.

When the time coordinates $\{u_i\}$ are drawn from probability spaces $U_i = (\Omega_i, B_i, P_i)$ then any time state ω can be regarded as an elementary sample point of the probability space (Ω_g, B, P) where Ω_g is the countably infinite cartesian product of the Ω_i, B is the σ-field of subsets of Ω_g derived from the σ-fields B_i, and P is the induced probability measure. We now show that $t_R(\omega)$ is a random variable on (Ω_g, B, P).

Theorem 2.1.3. $t_R(\omega)$ *is a measurable function on* (Ω_g, B, P).

Proof. Let r be any path of R: say $r \equiv (l_{i_1}, l_{i_2}, \ldots, l_{i_k})$.

Then $t(r, \omega) = u_{i_1} + u_{i_2} + \ldots + u_{i_k}$ is measurable on (Ω_g, B, P). Since $\{l_i\}$ is a countable set, the set of paths with exactly n arcs is a countable set R_n; and hence R, a subset of the union of $R_1, R_2, \ldots$ is also countable. Hence $t_R(\omega) = \inf_{r \in R} t(r, \omega)$ is also measurable on (Ω_g, B, P).

2.2. *Equivalence under lateral shift.* Two sets of paths R_1 and R_2 are said to be *equivalent under lateral shift* if there exists a one-to-one mapping f of R_1 onto R_2 such that for every subset of paths $r_1, r_2, \ldots, r_k$ belonging to R_1, the joint distribution of $t(r_1, \omega), t(r_2, \omega), \ldots, t(r_k, \omega)$ is identical with the joint distribution of $t(f(r_1), \omega), t(f(r_2), \omega), \ldots, t(f(r_k), \omega)$.

Theorem 2.2.1. *If R_1 and R_2 are sets of paths equivalent under lateral shift, $t_{R_1}(\omega)$ and $t_{R_2}(\omega)$ are identically distributed random variables.*

Proof. Let $r_1, r_2, \ldots$ be an enumeration of R_1. For prescribed $u \geq 0$, define

$$\Omega_k = \left\{\omega: \inf_{1 \leq i \leq k} t(r_i, \omega) \leq u\right\}$$
$$\Omega_k^f = \left\{\omega: \inf_{1 \leq i \leq k} t[f(r_i), \omega] \leq u\right\}. \tag{2.2.2}$$

Then Ω_k and Ω_k^f are nondecreasing set functions of k for which $P(\Omega_k) = P(\Omega_k^f)$ by the invariance hypothesis. Hence

$$P\left(\lim_{k \to \infty} \Omega_k\right) = P\left(\lim_{k \to \infty} \Omega_k^f\right) \tag{2.2.3}$$

and as u varies these two quantities are the distribution functions of $t_{R_1}(\omega)$ and $t_{R_2}(\omega)$.

Corollary 2.2.4. *Under the hypothesis of Theorem 2.2.1,* $Et_{R_1}(\omega) = Et_{R_2}(\omega)$.

Example 2.2.5. Let g be the square lattice on the Euclidean plane, R_1 the set of paths connecting the origin to $(m, 0)$, and R_2 the set of paths connecting $(a, 0)$ to $(a + m, 0)$. Suppose that the U_i-distribution associated with any arc is independent of that arc. Then R_1 and R_2 are equivalent under lateral shift.

2.3. *Inclusion and connection.* In this section we mention some rather obvious lemmas used in the comparison of first-passage times over different subsets of paths of g.

Lemma 2.3.1. (The inclusion lemma.) *If R_1 and R_2 are two sets of paths on g such that $R_1 \subset R_2$, where $\subset$ has its usual set theoretic meaning, then $t_{R_1}(\omega) \geq t_{R_1}(\omega)$.*

Proof. When $R_1 \subset R_2$, $\inf_{r \in R_1} t(r, \omega) \geq \inf_{r \in R_2} t(r, \omega)$.

Two paths r_1, r_2 are said to be *connected* if they have a common end point. If $r_1 \equiv (P_1, P_2, \ldots, P_n = Q)$ (that is, if $P_1, P_2, \ldots, P_n$, are the successive nodes on r_1), and $r_2 \equiv (Q = Q_1, Q_2, \ldots, Q_m)$ are connected, their *connection* $r_1 * r_2$ is defined to be the path $(P_1, P_2, \ldots, P_n = Q = Q_1, Q_2, \ldots, Q_m)$. Similarly, two sets of paths R_1, R_2, are said to be connected with connection $R_3 \equiv R_1 * R_2$ if for arbitrary $r_1 \in R_1$, $r_2 \in R_2$, we have $r_3 = r_1 * r_2 \in R_3$ and R_3 is made up only of paths of form $r_1 * r_2$. With this terminology we have

Lemma 2.3.2 (The connection lemma). *If R_1, R_2, are connected sets of paths, then*

$$t_{R_1}(\omega) + t_{R_2}(\omega) \geq t_{R_1 * R_2}(\omega).$$

The proof of the lemma is quite trivial; for, if r^1 and r^2 are the routes of $t_{R_1}(\omega)$ and $t_{R_2}(\omega)$ then $r^1 * r^2 \in R_1 * R_2$ and hence

$$t_{R_1 * R_2}(\omega) \leq t(r^1 * r^2, \omega) = t(r^1, \omega) + t(r^2, \omega) = t_{R_1}(\omega) + t_{R_2}(\omega).$$

This lemma is used mostly where $R_1 * R_2 \subset R_3$, for then by a combination of the above two lemmas

$$t_{R_1}(\omega) + t_{R_2}(\omega) \geq t_{R_3}(\omega). \tag{2.3.3}$$

3. Subadditive stochastic processes

3.1. *Definitions.* In our study of first-passage percolation we shall need a new concept — subadditive stochastic processes. We deal with a fixed probability space (Ω, B, P) in which ω denotes a typical sample point. On this space we define a family of real random variables $\{x_{st}(\omega)\}$ indexed by the pair of nonnegative integers s and t, where $s \leq t$. We call such a family a *subadditive process* if conditions (3.1.1), (3.1.2) and (3.1.3) hold:

$$x_{rt}(\omega) \leq x_{rs}(\omega) + x_{st}(\omega), \qquad (r \leq s \leq t); \tag{3.1.1}$$

$x_{st}(\omega)$ is *stationary* in the sense that its distribution depends only on the difference $t - s$; $\hspace{5cm}$ (3.1.2)

and

$M + Nt \leq g_t = Ex_{0t}(\omega) \leq E \mid x_{0t}(\omega) \mid < \infty$, where g_t is defined by this relation, and M and N are constants (possibly negative). (3.1.3)

Associated with this subadditive process there is a constant γ defined by

$$\gamma = \inf_t g_t/t. \tag{3.1.4}$$

This constant is finite by virtue of the left side of (3.1.3). We call γ the *time constant* of the process.

3.2. *Examples.* Let X_0, X_1, ... be a sequence of independent identically distributed random variables, each with the same finite mean γ. Then

$$x_{st} = \begin{cases} 0 & \text{if } s = t \\ X_s + X_{s+1} + \ldots + X_{t-1} & \text{if } s < t \end{cases} \qquad (3.2.1)$$

is a subadditive process with time constant γ.

The second example is less trivial. Distribute straight lines on the Euclidean plane uniformly and independently at random (that is, their directions are uniformly and independently distributed between 0 and 2π, and their perpendicular distances from the origin are the points of a Poisson process on the positive reals). These lines dissect the plane into convex polygons. Let K be some fixed positive constant, and let R_{st} be the rectangle with vertices $(s, \pm K)$ and $(t, \pm K)$. Let the number of polygons of some given class (say hexagons whose area is at least A) which intersect R_{st} be x_{st}. Then $\{x_{st}\}$ is a subadditive process.

For the third example let $X(\omega)$ be some given random variable with a nondegenerate distribution and a finite mean. Then $x_{st}(\omega) = (t-s)\,X(\omega)$ is a subadditive process.

3.3. *Blankets and smotherability.* In the first of the foregoing examples

$$x_{st}(\omega)/t \to \gamma \text{ with probability 1 as } t \to \infty \text{ (s fixed)}. \qquad (3.3.1)$$

In the third example this is false: indeed we do not even have convergence in probability. Thus to insure the convergence of $x_{st}(\omega)/t$ we need some ancillary conditions; and these are provided by the definitions of blankets and smotherability below. At first sight these definitions appear artificial. Their justification lies in the fact that smotherability is a necessary condition for convergence with probability 1 and a sufficient condition for convergence in probability. Whether or not smotherability is also sufficient for convergence with probability 1 remains an open question at present. Clearly, in view of the stationarity condition (3.1.2), we need only consider the convergence of $x_{0t}(\omega)/t$ as $t \to \infty$.

We shall say that the subadditive process $x_{st}(\omega)$ has a δ-*blanket* if, for some positive integer n, there exists a sequence $\{y_i(\omega)\}_{i=1}^{\infty}$ of identically distributed mutually independent random variables with common mean $E y_i(\omega) = n\delta$, such that with probability 1

$$x_{0,\,jn}(\omega) \leq y_1(\omega) + y_2(\omega) + \ldots + y_j(\omega), \qquad [j \geq j_0(\omega)]. \qquad (3.3.2)$$

We shall see presently that no δ-blanket can exist with $\delta < \gamma$. On the other hand, it may happen that a sequence of δ_i-blankets exist with $\inf_i \delta_i = \gamma$; and in this case we shall say that the process is *smotherable*.

If each $y_i(\omega)$ in (3.3.2) has the same distribution as $x_{0n}(\omega)$, we shall say that this blanket, for which clearly $\delta = g_n/n$, is a (g_n/n)-*self-blanket* of $x_{st}(\omega)$. If further the process $x_{st}(\omega)$ has (g_n/n)-self-blankets for infinitely many n, we say that the process is *self-smothering*. The following theorem shows that a self-smothering process is indeed smotherable.

Theorem 3.3.3.

$$g_t/t \geq \lim_{t\to\infty} g_t/t = \gamma \; .$$

Proof. If we take expectations in (3.1.1) and use (3.1.2), we get

$$g_{t-r} \leq g_s - r + g_{t-s} \; . \tag{3.3.4}$$

Thus g_t is a subadditive function of t; and the result (including the existence of the limit) follows from the fundamental theorem on subadditive functions [HILLE (1957), CHAPTER 6].

3.4. *Nonnegative vectorial cone of subadditive processes.* Subadditive processes generate a nonnegative vectorial cone in the sense that, if $\{x_{st}(\omega)\}$ and $\{x_{st}^*(\omega)\}$ are any two subadditive processes on (Ω, B, P), then so is $\{\alpha x_{st}(\omega) + \alpha^* x_{st}^*(\omega)\}$ for any pair of nonnegative numbers α and α^*. Further the time constant of a process, $\gamma = \gamma[\{x_{st}(\omega)\}]$, is clearly a linear functional from this nonnegative cone to the real line. Presumably, there is scope for applying standard theorems of functional analysis to this situation; but we shall not do so here.

We say that a subadditive process is *strictly positive* if $x_{st}(\omega) > 0$ for all s, t, ω. The strictly positive subadditive processes generate a positive subcone of the cone of all subadditive processes. Similar considerations apply to *nonnegative* subadditive processes.

3.5. *Convergence theorems.*

Theorem 3.5.1. *If the subadditive process $\{x_{st}(\omega)\}$ satisfies*

$$P\left[\limsup_{t\to\infty} x_{0t}(\omega)/t \leq \gamma\right] = 1 \, , \tag{3.5.2}$$

where γ is the time constant of the process, then

$$P\left[\limsup_{t\to\infty} x_{0t}(\omega)/t = \gamma\right] = 1 \tag{3.5.3}$$

and $x_{0t}(\omega)/t$ converges in probability to γ as $t \to \infty$.

Proof. The Heaviside function, $H(a) = 0$ or 1 according as $a < 0$ or $a \geq 0$, has the properties

$$H(a) \leq H(b) \text{ if } a \leq b \; ; \tag{3.5.4}$$

$$H(a + b) \leq H(a) + H(b) \; ; \tag{3.5.5}$$

and

$$xH(x - \alpha + b) \leq xH(x - \alpha) + \alpha H(b) \text{ if } \alpha \geq 0 \; . \tag{3.5.6}$$

The first two properties are obvious. For the third property, when $0 \leq \alpha \leq x$, we have $xH(x - \alpha + b) \leq x = xH(x - \alpha) \leq xH(x - \alpha) + \alpha H(b)$; and in the remaining case $x < \alpha \geq 0$ we have $xH(x - \alpha + b) \leq \alpha H(x - \alpha + b) \leq \alpha H(b) = xH(x - \alpha) + \alpha H(b)$.

For brevity hereafter we write $x_{st} = x_{st}(\omega)$. Repeated application of (3.1.1) yields

$$x_{0t} \leqq x_{0r} + x_{r,\,r+1} + x_{r+1,\,t} \leqq \sum_{r=0}^{t-1} x_{r,\,r+1} \qquad (0 \leqq r < t) \qquad (3.5.7)$$

with the convention that, in the central expression of (3.5.7), the terms x_{00} and x_{tt} are to be omitted in the cases $r = 0$ and $r = t - 1$ respectively. From (3.5.4) through (3.5.7), we now obtain for any $\alpha \geqq 0$

$$
\begin{aligned}
x_{0t}\,H\,(x_{0t} - t\alpha) &\leqq \sum_{r=0}^{t-1} x_{r,\,r+1}\,H\,(x_{0t} - t\alpha) \\
&\leqq \sum_{r=0}^{t-1} x_{r,\,r+1}\,H\,(x_{0r} + x_{r,\,r+1} + x_{r+1,\,t} - t\alpha) \\
&\leqq \sum_{r=0}^{t-1} \left\{ x_{r,\,r+1}\,H\,(x_{r,\,r+1} - \alpha) + \alpha H\,[x_{0r} + x_{r+1,\,t} - (t-1)\alpha] \right\} \\
&\leqq \sum_{r=0}^{t-1} \left\{ x_{r,\,r+1}\,H\,(x_{r,\,r+1} - \alpha) + \alpha H\,(x_{0r} - r\alpha) + \right. \\
&\qquad \left. + \alpha H\,[x_{r+1,\,t} - (t - r - 1)\,\alpha] \right\}
\end{aligned}
\qquad (3.5.8)
$$

with the understanding that $\alpha H\,(x_{0r} - r\alpha)$ is omitted when $r = 0$ and $\alpha H\,[x_{r+1,\,t} - (t - r - 1)\alpha]$ is omitted when $r = t - 1$.

Next let $G_r\,(x)$ denote the cumulative distribution function of $x_{0r}\,(\omega)$; and define

$$I_r\,(z) = \int_{rz}^{\infty} x\,dG_r\,(x) - E\,[x_{0r}\,H\,(x_{0r} - rz)]\,, \qquad (3.5.9)$$

$$J_r\,(z) = \int_{rz}^{\infty} dG_r\,(x) = EH\,(x_{0r} - rz)\,. \qquad (3.5.10)$$

The stationarity property (3.1.2) allows us to replace x_{0r} by $x_{s,\,r+s}$ in (3.5.9) and (3.5.10) when we wish. Taking expectations of (3.5.8) and recalling the understanding on the two omitted terms, we have

$$I_t\,(\alpha) \leqq t I_1\,(\alpha) + 2\,\alpha \sum_{r=1}^{t-1} J_r\,(\alpha)\,, \qquad (\alpha \geqq 0)\,. \qquad (3.5.11)$$

Now prescribe α, β, ε such that

$$\alpha \geqq 0, \quad \alpha > \beta > \gamma > \varepsilon\,, \qquad (3.5.12)$$

where γ is the time constant of $\{x_{st}\,(\omega)\}$. By (3.5.2), (3.5.10) and (3.5.12) we have

$$J_t\,(\alpha) \to 0 \text{ and } J_t\,(\beta) \to 0 \text{ as } t \to \infty\,. \qquad (3.5.13)$$

Using (3.5.11) we have

$$
\begin{aligned}
t^{-1}\,I_t\,(\beta) &= \left\{ \int_{t\beta}^{t\alpha} + \int_{t\alpha}^{\infty} \right\} t^{-1}\,x\,dG_t\,(x) \\
&\leqq \alpha \int_{t\beta}^{t\alpha} dG_t\,(x) + t^{-1}\,I_t\,(\alpha) \qquad (3.5.14) \\
&\leqq \alpha J_t\,(\beta) + I_1\,(\alpha) + 2\,\alpha t^{-1} \sum_{r=1}^{t-1} J_r\,(\alpha)\,;
\end{aligned}
$$

and now (3.5.13) gives

$$\limsup_{t\to\infty} t^{-1} I_t(\beta) \leqq I_1(\alpha) .\tag{3.5.15}$$

By Theorem 3.3.3

$$\gamma \leqq t^{-1} g_t = t^{-1} \left\{ \int_{-\infty}^{\varepsilon t} + \int_{\varepsilon t}^{\beta t} + \int_{\beta t}^{\infty} \right\} x dG_t(x)$$

$$\leqq \varepsilon \int_{-\infty}^{\varepsilon t} dG_t(x) + \beta \int_{\varepsilon t}^{\beta t} dG_t(x) + t^{-1} I_t(\beta)\tag{3.5.16}$$

$$= (\varepsilon - \beta)[1 - J_t(\varepsilon)] + \beta[1 - J_t(\beta)] + t^{-1} I_t(\beta) ;$$

and hence

$$\gamma - \varepsilon \leqq (\beta - \varepsilon) J_t(\varepsilon) - \beta J_t(\beta) + t^{-1} I_t(\beta) .\tag{3.5.17}$$

In (3.5.17) let $t \to \infty$ through a subsequence of values of t such that $J_t(\varepsilon) \to \liminf_{t\to\infty} J_t(\varepsilon)$. From (3.5.13) and (3.5.15) we get

$$\gamma - \varepsilon \leqq (\beta - \varepsilon) \liminf_{t\to\infty} J_t(\varepsilon) + I_1(\alpha) .\tag{3.5.18}$$

Now let $\alpha \to \infty$, so that $I_1(\alpha) \to 0$ in view of the existence of $E \mid x_1(\omega) \mid$. This gives

$$\liminf_{t\to\infty} J_t(\varepsilon) \geqq (\gamma - \varepsilon)/(\beta - \varepsilon) .\tag{3.5.19}$$

Let $\beta \to \gamma$ from above. Since $J_t(\varepsilon) \leqq 1$ by (3.5.10) we get

$$\lim_{t\to\infty} J_t(\varepsilon) = 1 .\tag{3.5.20}$$

Since (3.5.20) holds for arbitrary $\varepsilon < \gamma$, the convergence of x_{0t}/t to γ in probability follows from (3.5.2). Any sequence which converges in probability contains a subsequence which converges with probability 1. Hence (3.5.3) follows from (3.5.2). This completes the proof.

Theorem 3.5.21. *If $\{x_{st}(\omega)\}$ is a subadditive process (with time constant γ) satisfying*

$$P[\limsup_{t\to\infty} x_{0t}(\omega)/t = \gamma] = 1\tag{3.5.22}$$

and if $\{z_{st}(\omega)\}$ is any strictly positive subadditive process satisfying

$$P[\limsup_{t\to\infty} z_{0t}(\omega)/t = 0] = 1 ,\tag{3.5.23}$$

then $\{x_{st}(\omega) + z_{st}(\omega)\}$ is a self-smothering subadditive process.

Proof. Write $h_t = E z_{0t}(\omega) > 0$, because $z_{st}(\omega)$ is strictly positive. For any fixed positive integer n, let $\{y_i(\omega)\}_{i=1}^{\infty}$ be a sequence of mutually independent random variables each distributed with the distribution of $x_{0n}(\omega) + z_{0n}(\omega)$. Then with probability 1

$$\lim_{j\to\infty} j^{-1} \sum_{i=1}^{j} y_i(\omega) = E\{x_{0n}(\omega) + z_{0n}(\omega)\} = g_n + h_n\tag{3.5.24}$$

$$\geqq \gamma n + h_n > \gamma n \geqq \limsup_{j\to\infty} j^{-1}\{x_{0,jn}(\omega) + z_{0,jn}(\omega)\} .$$

Here we have used the strong law of large numbers, Theorem 3.3.3, and (3.5.22) and (3.5.23). Since strict inequality holds in the middle of (3.5.24), we have with probability 1

$$x_{0,jn}(\omega) + z_{0,jn}(\omega) \leq \sum_{i=1}^{j} y_i(\omega), \qquad [j \geq j_0(\omega)] . \qquad (3.5.25)$$

This affords a self-blanket. Since n is an arbitrary positive integer, $\{x_{st}(\omega) + z_{st}(\omega)\}$ is self-smothering.

Theorem 3.5.26. *If a subadditive process $\{x_{st}(\omega)\}$ has a δ-blanket, then*

$$P\,[\limsup_{t\to\infty} x_{0t}(\omega)/t \leq \delta] = 1 . \qquad (3.5.27)$$

Proof. Fix a nonnegative integer r and let $F(x)$ be the cumulative distribution function of $|x_{0r}(\omega)|$. Fix $\varepsilon > 0$, and with the aid of the stationarity condition (3.1.2) write

$$Q_s = P\,[\,|x_{s,\,s+r}(\omega)|\geq s\varepsilon] = \int_{s\varepsilon}^{\infty} dF(x) . \qquad (3.5.28)$$

Then

$$E\,|x_{0r}(\omega)| = \int_0^{\infty} x\,dF(x) = \sum_{t=0}^{s}\int_{t\varepsilon}^{(t+1)\varepsilon} x\,dF(x) + \int_{(s+1)\varepsilon}^{\infty} x\,dF(x) \qquad (3.5.29)$$

$$\geq \sum_{t=0}^{s} t\,\varepsilon\,(Q_t - Q_{t+1}) + (s+1)\,\varepsilon\,Q_{s+1} = \varepsilon\sum_{t=1}^{s+1} Q_t .$$

The left side of (3.5.29) is finite by (3.1.3) and independent of s. So the series of positive terms $\sum_{t=0}^{\infty} Q_t$ converges; and the Borel-Cantelli lemma applied to (3.5.28) yields

$$P\,[\lim_{s\to\infty} s^{-1}\,|x_{s,\,s+r}(\omega)| = 0] = 1 , \qquad (r \text{ fixed}). \qquad (3.5.30)$$

Now let the δ-blanket be defined as in (3.3.2); and write $t = jn + r$, where j and r are nonnegative integers and $0 \leq r < n$. By (3.1.1) and (3.3.2), we have

$$t^{-1} x_{0t}(\omega) \leq t^{-1} x_{0,jn}(\omega) + t^{-1} x_{jn,\,jn+r}(\omega)$$

$$\leq (j/t)\,j^{-1}\sum_{i=1}^{j} y_i(\omega) + (jn)^{-1}\,|x_{jn,\,jn+r}(\omega)| . \qquad (3.5.31)$$

When $t \to \infty$, $j/t \to 1/n$ and, by the strong law of large numbers, $j^{-1}\sum_{i=1}^{j} y_i(\omega) \to Ey_i(\omega) = n\delta$ with probability 1. The last term in (3.5.31) tends to zero with probability 1 by (3.5.30). [Strictly speaking, r is not fixed in (3.5.31); but r has one of the fixed values $0, 1, \ldots, n-1$, and (3.5.30) can be applied to each of these n values.] This completes the proof.

Theorem 3.5.32. *A smotherable subadditive process $\{x_{st}(\omega)\}$ with a time constant γ satisfies*

$$P\,[\limsup_{t\to\infty} x_{0t}(\omega)/t = \gamma] = 1 . \qquad (3.5.33)$$

Proof. Since the process is smotherable it has δ_i-blankets with $\inf_i \delta_i = \gamma$. Hence (3.5.27) holds for all such δ_i, which implies (3.5.2). The result follows from Theorem 3.5.1.

Theorem 3.5.34. *A subadditive process with time constant γ cannot have a δ-blanket with $\delta < \gamma$.*

Proof. If such a δ-blanket exists, (3.5.27) holds with $\delta < \gamma$. A fortiori, (3.5.2) holds; and this implies (3.5.3) in contradiction to (3.5.27).

Theorem 3.5.35. *Let $\{x_{st}(\omega)\}$ be a subadditive process; and let $\{z_{st}(\omega)\}$ be a strictly positive subadditive process with time constant zero. Then, if $\{x_{st}(\omega) + z_{st}(\omega)\}$ is self-smothering, $\{x_{st}(\omega)\}$ is smotherable.*

Proof. Since $\{z_{st}(\omega)\}$ is strictly positive

$$x_{st}(\omega) < x_{st}(\omega) + z_{st}(\omega) \, . \tag{3.5.36}$$

Therefore every δ-blanket of $\{x_{st}(\omega) + z_{st}(\omega)\}$ is also a δ-blanket of $\{x_{st}(\omega)\}$. However, since $\{x_{st}(\omega) + z_{st}(\omega)\}$ is self-smothering, it has for infinitely many values of n, a δ_n-blanket with

$$\delta_n = (g_n + h_n)/n \to \gamma \text{ as } n \to \infty \, , \tag{3.5.37}$$

where g_n, h_n, and γ have the meanings assigned in the proof of Theorem 3.5.21. These blankets therefore smother $\{x_{st}(\omega)\}$.

We now assemble the foregoing theorems into a single main theorem.

Theorem 3.5.38. *Let $\{x_{st}(\omega)\}$ be any subadditive process with time constant γ; and let $\{z_{st}(\omega)\}$ be any strictly positive smotherable subadditive process with time constant zero. (In particular, $z_{st}(\omega)$ may degenerate to a subadditive function $h_{t-s} = o(t-s)$ as $t-s \to \infty$.) Then the following four statements are equivalent in the sense that any one of them implies the other three:*

$$\{x_{st}(\omega)\} \qquad \text{is smotherable}; \tag{3.5.39}$$

$$\{x_{st}(\omega) + z_{st}(\omega)\} \qquad \text{is self-smothering}; \tag{3.5.40}$$

$$P\,[\limsup_{t \to \infty} x_{0t}(\omega)/t \leqq \gamma] = 1 \, ; \tag{3.5.41}$$

and

$$P\,[\limsup_{t \to \infty} x_{0t}(\omega)/t = \gamma] = 1 \, . \tag{3.5.42}$$

Each of the four is a sufficient condition that $x_{0t}(\omega)/t$ should converge to γ in probability as $t \to \infty$, and a necessary condition for convergence with probability 1. In any case, whatever the behavior of $\{x_{st}(\omega)\}$, we have with probability 1

$$\lim_{t \to \infty} z_{0t}(\omega)/t = 0 \, . \tag{3.5.43}$$

Proof. Since $\{z_{st}(\omega)\}$ is smotherable and has time constant zero, Theorem 3.5.32 shows that (3.5.23) holds. Hence (3.5.42) implies (3.5.40), by Theorem 3.5.21; (3.5.40) implies (3.5.39), by Theorem 3.5.35; and (3.5.39) implies (3.5.42), by Theorem 3.5.32. Also (3.5.42) implies (3.5.41) trivially; and (3.5.41) implies (3.5.42), by Theorem 3.5.1. This proves the equivalence of the four statements. Further (3.5.41) implies the convergence of $x_{0t}(\omega)/t$ in probability, by Theorem 3.5.1; and convergence with probability 1 trivially implies (3.5.42). Finally (3.5.43) follows from (3.5.23) and the fact that $\{z_{st}(\omega)\}$ is strictly positive.

3.6. *Second moments and convergence in quadratic mean.* In general, there is little to say about the properties of the second moment of a subadditive process: for instance, the existence of $E(x_{01}^2)$ does not necessarily imply the existence of $E(x_{0t}^2)$ for $t > 1$, since nothing prevents x_{0t} assuming large negative values too frequently. For nonnegative subadditive processes we can assert something, however. More generally, these same assertions hold for subadditive processes which are bounded below by a linear function of t; since, by adding a linear function of t to such a process we can convert it to a nonnegative subadditive process without altering its variance. In what follows, we shall write

$$v_t = E\{[x_{0t}(\omega)]^2\} - \{E[x_{0t}(\omega)]\}^2 \tag{3.6.1}$$

for the variance of $x_{0t}(\omega)$. By the stationarity condition (3.1.2), v_t is also the variance of $x_{s,\,s+t}(\omega)$. We also write

$$\gamma_t = g_t/t = Ex_{0t}(\omega)/t . \tag{3.6.2}$$

Theorem 3.3.3 shows that $\gamma_t \to \nu$ from above as $t \to \infty$.

Theorem 3.6.3. *If $\{x_{st}(\omega)\}$ is a nonnegative subadditive process for which Ex_{01}^2 exists, then Ex_{0t}^2 exists for $t \geq 1$ and*

$$v_t \leqq (v_1 + \gamma_1^2 - \gamma_t^2)\, t^2 \leqq (v_1 + \gamma_1^2 - \gamma^2)\, t^2 . \tag{3.6.4}$$

Proof. From (3.1.1)

$$x_{0t}(\omega) \leqq \sum_{s=0}^{t-1} x_{s,\,s+1}(\omega) . \tag{3.6.5}$$

Since the process is nonnegative, we may square this inequality and take expected values. This gives

$$v_t + g_t^2 \leqq (v_1 + g_1^2)\, t^2 \tag{3.6.6}$$

because the covariance of two variables with equal variances cannot exceed that common variance. This leads to (3.6.4) on using (3.6.2).

Theorem 3.6.3 shows that $v_t = O(t^2)$ as $t \to \infty$ whenever a nonnegative subadditive process possesses variances. The third example of Section 3.2 shows that no sharper result holds in general. To make further progress we define an *uncorrelated process* to be one in which $x_{qr}(\omega)$ and $x_{st}(\omega)$ are uncorrelated whenever (q, r) and (s, t) are disjoint open intervals. We also say that a process $\{x_{st}^*(\omega)\}$ *dominates* $\{x_{st}(\omega)\}$ if $x_{st}(\omega) \leqq x_{st}^*(\omega)$ for all s, t, ω. Since it only makes sense to talk about an uncorrelated process when the covariances and variances of x_{st} exist, we shall automatically assume that they do when speaking of uncorrelated processes. As Theorem 3.6.3 shows, the existence of Ex_{01}^2 is a sufficient condition for this in a nonnegative process.

We then have the following theorem, in which the constant 16 could be replaced by a somewhat smaller constant at the expense of added complications in the proof.

Theorem 3.6.7. *Let $\{x_{st}(\omega)\}$ and $\{x_{st}^*(\omega)\}$ be nonnegative subadditive processes with the same time constant γ. Suppose that $\{x_{st}(\omega)\}$ is an uncorrelated process and that it dominates $\{x_{st}^*(\omega)\}$. Then the variance v_t of $x_{0t}(\omega)$ satisfies*

$$v_t \leq tv_1 + 16\, t\gamma_1 \sum_{i=1}^{t} (\gamma_i - \gamma). \tag{3.6.8}$$

The variance v_t^* of $x_{0t}^*(\omega)$ satisfies $v_t^* = o\,(t^2)$ as $t \to \infty$. Finally, both $x_{0t}(\omega)/t$ and $x_{0t}^*(\omega)/t$ converge to γ in quadratic mean as $t \to \infty$.

Proof. Let $n = a + b + \ldots + j + k$, where $a, b, \ldots, j, k$ are positive integers. From (3.1.1) we have

$$x_{0n}(\omega) \leq x_{0a}(\omega) + x_{a,a+b}(\omega) + \ldots + x_{a+b+\ldots+j,a+b+\ldots+j+k}(\omega). \tag{3.6.9}$$

If we square this inequality, take expectations, and remember that the terms on the right are uncorrelated, we get

$$v_n + g_n^2 \leq v_a + v_b + \cdots + v_k + (g_a + g_b + \cdots + g_k)^2. \tag{3.6.10}$$

Now $g_t \leq tg_1 = t\gamma_1$ by Theorem 3.3.3. So (3.6.10) yields

$$v_n \leq v_a + v_b + \cdots + v_k + 2\,n\gamma_1\,(g_a + g_b + \cdots + g_k - g_n). \tag{3.6.11}$$

We now define

$$W(n) = v_n/2\,n\gamma_1, \qquad Q(n) = g_n - n\gamma. \tag{3.6.12}$$

We find from (3.6.11)

$$W(n) \leq \frac{a}{n}\,W(a) + \frac{b}{n}\,W(b) + \cdots + \frac{k}{n}\,W(k) + Q(a) + \tag{3.6.13}$$
$$+ Q(b) + \cdots + Q(k) - Q(n).$$

In particular when $n = a + b$ and $a = b = 2^i$ we have

$$W(2^{i+1}) \leq W(2^i) + 2\,Q(2^i) - Q(2^{i+1}); \tag{3.6.14}$$

and, summing (3.6.14) over $i = 0, 1, \ldots, h-1$, we obtain

$$W(2^h) \leq W(1) + Q(1) - Q(2^h) + \sum_{i=0}^{h-1} Q(2^i). \tag{3.6.15}$$

Any positive integer t can be expressed in the binary form

$$t = \sum_{j=1}^{l} 2^{h(j)}, \qquad 0 \leq h(1) < h(2) < \cdots < h(l). \tag{3.6.16}$$

Putting $a = 2^{h(1)}, b = 2^{h(2)}, \ldots, k = 2^{h(l)}$ and $n = t$ in (3.6.13) and substituting from (3.6.15), we find

$$W(t) \leq W(1) + Q(1) - Q(t) +$$
$$+ \sum_{j=1}^{l} \left\{ \left[1 - \frac{2^{h(j)}}{t}\right] Q[2^{h(j)}] + \frac{2^{h(j)}}{t} \sum_{i=0}^{h(j)-1} Q(2^i) \right\}. \tag{3.6.17}$$

In this expression each Q is nonnegative, because of Theorem 3.3.3; and the coefficient of $Q(2^i)$ consists of two parts. The first part is the sum of terms $2^{h(j)}/t$ taken over those j for which $i < h(j) \leq h(l)$; and this sum cannot exceed 1 because of (3.6.16). The second part vanishes unless $i = h(j)$ for some j; and in this exceptional case it has the value $1 - 2^{h(j)}/t \leq 1$. Thus the coefficient of $Q(2^i)$ is at most 2; and we have, omitting (as we may) $-Q(t)$,

$$W(t) \leq W(1) + Q(1) + 2 \sum_{i=0}^{h(l)} Q(2^i). \tag{3.6.18}$$

Now g_n is a subadditive function; and so, by (3.6.12), $Q(n)$ is also sub-additive. Hence for $i \geq 2$

$$Q(2^i) \leq Q(m) + Q(2^i - m), \qquad (m = 1, 2, \ldots, 2^{i-1} - 1), \tag{3.6.19}$$

and therefore

$$Q(2^i) \leq \frac{1}{2^{i-1}} \left[\sum_{m=1}^{2^{i-1}-1} + \sum_{m=2^{i-1}+1}^{2^i} \right] Q(m), \qquad (i \geq 2). \tag{3.6.20}$$

Inserting (3.6.20) into (3.6.18), we have

$$\begin{aligned}
W(t) \leq \; & W(1) + 3Q(1) + 2Q(2) + \\
& + 2\Big\{ Q(1) + \tfrac{1}{2}Q(2) + Q(3) + \tfrac{3}{4}Q(4) + \\
& + \tfrac{1}{2}[Q(5) + \cdots + Q(7)] + \tfrac{3}{8}Q(8) + \\
& + \tfrac{1}{4}[Q(9) + \cdots + Q(15)] + \tfrac{3}{16}Q(16) + \cdots + \\
& + \frac{1}{2^{h(l)-2}}[Q(2^{h(l)-1} + 1) + \cdots + Q(2^{h(l)})]\Big\} \leq
\end{aligned} \tag{3.6.21}$$

$$\leq W(1) + 8 \sum_{m=1}^{n} Q(m)/m = W(1) + 8 \sum_{m=1}^{n} (\gamma_m - \gamma).$$

Now (3.6.12) and (3.6.21) yield (3.6.8).

If we square the inequality $x_{0t}^*(\omega) \leq x_{0t}(\omega)$ and then take expected values, we get

$$v_t^* \leq v_t + t^2 (\gamma_t^2 - \gamma_t^{*2}). \tag{3.6.22}$$

Since both processes have the same time constant, $\gamma_t^2 - \gamma_t^{*2} \to 0$ as $t \to \infty$. Also $v_t = o(t^2)$ as $t \to \infty$, by (3.6.8) since $\gamma_i \to \gamma$ as $i \to \infty$; and $v_t^* = o(t^2)$ as $t \to \infty$, by (3.6.22). Finally

$$E\big\{x_{0t}(\omega)/t - \gamma\big\}^2 = v_t/t^2 + \gamma_t^2 - 2\gamma\gamma_t + \gamma^2 \to 0 \text{ as } t \to \infty, \tag{3.6.23}$$

which shows that $x_{0t}(\omega)/t$ converges to γ in quadratic mean as $t \to \infty$; and a similar result holds for $x_{0t}^*(\omega)/t$.

3.7. *Convergence with probability* 1. *Independent processes.*

Theorem 3.3.3 asserts that $\gamma_t \to \gamma$ as $t \to \infty$, but it gives no information on the rate of convergence. In a sense, this is inevitable because it relies only upon the properties of the subadditive function g_t. If γ_t is any nonincreasing function of t (which therefore tends to a limit γ as $t \to \infty$)

then it is easy to see that $g_t = t\gamma_t$ is subadditive: in fact, $g_{a+b} = (a + b)\gamma_{a+b} = a\gamma_{a+b} + b\gamma_{a+b} \leq a\gamma_a + b\gamma_b = g_a + g_b$. It may however happen that a subadditive process satisfies certain further conditions, say some conditions inherent in the physical situation it represents, which suffice to show that $\gamma_t \to \gamma$ not too slowly. The following theorem may then apply. Nevertheless the condition (3.7.2) is an artificial one, which is introduced to implement the proof of the theorem and seems to bear little relation to either the conclusions or the hypothesis of the theorem. To state this theorem we need to introduce the idea of an *independent subadditive process*. The subadditive process $\{ x_{st}(w) \}$ is called *independent* if the random variables $x_{s_i\,t_i}(w)$, $i = 1,2,\ldots$, are mutually independent whenever (s_i, t_i) are mutually disjoint *open* intervals. Clearly an independent subadditive process, which possesses second moments, is an uncorrelated process; though the converse is generally false. An independent subadditive process is obviously self-smothering.

Theorem 3.7.1. *Let* $\{ x_{st}(\omega) \}$ *and* $\{ x^*_{st}(\omega) \}$ *be non-negative subadditive processes with the same time constant* γ. *Suppose that* Ex^2_{01} *exists, and that* $\{ x_{st}(\omega) \}$ *is an independent process which dominates* $\{ x^*_{st}(\omega) \}$. *Then the condition*

$$\sum_{t=1}^{\infty} (\gamma_t - \gamma)/t < \infty \tag{3.7.2}$$

is sufficient for the convergence of $x_{ot}(\omega)/t$ *and of* $x^*_{ot}(\omega)/t$ *to* γ *with probability 1 as* $t \to \infty$.

Proof: Since g_t is subadditive we have

$$\gamma_t - \gamma = \frac{tg_t}{t^2} - \gamma \leq \frac{(g_1 + g_{t-1}) + (g_2 + g_{t-2}) + \ldots + (g_{t-1} + g_1) + g_t}{t^2} - \gamma$$

$$= \frac{2(g_1 + g_2 + \ldots + g_{t-1}) + g_t}{t^2} - \gamma = \frac{2\sum_{i=1}^{t-1}(g_i - i\gamma) + (g_t - t\gamma)}{t^2}$$

$$\leq \frac{2}{t} \sum_{i=1}^{t} (\gamma_i - \gamma). \tag{3.7.3}$$

Hence

$$\gamma_t^2 - \gamma_t^{*2} = (\gamma_t + \gamma_t^*)(\gamma_t - \gamma_t^*) \leq (\gamma_1 + \gamma_1^*)(\gamma_t - \gamma_t^*) \leq (\gamma_1 + \gamma_1^*)(\gamma_t - \gamma)$$

$$\leq \frac{2(\gamma_1 + \gamma_1^*)}{t} \sum_{i=1}^{t} (\gamma_i - \gamma). \tag{3.7.4}$$

Prescribe $\varepsilon > 0$. Chebyshev's inequality gives

$$P\left[\,|\, x_{ot}^*(\omega)/t - \gamma_t^* \,| \geq \varepsilon \right] \leq \mathrm{var}\,[x_{ot}^*(\omega)/t]/\varepsilon^2 = v_t^*/t^2\,\varepsilon^2$$

$$\leq \frac{1}{\varepsilon^2} \left\{ \frac{v_1}{t} + \frac{18\,\gamma_1 + 2\gamma_1^*}{t} \sum_{i=1}^{t} (\gamma_i - \gamma) \right\} \tag{3.7.5}$$

on using (3.7.4), (3.6.22), and (3.6.8). Prescribe $\lambda > 1$; and let t_j denote the smallest integer not less than λ^j. Clearly

$$\sum_{j=1}^{\infty} 1/t_j \leq \sum_{j=1}^{\infty} \lambda^{-j} = (\lambda - 1)^{-1} < \infty \tag{3.7.6}$$

Also let $j(i)$ be the smallest positive integer such that $\lambda^{j(i)} > i - 1$. If $i > 1$, then $\lambda^{j(i)} \geq \frac{1}{2}i$; while if $i = 1$, then $j(i) = 1$ and $\lambda^{j(i)} = \lambda > 1$, so that $\lambda^{j(i)} \geq \frac{1}{2}i$ is true in this case as well. We have

$$\sum_{j=1}^{\infty} \frac{1}{t_j} \sum_{i=1}^{t_j} (\gamma_i - \gamma) \leq \sum_{j=1}^{\infty} \frac{1}{\lambda^j} \sum_{1 \leq i < \lambda^j + 1} (\gamma_i - \gamma)$$

$$= \sum_{i=1}^{\infty} (\gamma_i - \gamma) \sum_{j=j(i)}^{\infty} \lambda^{-j} = \frac{\lambda}{\lambda - 1} \sum_{i=1}^{\infty} \frac{(\gamma_i - \gamma)}{\lambda^{j(i)}} \leq \frac{2\lambda}{\lambda - 1} \sum_{i=1}^{\infty} \frac{(\gamma_i - \gamma)}{i} < \infty \quad (3.7.7)$$

by (3.7.2). Consequently (3.7.5), (3.7.6), and (3.7.7) show that

$$\sum_{j=1}^{\infty} P\left[\, |\, x_{0t_j}^{*}(\omega)/t_j - \gamma_{t_j}^{*}\,| \geq \varepsilon \right] < \infty \; ; \quad (3.7.8)$$

and since this holds for arbitrary $\varepsilon > 0$, the Borel-Cantelli lemma shows that with probability 1

$$\lim_{j \to \infty} \left\{ x_{0t_j}^{*}(\omega)/t_j - \gamma_{t_j}^{*} \right\} = 0 \, . \quad (3.7.9)$$

Hence

$$\lim x_{0t_j}^{*}(\omega)/t_j = \gamma \quad (3.7.10)$$

with probability 1, since $\gamma_{t_j}^{*} \to \gamma$ as $j \to \infty$.

Next prescribe a positive integer n. To each positive integer t, there corresponds an integer $j = j(t)$ such that $t_{j-1} \leq t < t_j$. Then write $t_{j+1} - t = kn + r$, where k and r are integers and $0 < r \leq n$. This is legitimate since $t_{j+1} - t$ is an integer and $t_{j+1} - t \geq t_j - t > 0$. Both $k = k(t)$ and $r = r(t)$ are functions of t. Since $t < t_j$, we must have $j \to \infty$ when $t \to \infty$. Also

$$t_{j+1} - t \geq t_{j+1} - t_j + 1 \geq \lambda^{j+1} - \lambda^j = \lambda^j (\lambda - 1) \to \infty \text{ as } j \to \infty. \quad (3.7.11)$$

Hence $t_{j+1} - t \to \infty$ as $t \to \infty$. Further, by (3.7.11),

$$n(k + 1) \geq kn + r = t_{j+1} - t \geq \lambda^j (\lambda - 1); \quad (3.7.12)$$

and $kn \leq kn + r - 1 = t_{j+1} - t - 1 \leq t_{j+1} - t_{j-1} - 1 \leq \lambda^{j+1} - \lambda^{j-1}$

$$\leq \lambda^{j-1} (\lambda^2 - 1). \quad (3.7.13)$$

It follows from (3.7.12) that $k \to \infty$ when $t \to \infty$. From (3.7.12) and (3.7.13)

$$\frac{\log kn - \log (\lambda^2 - 1) + \log \lambda}{\log \lambda} \leq j \leq \frac{\log (k + 1) n - \log (\lambda - 1)}{\log \lambda} \quad (3.7.14)$$

and hence, if k has a given value, the number of different values of j which could give this value of k is at most

$$\frac{\log (k + 1) n - \log (\lambda - 1) - \log kn + \log (\lambda^2 - 1) - \log \lambda}{\log \lambda} + 1$$

$$= \frac{\log \left(1 + \dfrac{1}{k}\right) + \log (1 + \lambda)}{\log \lambda} \leq \frac{\log [2(1 + \lambda)]}{\log \lambda} \quad (3.7.15)$$

If both j and k have given values, then $t_{j+1} - t = kn + r$ shows that t can have at most n different values. Hence, by (3.7.15), the number of distinct values of t which can give rise to a given value of $k = k(t)$ is at most

$$n \log [2 (1 + \lambda)]/\log \lambda . \tag{3.7.16}$$

Let $z_1, z_2, \ldots, z_m$ be m independent observations each distributed with the same distribution as $x_{on}(\omega) - g_n$. For prescribed $\varepsilon > 0$, write

$$P_m = P \left[\, | \sum_{i=1}^{m} z_i | > m\varepsilon \right] . \tag{3.7.17}$$

Since the z_i are distributed with zero mean and finite variance, a theorem of Erdös (1950) asserts that

$$\sum_{m=0}^{\infty} P_m < \infty . \tag{3.7.18}$$

Since $\{ x_{st}(\omega) \}$ is an independent subadditive process, we have

$$\sum_{t=1}^{\infty} P\left[\, | \frac{1}{k(t)} \sum_{i=0}^{k(t)-1} x_{t+in,\, t+(i+1)n}(\omega) - g_n | > \varepsilon \right] = \sum_{t=1}^{\infty} P_{k(t)} \leq$$

$$\leq \frac{n \log [2 (1 + \lambda)]}{\log \lambda} \sum_{k=0}^{\infty} P_k < \infty, \tag{3.7.19}$$

by (3.7.16) and (3.7.18). Hence, by the Borel-Cantelli lemma and the fact that $k \to \infty$ when $t \to \infty$, we have

$$\frac{1}{k(t)} \sum_{i=0}^{k(t)-1} x_{t+in,\, t+(i+1)n}(\omega) \to g_n \text{ as } t \to \infty \tag{3.7.20}$$

with probability 1.

When $t_{j-1} \leq t < t_j$ we have

$$x^*_{ot_{j+1}}(\omega) \leq x^*_{ot}(\omega) + x^*_{t, t_{j+1}}(\omega) \leq x^*_{ot}(\omega) + x_{t, t_{j+1}}(\omega) ; \tag{3.7.21}$$

and hence

$$\frac{x^*_{ot}(\omega)}{t} \geq \frac{x^*_{ot_{j+1}}(\omega)}{t_{j+1}} - \frac{t_{j+1}-t}{t} \cdot \frac{x_{t, t_{j+1}}(\omega)}{t_{j+1}-t} . \tag{3.7.22}$$

Further

$$\frac{x_{t, t_{j+1}}(\omega)}{t_{j+1}-t} \leq \frac{1}{k(t)\,n} \sum_{i=0}^{k(t)-1} x_{t+in,\, t+(i+1)n}(\omega) + \frac{x_{t_{j+1}-r, t_{j+1}}(\omega)}{t_{j+1}-t} \tag{3.7.23}$$

When $t \to \infty$, the final term in (3.7.23) tends to zero with probability 1, as we may see by a simple change of notation in the argument leading from (3.5.28) to (3.5.30). Also the first term on the right of (3.7.23) tends with probability 1 to $g_n/n = \gamma_n$ by virtue of (3.7.20). Hence

$$\limsup_{t \to \infty} \frac{x_{t, t_{j+1}}(\omega)}{t_{j+1}-t} \leq \gamma_n \tag{3.7.24}$$

with probability 1. Moreover as $t \to \infty$

$$\frac{t_{j+1}-t}{t} \leq \frac{t_{j+1}}{t_{j-1}} - 1 \to \lambda^2 - 1 . \tag{3.7.25}$$

It now follows from (3.7.10), (3.7.22), (3.7.24), and (3.7.25) that with probability 1

$$\liminf_{t \to \infty} \frac{x_{ot}^*(\omega)}{t} \geq \gamma - (\lambda^2 - 1)\gamma_n . \tag{3.7.26}$$

However, the left side of (3.7.26) is independent of λ; so, on letting $\lambda \to 1$ from above, we have

$$P\left[\liminf_{t \to \infty} x_{ot}^*(\omega)/t \geq \gamma\right] = 1 . \tag{3.7.27}$$

Finally, $x_{st}(\omega)$ is self-smothering since it is an independent subadditive process. Thus Theorem 3.5.38 gives

$$P\left[\limsup_{t \to \infty} x_{ot}(\omega)/t = \gamma\right] = 1 . \tag{3.7.28}$$

But $x_{ot}(\omega) \geq x_{ot}^*(\omega)$, since $\{ x_{st}(\omega) \}$ dominates $\{ x_{st}^*(\omega) \}$. Therefore (3.7.27) and (3.7.28) yield

$$P\left[\lim_{t \to \infty} x_{ot}(\omega)/t = \gamma\right] = P\left[\lim_{t \to \infty} x_{ot}^*(\omega)/t = \gamma\right] = 1 \tag{3.7.29}$$

which completes the proof of Theorem 3.7.1.

3.8 *Processes with subadditive means.* In the following part of the paper we shall have to deal with processes that are not quite subadditive. Specifically, these will be processes which satisfy conditions (3.1.2) and (3.1.3) but, instead of necessarily satisfying (3.1.1), they merely satisfy the weaker condition

$$g_{t-r} \leq g_{s-r} + g_{t-s} , \qquad (r \leq s \leq t) . \tag{3.8.1}$$

We shall call these *processes with subadditive means.* Such a process, of course, satisfies Theorem 3.3.3, and possesses a time constant γ.

Theorem 3.8.1. *Let $\{x_{st}^*(\omega)\}$ be a process with subadditive means and a time constant γ; and let $\{x_{st}(\omega)\}$ be a subadditive process, also having time constant γ, which dominates $\{x_{st}^*(\omega)\}$. If*

$$P\left[\limsup_{t \to \infty} x_{0t}(\omega)/t \leq \gamma\right] = 1 , \tag{3.8.2}$$

then

$$P\left[\limsup_{t \to \infty} x_{0t}^*(\omega)/t = \gamma\right] = 1 , \tag{3.8.3}$$

and $x_{0t}^*(\omega)/t$ converges in probability to γ as $t \to \infty$.

Proof. The proof is similar to that of Theorem 3.5.1; and we use the same symbolism, adding asterisks where necessary to refer to the corresponding quantities defined in terms of $\{x_{st}^*(\omega)\}$ instead of $\{x_{st}(\omega)\}$. Since $\{x_{st}(\omega)\}$ dominates $\{x_{st}^*(\omega)\}$ we have

$$P\left[\limsup_{t \to \infty} x_{0t}^*(\omega)/t \leq \gamma\right] = 1 ; \tag{3.8.4}$$

and hence for $\beta > \gamma$

$$J_t^*(\beta) \to 0 \text{ as } t \to \infty . \tag{3.8.5}$$

Instead of (3.5.16) we have

$$\gamma \leq t^{-1} g_t^* \leq (\varepsilon - \beta) [1 - J_t^* (\varepsilon)] + \beta [1 - J_t^* (\beta)] + t^{-1} I_t^* (\beta) \leq$$
$$\leq (\varepsilon - \beta) [1 - J_t^* (\varepsilon)] + \beta [1 - J_t^* (\beta)] + t^{-1} I_t (\beta) , \qquad (3.8.6)$$

and hence

$$\gamma - \varepsilon \leq (\beta - \varepsilon) \liminf_{t \to \infty} J_t^* (\varepsilon) + I_1 (\alpha) . \qquad (3.8.7)$$

The remainder of the proof now goes as in Theorem 3.5.1.

Theorem 3.8.2. *Theorem 3.6.7 remains true if $\{x_{st}^* (\omega)\}$ is merely a nonnegative process with subadditive means and time constant γ.*

Proof. The proof used for Theorem 3.6.7 remains valid without alteration.

4. First-passage percolation on the square lattice

In Section 2 we outlined the general theory of first-passage theory on an arbitrary graph g. Here we shall study the problem in detail for the case where g is the square lattice, that is, the lattice of integer points (x, y), the arcs (all of unit length) being parallel to the x- and y-axes. The phase space (Ω, B, P) on this lattice is induced by a distribution U of nonnegative random variables u with finite mean $\bar{u}$.

A standard principle in first-passage theory is that the more restricted the set of paths R, the more tractable is $t_R (\omega)$. Therefore in Section 4.1 through Section 4.3 we study first-passage times between nodes of the lattice over paths which are subject to a *cylinder restriction* (which will be specified below). Then in subsequent sections we use the results obtained to determine the first-passage times between nodes of the lattice over paths which are subject to no restrictions whatsoever. Such first-passage times are termed *absolute* first-passage times.

The main results of this section will show that these first-passage times are subadditive stochastic processes with a time constant $\mu = \mu (U)$, which is the same for both cylinder and absolute times. The results of Section 3 may then be applied to these processes.

4.1. *The cylinder process $t_{mn} (\omega)$ — An independent subadditive process.* The *cylinder* defined by two nodes (m_1, m_2), (n_1, n_2) of the lattice is the strip enclosed between the lines $x = m_1$, and $x = m_2$. $t [(m_1, m_2),$ $(n_1, n_2); \omega]$ is defined to be the first-passage time under ω between (m_1, m_2), (n_1, n_2) over paths on the lattice lying *strictly* (save for the first endpoint) inside the cylinder defined by $m_1 < X \leq m_2$. Such a first-passage time is called a *cylinder time*.

By Theorem 2.1.3, $t [(m_1, m_2), (n_1, n_2); \omega]$ is a random variable on (Ω, B, P). We denote $t [(m, 0), (n, 0); \omega]$ by $t_{mn} (\omega)$ where $m \leq n$ and now we may state

Theorem 4.1.1. $\{t_{mn} (\omega)\}$ *is an independent nonnegative subadditive process on (Ω, B, P).*

Proof. Let m be integer $\leq n$. Since the time coordinates of the arcs of the lattice are nonnegative,

$$t_{mn}(\omega) \geq 0, \qquad (m \leq n, \omega \in \Omega). \tag{4.1.2}$$

By a simple application of the connection lemma 2.3.2 we have

$$t_{mn}(\omega) + t_{np}(\omega) \geq t_{mp}(\omega), \qquad (m \leq n \leq p). \tag{4.1.3}$$

Also if $l_{k_1}, l_{k_2}, \ldots, l_{k_{n-m}}$ are the arcs of the lattice which make up the straight line path from $(m, 0)$ to $(n, 0)$ and u_i is the time coordinate of l_i under ω

$$t_{mn}(\omega) \leq u_{k_1} + u_{k_2} + \ldots + u_{k_{n-m}}. \tag{4.1.4}$$

So that taking expectations of (4.1.4)

$$\tau(m, n) = E t_{mn}(\omega) \leq (n - m)\,\overline{u}, \qquad (m \leq n). \tag{4.1.5}$$

Also (Example 2.2.5) the set of cylinder paths from $(m, 0)$ to $(n, 0)$ is equivalent under lateral shift with the set of paths from $(m + a, 0)$ to $(n + a, 0)$ for any integer a. Hence the distribution of $t_{mn}(\omega)$ depends only on the difference $(n - m)$.

Thus we see that $\{t_{mn}(\omega)\}$ is a nonnegative subadditive process. Also since the distribution of $t_{mn}(\omega)$ depends *only* on the time coordinates of the arcs of the lattice which lie strictly inside the ordinates $x = n$, $x = m$, we see that $\{t_{mn}(\omega)\}$ is an independent subadditive stochastic process.

Hence, Theorem 3.1.3 shows that there exists a constant $\mu = \mu(U)$ such that

$$\tau(m, n)/(n - m) \geq \mu = \lim_{n \to \infty} \tau(m, n)/(n - m). \tag{4.1.6}$$

Notice that by (4.1.5), the time constant $\mu(U)$ satisfies

$$0 \leq \mu(U) \leq \overline{u}. \tag{4.1.7}$$

That strict inequality does not always hold in (4.1.7) is seen by

Example 4.1.8. Let the distribution U be such that each arc of the lattice has a constant time coordinate k with probability 1. Then for this, the *constant distribution*, we see that $\mu(U) = k = \overline{u}$.

However, we do have

Theorem 4.1.9. *Provided the underlying distribution U is not the trivial distribution $u = 0$ or 1 according as $u < \overline{u}$ or $u \geq \overline{u}$ then $\mu(U) < \overline{u}$.*

The proof of Theorem 4.1.9 depends upon results in Section 4.2 and we will give it at the end of Section 4.2.

Since $\{t_{mn}(\omega)\}$ is an independent subadditive process and hence, a fortiori, smotherable, Theorem 3.5.38 gives

Theorem 4.1.10. *As $n \to \infty$ the random variable $t_{0n}(\omega)/n$ converges in probability to μ and*

$$P\,[\limsup_{n \to \infty} t_{0n}(\omega)/n = \mu] = 1. \tag{4.1.11}$$

Also we may make further remarks if we stipulate that the underlying distribution U has a finite variance. For, let U have variance σ^2. Then $Et_{01}^2 \leqq Eu^2 = \bar{u}^2 + \sigma^2$. Thus the conditions of Theorem 3.6.7 are satisfied, and as a result we have

Theorem 4.1.12. *If the U distribution has a finite variance σ^2, the random variable $t_{mn}(\omega)$ satisfies*

$$\lim_{n\to\infty} \frac{\operatorname{var} t_{mn}(\omega)}{(n-m)^2} = 0, \qquad (m \text{ fixed}), \qquad (4.1.13)$$

and $\{t_{mn}(\omega)/(n-m)\}$ converges in mean square to the time constant $\mu(U)$ as $n \to \infty$.

Thus it may be seen that the cylinder process $\{t_{mn}(\omega)\}$ is a comparatively well-behaved subadditive process. However it is not always easy to handle this process. For example:

Conjecture 4.1.14. While many distributions U exist which include time states ω on the lattice for which $t_{mn}(\omega)$ is *not* monotonic in n for fixed m, it nevertheless seems a reasonable conjecture (or even intuitively obvious), that

$$\tau(0, n) \leqq \tau(0, n+1), \quad (n \geqq 0). \qquad (4.1.15)$$

This result we *cannot* prove.

In later sections we shall study the process $\{t_{mn}(\omega)\}$ more closely for specified U distributions (the uniform rectangular, the exponential, and others). In Section 8 we shall study some "geometrical" properties of the route of $t_{mn}(\omega)$ (for example, its existence, its expected number of arcs).

4.2. *The cylinder process* $\{s_{mn}(\omega)\}$. Apart from its own interest, the study of the stochastic process $\{s_{mn}(\omega)\}$ (defined below) is essential if the problem of absolute first-passage theory is to be solved. It will be seen that the process, although *not* subadditive, is a process with subadditive means.

The cylinder time $s[(m, y), X = n; \omega]$ is defined for y any positive, negative or zero integer by

$$s[(m, y), X = n; \omega] = \inf_k t[(m, y), (n, k); \omega] \qquad (4.2.1)$$

where k runs through the integers $(-\infty, \infty)$.

More loosely, $s[(m, y), X = n; \omega]$ is the cylinder time between (m, y) and the line $X = n$. By Theorem 2.1.3 this cylinder time is a random variable on (Ω, B, P).

Define $s_{mn}(\omega)$ to be $s[(m, 0), X = n; \omega]$. Then $\{s_{mn}(\omega)\}$ is a 2-parameter stochastic process on (Ω, B, P). By definition, for all $\omega \in \Omega$

$$0 \leqq s_{mn}(\omega) \leqq t_{mn}(\omega), \qquad (m \leqq n). \qquad (4.2.2)$$

Thus $\Psi(m, n) = Es_{mn}(\omega)$ exists, and satisfies

$$0 \leq \Psi(m, n) \leq \tau(m, n) \leq (n - m)\,\overline{u}\,. \qquad (4.2.3)$$

By the principle of equivalence under lateral shift, Theorem 2.2.1, we see that $s_{mn}(\omega)$ and $s[(m, y), X = n; \omega]$ are identically distributed. In particular the distribution of $s_{mn}(\omega)$ depends only on $(n - m)$. However, although conditions (3.1.2) and (3.1.3) are satisfied, it is *not* possible to say in general that

$$s_{mn}(\omega) + s_{np}(\omega) \geq s_{m,\,p}(\omega) \qquad (4.2.4)$$

and hence $\{s_{mn}(\omega)\}$ is *not* a subadditive process. However, we do have

Theorem 4.2.5. *For any distribution* U, *the function* $\Psi(m, n)$ *satisfies*

$$\Psi(m, n) + \Psi(n, p) \geq \Psi(m, p)\,, \qquad (m \leq n \leq p)\,. \qquad (4.2.6)$$

Proof. Let r_1 be the route of $s_{mn}(\omega)$ and let it meet $X = n$ at $P \equiv (n, y_1)$. [The existence of r_1 will be proved in Section 8.] Let $f(\omega)$ be the first-passage time from P to $X = p$ over cylinder paths whose first arc is from (n, y_1) to $(n + 1, y_1)$. $f(\omega)$ is a random variable on (Ω, B, P). Its distribution depends on the distribution of time coordinates of the arcs in the strip bounded by $X = n$, $X = p$.

Hence $f(\omega)$ has the distribution of $s_{np}(\omega)$. If r_2 is the route of $f(\omega)$, by a simple application of the connection lemma 2.3.2

$$s_{mn}(\omega) + f(\omega) \geq s_{mp}(\omega)\,, \qquad (m \leq n \leq p)\,. \qquad (4.2.7)$$

Hence taking expected values of (4.2.7), since $Ef(\omega) = \Psi(n, p)$, we have the required result (4.2.6) and this completes the proof of Theorem 4.2.5.

Since $\{s_{mn}(\omega)\}$ is a stationary process with subadditive means

$$\inf_{n} \frac{\Psi(m, n)}{(n - m)} = \mu^*(U) = \lim_{n \to \infty} \frac{\Psi(m, n)}{(n - m)}\,, \qquad (m \text{ fixed})\,. \qquad (4.2.8)$$

The time constant $\mu^*(U)$ depends only on U. From (4.2.3) it satisfies

$$0 \leq \mu^*(U) \leq \mu(U) \leq u < \infty\,. \qquad (4.2.9)$$

The main result of this section is

Theorem 4.2.10. *For any distribution* U, *the time constants* $\mu(U)$, $\mu^*(U)$ *are equal.*

Theorem 4.2.10 has important mathematical consequences in this work; and physically it has the following interpretation. If fluid is supplied at r collinear nodes of the lattice and the fluid can only flow along arcs of the lattice, the time of flow along any arc being a random variable, then the expected time to "wet" a specified node is asymptotically independent of r.

Proof of Theorem 4.2.10. Let x_0 be a prescribed integer. Let $s_{x_0}^1(\omega)$ be $s_{0,\,x_0}(\omega)$. Let its route r_1 meet the line $X = x_0$ at $P_1 \equiv (x_0, h_1(\omega))$. Let $s_{x_0}^2(\omega)$ be the cylinder first-passage time from P_1 to the line $X = 2x_0$ over paths whose first arc links P_1 to $(x_0 + 1, h_1(\omega))$. Let r_2 be the route of $s_{x_0}^2(\omega)$, and let r_2 meet $X = 2 x_0$ at $P_2 \equiv (2 x_0, h_1(\omega) + h_2(\omega))$. Similarly define $s_{x_0}^3(\omega)$ to be the first-passage time, under the same conditions, from P_2 to $X = 3 x_0$.

Continuing in this way it is possible to define sequences $\{s_{x_0}^i(\omega)\}_{i=1}^n$, $\{r_i\}_{i=1}^n$, $h_i(\omega)\}_{i=1}^n$ for n, any positive integer, such that

a) $\{s_{x_0}^i(\omega)\}_{i=1}^n$ forms a sequence of independent, identically distributed random variables having the distribution of $s_{0,\,x_0}(\omega)$.

b) $\{r_i\}_{i=1}^n$ is a sequence of paths on the lattice such that $r_1 * r_2 * \ldots * r_n$ is a connected path from the origin to $X = n x_0$.

c) $\{h_i(\omega)\}_{i=1}^n$ is a sequence of integer-valued, symmetric, identically distributed random variables.

Define $N(\omega) \equiv \sum_{i=1}^n h_i(\omega)$. Let r_0 be the straight line path on the lattice from $(n x_0, N(\omega))$ to $(n x_0, 0)$. Since the time coordinates of the arcs of r_0 are independent of the value of $N(\omega)$ we have by the theory of a random number of random variables (Feller 1957, p. 268)

$$Et(r_0, \omega) = \overline{u}E \mid N(\omega) \mid \qquad (4.2.11)$$

provided that $E \mid N(\omega) \mid$ exists, and we now show that it in fact does.

We consider the cylinder $0 < X \leq x_0$. Let B_y denote the set of all links in this cylinder and also in the orthogonal cylinder $y \leq Y \leq y + x_0$. We say that B_y forms a barrier if every link in $(0 < X \leq x_0, Y = y)$ has a time coordinate at least as great as the greatest time coordinate in the rest of B_y. Clearly the route r_1 cannot cross B_y vertically if B_y is a barrier, because the passage time of such a path would be at least as great as the passage time along some path in $(0 < X \leq x_0, Y \leq y)$. If y is a positive integer, the probability that B_y is a barrier is strictly positive, say π_0. [In fact $\pi_0 \geq 2^{-x_0(1+2x_0)}$, but the actual value of π_0 is immaterial aside from the fact that $\pi_0 > 0$.] If $h_1(\omega) \geq (x_0 + 1)k$, then none of the disjoint sets $B_1, B_{1+(x_0+1)}, \ldots, B_{1+(k-1)(x_0+1)}$ can have been a barrier; and since they are disjoint sets, the probability of this does not exceed $(1 - \pi_0)^k$. Hence

$$P[h_1(\omega) \geq (x_0 + 1)k] \leq (1 - \pi_0)^k. \qquad (4.2.12)$$

Thus the distribution of $h_1(\omega)$ falls off at least exponentially, and therefore

$$E \mid h_1(\omega) \mid < \infty, \qquad (4.2.13)$$

by the symmetry of the distribution. Finally

$$E \mid N(\omega) \mid = E \left| \sum_{i=1}^n h_i(\omega) \right| \leq E \sum_{i=1}^n \mid h_i(\omega) \mid = nE \mid h_1(\omega) \mid < \infty. \qquad (4.2.14)$$

It also follows from (4.2.12) that

$$E[h_1(\omega)]^2 < \infty. \qquad (4.2.15)$$

Therefore by Schwarz's inequality

$$[E \mid N\,(\omega)\mid]^2 \leqq E\,[N\,(\omega)]^2 = \sum_{i=1}^{n} E\,[h_i\,(\omega)]^2 = nE\,[h_1\,(\omega)]^2 \,. \qquad (4.2.16)$$

Now, using the inclusion lemma and b) above

$$t_{0,\,nx_0}\,(\omega) \leqq t\,(r_1 * r_2 * \ldots * r_n * r_0;\,\omega)$$

$$= \sum_{i=1}^{n} t\,(r_i;\,\omega) + t\,(r_0,\,\omega) \qquad (4.2.17)$$

$$= \sum_{i=1}^{n} s_{x_0}^{i}\,(\omega) + t\,(r_0;\,\omega) \,.$$

Taking expected values of (4.2.17) and using (4.2.11) and (4.2.16) we have

$$\tau\,(0,\,nx_0) \leqq n\Psi\,(0,\,x_0) + \overline{u}\,n^{\frac{1}{2}}\{E\,[h_1\,(\omega)]^2\}^{\frac{1}{2}} \,. \qquad (4.2.18)$$

If we divide this equation by nx_0 and let $n \to \infty$ we get

$$\mu\,(U) \leqq \Psi\,(0,\,x_0)/x_0 \,; \qquad (4.2.19)$$

and if we now let $x_0 \to \infty$, we get in combination with (4.2.9)

$$\mu\,(U) \leqq \mu^*\,(U) \leqq \mu\,(U) \,. \qquad (4.2.20)$$

This proves the theorem.

Since $\{s_{mn}\,(\omega)\}$ is a process with subadditive means, even though it is not a subadditive process, Theorems 3.8.1 and 3.8.2 give

$$P\,[\limsup_{n \to \infty} s_{0n}\,(\omega)/n = \mu] = 1 \qquad (4.2.21)$$

and

Theorem 4.2.22. *As $n \to \infty$ the random variable $s_{0n}\,(\omega)/n$ converges in probability to the time constant μ.*

Theorem 4.2.23. *If the U distribution has, in addition to a finite mean $\overline{u}$, a finite variance σ^2 then var $s_{0n}\,(\omega)$ exists and satisfies*

$$\lim_{n \to \infty} \text{var } s_{0n}\,(\omega)/n^2 = 0 \qquad (4.2.24)$$

and as a result $s_{0n}\,(\omega)/n$ converges to μ in mean square as $n \to \infty$.

Because of stationarity the results (4.2.21) through (4.2.24) may be extended to $s_{mn}\,(\omega)/(n - m)$.

The results of this section exhibit the close relationship between the s- and t-processes; and we shall explore this further in later sections.

Proof of Theorem 4.1.9. Let $p\,(x) = P\,(u \leqq x)$. Choose x such that $0 < x < \overline{u}$ and $p = p\,(x) > 0$. By hypothesis, such a choice is possible. Then choose n to satisfy

$$(n + 1)\,x < n\,\overline{u} \,. \qquad (4.2.25)$$

Now let $l_{k_1}, l_{k_2}, \ldots, l_{k_n}$ be the arcs making up the straight line path on the lattice from $(0, 0)$ to $(n, 0)$. Let $l_{j_1}, l_{j_2}, \ldots, l_{j_n}$ be the arcs making up the straight line lattice path from $(0, 1)$ to $(n, 1)$. Finally let l_{j_0} be the arc from the origin to $(0, 1)$. Then, if u_i denotes the time coordinate of l_i under ω, define

$$y(\omega) = \min\left[\sum_{i=1}^{n} u_{k_i}, \sum_{i=0}^{n} u_{j_i}\right] \qquad (4.2.26)$$

and

$$y^*(\omega) = \sum_{i=0}^{n} u_{j_i} \qquad \text{if and only if each } u_{j_i} \leqq x$$
$$= \sum_{i=1}^{n} u_{k_i}, \text{ otherwise .} \qquad (4.2.27)$$

Then

$$Ey^*(\omega) = p^{n+1} E\left[\sum_{i=0}^{n} u_{j_i} \mid u_{j_i} \leqq x \text{ for all } i\right] + (1 - p^{n+1}) E\left[\sum_{i=1}^{n} u_{k_i}\right] \leqq$$
$$\leqq p^{n+1}(n+1) x + (1 - p^{n+1}) n \bar{u} < n \bar{u} . \qquad (4.2.28)$$

But with the notation and results of Section 4.2 we shall have $s_{0n}(\omega) \leqq$
$\leqq y(\omega) \leqq y^*(\omega)$ and hence

$$n\mu(U) \leqq \Psi(0, n) < n \bar{u}, \qquad (4.2.29)$$

which implies Theorem 4.1.9.

4.3. *The absolute first-passage time* $a_{mn}(\omega)$: *A smotherable but not independent subadditive process.* The previous sections have dealt with first-passage times on the square lattice over cylinder paths. For most practical purposes, more important quantities *are absolute first-passage times*, that is, first-passage times over paths which are subject to no restriction whatsoever. Let $a_{mn}(\omega)$ denote the absolute first-passage time between $(m, 0)$ and $(n, 0)$ under ω. In this section we shall show that $a_{mn}(\omega)$ is "asymptotically equivalent" to the cylinder process $t_{mn}(\omega)$, or in other words, we shall prove quantitatively the intuitively appealing idea that "the average time spent outside the fundamental cylinder when traveling as quickly as possible from $(m, 0)$ to $(n, 0)$ is relatively small.

From Theorem 2.1.3 we have that $\{a_{mn}(\omega), \omega \in \Omega, m, n \text{ integers}\}$ is a 2-parameter stochastic process. By the inclusion lemma 2.3.1 it is immediate that

$$0 \leqq a_{mn}(\omega) \leqq t_{mn}(\omega) , \qquad (\omega \in \Omega, m \leqq n) . \qquad (4.3.1)$$

Theorem 4.3.2. $\{a_{mn}(\omega)\}$ *is a nonnegative subadditive process on* (Ω, B, P).

Proof. Trivially from (4.3.1), $a_{mn}(\omega) \geqq 0$, while $\alpha(m, n) = E a_{mn}(\omega)$ exists and satisfies

$$\alpha(m, n) \leqq \tau(m, n) \leqq (n - m) \bar{u} . \qquad (4.3.3)$$

By the principle of equivalence under lateral shift (2.2.1), the distribution of $a_{mn}(\omega)$ depends only on the difference $(n - m)$, while the connection lemma (2.3.2) proves that

$$a_{mn}(\omega) + a_{np}(\omega) \geqq a_{mp}(\omega) , \qquad (m \leqq n \leqq p) . \qquad (4.3.4)$$

Hence $\{a_{mn}(\omega)\}$ is a subadditive stochastic process. However, since the paths over which these first-passage times are taken are not restricted to being inside a cylinder it is not true that $a_{mn}(\omega)$ and $a_{np}(\omega)$ are independent random variables. Hence, unlike $\{t_{mn}(\omega)\}$, $\{a_{mn}(\omega)\}$ is *not* an independent subadditive process.

By Theorem **3.3.3** there exists a constant $\mu_A(U)$ such that

$$\alpha(m, n)/(n - m) \geq \mu_A(U) = \lim_{n \to \infty} \alpha(m, n)/(n - m) . \qquad (4.3.5)$$

From (4.3.3) $\mu_A(U)$ satisfies

$$0 \leq \mu_A(U) \leq \mu(U) \leq \bar{u} < \infty . \qquad (4.3.6)$$

Intuitively one would expect the difference $[\tau(m, n) - \alpha(m, n)]$ to be relatively small. In the case where U is the constant distribution (Example 4.1.8) it is obvious that for all $m, n, \alpha(m, n)$ and $\tau(m, n)$ are equal.

Theorem 4.3.7. *The time constants $\mu_A(U)$ and $\mu(U)$ are equal for any distribution U.*

Proof. Define $q_{mn}^k(\omega)$, for m, n and k positive integers, to be the first-passage time between $(m, 0)$ and $(n, 0)$ under ω, over paths which lie strictly inside the strip bounded by $X = m - k, X = n + k$. $q_{mn}^k(\omega)$ is a nonnegative random variable on (Ω, B, P). By the principle of equivalence under lateral shift, $q_{mn}^k(\omega)$ has a distribution which depends only on $(n - m)$ for fixed k. By the connection lemma (2.3.2)

$$q_{mn}^k(\omega) + q_{np}^k(\omega) \geq q_{mp}^k(\omega) , \qquad (m \leq n \leq p) . \qquad (4.3.8)$$

Hence for fixed k, $\{q_{mn}^k(\omega)\}$ is a subadditive stochastic process on (Ω, B, P).

Hence by Theorem **3.3.3** there exists a constant $\mu_k(U)$ such that $Q_k(n) = E q_{0n}^k(\omega)$ satisfies

$$\frac{Q_k(n)}{n} \geq \mu_k(U) = \lim_{n \to \infty} \frac{Q_k(n)}{n} . \qquad (4.3.9)$$

Now by the inclusion lemma (2.3.1)

$$a_{0n}(\omega) \leq q_{0n}^k(\omega) \leq q_{0n}^{k-1}(\omega) \leq t_{0n}(\omega) , \qquad (k \geq 2) . \qquad (4.3.10)$$

Hence we have

$$\mu_A(U) \leq \mu_k(U) \leq \mu_{k-1}(U) \leq \mu(U) , \qquad k \geq 2) . \qquad (4.3.11)$$

Let r_0 be the route of $t_{-k, 0}(\omega)$. Let r_1 be the route of $q_{0n}^k(\omega)$. Let r_2 be the route of $t_{n, n+k}(\omega)$. Then $r_0 * r_1 * r_2$ is a connected cylinder path from $(-k, 0)$ to $(n + k, 0)$ and hence

$$
\begin{aligned}
t_{-k, n+k}(\omega) &\leq t(r_0 * r_1 * r_2, \omega) \\
&= t(r_0, \omega) + t(r_1, \omega) + t(r_2, \omega) \qquad (4.3.12) \\
&= t_{-k, 0}(\omega) + q_{0n}^k(\omega) + t_{n, n+k}(\omega) .
\end{aligned}
$$

Hence taking expected values of (4.3.12), by stationarity we have

$$\tau(0, n + 2k) \leq 2\tau(k) + Q_k(n) . \tag{4.3.13}$$

Dividing (4.3.13) by n and taking the limit as $n \to \infty$ with k fixed, we therefore get

$$\mu(U) \leq \mu_k(U) \tag{4.3.14}$$

which together with (4.3.11) implies that for all fixed k

$$\mu(U) = \mu_k(U) . \tag{4.3.15}$$

Now consider the random variable $q_{0n}^k(\omega)$. This is monotonic decreasing in k for fixed n, ω and

$$\lim_{k \to \infty} q_{0n}^k(\omega) = a_{0n}(\omega) , \qquad (n, \omega \text{ fixed}) . \tag{4.3.16}$$

Hence by the Monotone Convergence Theorem

$$\lim_{k \to \infty} Q_k(n) = \alpha(0, n) \qquad (n \text{ fixed}) . \tag{4.3.17}$$

Now by (4.3.9) and (4.3.15)

$$Q_k(n)/n \geq \mu(U) \qquad \text{for all } k, n . \tag{4.3.18}$$

Hence by (4.3.17) and (4.3.18)

$$\alpha(0, n)/n \geq \mu(U) . \tag{4.3.19}$$

In (4.3.19) let $n \to \infty$ with the result that $\mu_A(U) \geq \mu(U)$. This with (4.3.6) completes the proof of Theorem 4.3.7.

We can now enter Theorems 3.6.7 and 3.7.1 with $x_{mn}(\omega) = t_{mn}(\omega)$ and $x_{mn}^*(\omega) = a_{mn}(\omega)$. Also we can use $\{t_{mn}(\omega)\}$ to smother $\{a_{mn}(\omega)\}$. In fact

$$a_{0, jn} \leq \sum_{i=0}^{j-1} a_{in, in+n}(\omega) \leq \sum_{i=0}^{j-1} t_{in, in+n}(\omega) = \sum_{i=0}^{j-1} y_i(\omega) \tag{4.3.20}$$

say; thus $\{a_{mn}(\omega)\}$ has a δ-blanket where δ is given by

$$\delta n = E t_{(i-1)n, in}(\omega) = \tau(0, n) . \tag{4.3.21}$$

Since $\lim_{n \to \infty} \tau(0, n)/n = \inf_n \tau(0, n)/n = \mu(U)$, $a_{mn}(\omega)$ is a smotherable subadditive process and hence by Theorem 3.5.1 we have the result

Theorem 4.3.22. *The absolute first-passage time $a_{0n}(\omega)$ satisfies*

$$P[\limsup_{n \to \infty} a_{0n}(\omega)/n = \mu(U)] = 1 \tag{4.3.23}$$

and as $n \to \infty$, $a_{0n}(\omega)/n$ converges in probability to $\mu(U)$.

Likewise Theorems 3.6.7 and 3.7.1 give

Theorem 4.3.23. *If the U distribution has finite variance then*

$$\lim_{n \to \infty} \operatorname{var} a_{0n}(\omega)/n^2 = 0 \tag{4.3.24}$$

and $a_{0n}(\omega)/n$ converges to γ in quadratic mean as $n \to \infty$. It also con-

verges with probability **1** *if further*

$$\sum_{n=1}^{\infty} \frac{1}{n} \left[\frac{\tau\,(0,n)}{n} - \mu\,(U) \right] < \infty \; ; \tag{4.3.25}$$

but it is not yet known whether or not (4.3.25) *is true. [See also section 8.3].*

This completes our study of $a_{mn}\,(\omega)$ for the time being. The close relationship between $a_{mn}\,(\omega)$ and $t_{mn}\,(\omega)$ is evident. In the next section we shall study the absolute analogue of $s_{mn}\,(\omega)$. This, it will be seen is a much more difficult process to handle.

4.4. *The absolute first-passage times between a point and a line.* For many practical purposes an important quantity is not the absolute first-passage time between two nodes of the lattice, but the first-passage times between a specified node and some linear barrier. This problem was first tackled in Section 4.2 when, however, we restricted ourselves to considering first-passage time over cylinder paths. This problem was not too difficult to deal with, even though it was not subadditive. The corresponding analogue however is so difficult that our results are sparse and consist mainly of conjectures based on quite strong heuristic evidence.

Let $b_{mn}\,(\omega)$ be the absolute first-passage time from $(m, 0)$ to the line $X = n$, $(m \leq n)$. From (2.1.3) $\{b_{mn}\,(\omega)\}$ is a 2-parametered stochastic process on the phase space (Ω, B, P). Its expected value $\beta\,(m, n)$ exists, and the following inequalities are true for all m, n, ω,

$$0 \leq b_{mn}\,(\omega) \leq a_{mn}\,(\omega) \; , \tag{4.4.1}$$

$$0 \leq b_{mn}\,(\omega) \leq s_{mn}\,(\omega) \; , \tag{4.4.2}$$

$$0 \leq \beta\,(m, n) \leq \min\left\{\alpha\,(m, n), \Psi\,(m, n)\right\}. \tag{4.4.3}$$

Physically, $b_{mn}\,(\omega)$ bears the same relationship to $s_{mn}\,(\omega)$ as does $a_{mn}\,(\omega)$ to $t_{mn}\,(\omega)$. It is tempting to *conjecture* that

$$\beta\,(m, n) + \beta\,(n, p) \geq \beta\,(m, p) \; , \qquad (m \leq n \leq p) \; . \tag{4.4.4}$$

This cannot be proved by a straightforward application of the connection lemma.

We further *conjecture* that $\lim_{n \to \infty} \beta\,(0, n)/n$ exists and satisfies

$$\lim_{n \to \infty} \beta\,(0, n)/n = \mu\,(U) \tag{4.4.5}$$

where $\mu\,(U)$ is the time constant of the a-process and the t-process. If we could prove (4.4.4) and (4.4.5) we could use Theorems 3.8.1 and 3.8.2 to establish results on the convergence of $b_{0n}\,(\omega)/n$ to $\mu\,(U)$ as $n \to \infty$.

We sketch our reasons for conjecturing (4.4.5): Let $s_{mn}^{k}\,(\omega)$ be defined for $k \geq 0$ as the first-passage time from $(m, 0)$ to $X = n$ over paths which lie strictly between the lines $X = m - k$, $X = n$. Then, if $\Psi_k\,(n)$

$= E s_{0n}^{k} (\omega)$, it can be shown (as in the proof of Theorem 4.3.7) that

$$\lim_{n \to \infty} \Psi_k (n)/n = \mu (U) \qquad \text{for any } k . \tag{4.4.6}$$

By the inclusion lemma (2.3.1)

$$\beta (0, n) \leq \Psi_{k+1} (n) \leq \Psi_k (n) \leq \Psi (0, n) , \qquad (k \geq 0) . \tag{4.4.7}$$

Also by an application of the Monotone Convergence Theorem

$$\lim_{k \to \infty} \Psi_k (n) = \beta (0, n) \qquad \text{for any } n . \tag{4.4.8}$$

If it could be shown that $\Psi_k (n)/n \geq \mu (U)$ (which would be the case if $s_{mn}^{k} (\omega)$ were a subadditive process for fixed k) (4.4.5) would follow. However, we see no way of proving this result at the moment.

This completes our study of first-passage theory for the time being. We shall return to these processes in Sections 6 to 8.

5. Generalized renewal processes

In this section we demonstrate a relationship between first-passage percolation theory and renewal theory. Throughout this section the underlying graph g will be the square lattice, with phase space (Ω, B, P) induced by a distribution U. We introduce *reach functions* on the space Ω which are random variables whose expected values have properties analogous to the renewal function. The reach functions are random variables having an inverse relationship with the cylinder first-passage times. This relationship is explored below.

5.1. *The relation between first-passage percolation theory and renewal theory.* Following the notation of W. L. Smith (1958), let $\{X_i\}_{i=1}^{\infty}$ be a sequence of nonnegative, independent, identically distributed random variables with finite mean. Renewal theory is concerned with the distribution of the nth partial sum $S_n = X_1 + X_2 + \ldots + X_n$ and, more especially, with the distribution of $N (t)$, where $N (t)$ is the random variable defined as the maximum n such that $S_n \leq t$. The renewal function $H (t)$ is defined to be $E N (t)$.

More generally, let (Ω, B, P) be the phase space of the square lattice and define the x-reach function $x (m, t; \omega)$ for all $t \geq 0$, all integers m by

$$x (m, t; \omega) = \sup (m_1 - m) \mid t_{m, m_1} (\omega) \leq t . \tag{5.1.1}$$

[The notation here means the supremum of $(m_1 - m)$ subject to the condition $t_{m, m_1} (\omega) \leq t$.]

Similarly define the *y-reach function* $y (m, t; \omega)$ by

$$y (m, t; \omega) = \sup (m_1 - m) \mid s_{m, m_1} (\omega) \leq t . \tag{5.1.2}$$

[If, instead of the two-dimensional square lattice, we were to consider the one-dimensional line, then $x (m, t; \omega)$ and $y (m, t; \omega)$ would coincide with $N(t)$ in the usual renewal theory. For this reason the reach functions afford a generalization of ordinary renewal theory.]

Theorem 5.1.3. *The x and y reach functions are random variables on the phase space (Ω, B, P).*

The proof of this theorem is not difficult and will be found in Section 5.6.

It is easy to prove, by a method similar to the principle of equivalence under lateral shift, that the distributions of $x(m, t; \omega)$, $y(m, t; \omega)$ are independent of m. Hence, except where absolutely necessary, we will consider only $x_t(\omega) = x(0, t; \omega)$ and $y_t(\omega) = y(0, t; \omega)$. By their definition, we have, for all positive integers k,

$$P[x_t(\omega) < k] \leqq P[t_{0k}(\omega) > t], \tag{5.1.4}$$

$$P[y_t(\omega) < k] = P[s_{0k}(\omega) > t], \tag{5.1.5}$$

$$x_t(\omega) \leqq y_t(\omega), \qquad (t \geqq 0). \tag{5.1.6}$$

Note that equality holds in (5.1.5) but not necessarily in (5.1.4).

As their names suggest the reach functions may be loosely interpreted as the x-coordinates of the easternmost points which are attainable from the origin by cylinder paths in a time not exceeding t, when the time state of the lattice is ω. In a certain sense, therefore, the reach functions are (as already mentioned) two-dimensional analogues of $N(t)$. This analogy will be heightened by some of the results of this section. To prove these results we have unfortunately to impose a rather heavy bounding restriction on the U distribution. At the moment we see no way of removing this restriction.

5.2. *The reach function for bounded U.* Henceforth the phase space $[\Omega, B, P]$ is derived from a distribution U which is bounded: that is, the time coordinate u_i of l_i satisfies for all i

$$0 < U_0 \leqq u_i \leqq U_1 < \infty, \qquad (U_0, U_1 \text{ being constants}). \tag{5.2.1}$$

With this restriction it is easy to see that for all $\omega \in \Omega$

$$t/U_1 \leqq x_t(\omega) \leqq y_t(\omega) \leqq t/U_0. \tag{5.2.2}$$

We now state an extension of Theorem 5.1.3.

Theorem 5.2.3. *The reach functions $x_t(\omega)$, $y_t(\omega)$ are measurable functions on the product space $\Omega \times T$ where T is the interval $(0, \infty)$ of the real line.*

The proof of this is also postponed until Section 5.6. Thus it may be seen that $\{x_t(\omega)\}_{t \in T}$, $\{y_t(\omega)\}_{t \in T}$ are continuous-parametered, integer-valued, measurable stochastic processes on Ω. From (5.2.3), by Doob (1952, p. 67), we have

Corollary 5.2.4. $X(t) = Ex_t(\omega)$, $Y(t) = Ey_t(\omega)$, *both exist and are Lebesgue-measurable functions of t on the real line.*

Trivially in view of (5.2.2)

$$t/U_1 \leqq X(t) \leqq Y(t) \leqq t/U_0, \qquad (t \geqq 0) . \tag{5.2.5}$$

Finally we notice, for all $\omega \in \Omega$, $x_t(\omega)$, $y_t(\omega)$ are nondecreasing in t.

SMITH (1958) gives the elementary renewal theorem, which in the notation of Section 5.1 states

$$\lim_{t \to \infty} H(t)/t = 1/\overline{u} , \qquad (\overline{u} = E\,X_i) . \tag{5.2.6}$$

Clearly, if the analogy between the reach functions and $N(t)$ is to be of any standing, we would expect an "elementary reach theorem".

Theorem 5.2.7. $Y(t)/t$ *tends to a finite limit* $\lambda(U)$ *as* $t \to \infty$, *provided that the U-distribution is bounded.*

The proof of Theorem 5.2.7 will follow from some lemmas which we shall prove below. For each node A of the lattice define a y-reach function $y_t(A, \omega)$. In other words $y_t(A, \omega)$ has the value $y_t(\omega)$ would take if the lattice were translated horizontally and vertically so that A became the origin. By the principle of equivalence under lateral shift we therefore have

Lemma 5.2.8. *For fixed* A, $y_t(A, \omega)$ *obeys the same probability law as* $y_t(\omega)$; *and in particular* $E y_t(A, \omega) = Y(t)$.

Lemma 5.2.9. *The function* $Y(t)$ *satisfies*

$$Y(t_1) + Y(t_2) + 1 \leqq Y(t_1 + t_2 + U_1) , \qquad (t_1, t_2 \geqq 0) . \tag{5.2.10}$$

Proof. Let t_1, t_2 be fixed. Let $y_{t_1}(\omega) = m_1$. Then by definition

$$s_{0, m_1}(\omega) \leqq t_1 \quad \text{and} \quad s_{0, m_1 + 1}(\omega) > t_1 . \tag{5.2.11}$$

Let r_1 be the route of $s_{0, m_1}(\omega)$; (r_1 must exist since the U distribution is bounded). Let the endpoint of r_1 be $P \equiv (m_1, Z)$. Notice that $y_{t_1}(\omega)$ is determined *only* by the time coordinates of the arcs lying strictly inside the strip bounded by $X = 0$, $X = m_1 + 1$. Let l be the horizontal arc linking P to $P' \equiv (m_1 + 1, Z)$. Since the time coordinates of the arcs of the lattice do not exceed U_1, we may write

$$t(r_1 * l, \omega) = t(r_1, \omega) + t(l, \omega) \leqq t_1 + U_1 . \tag{5.2.12}$$

Consider now the random variable $y_{t_2}(P', \omega)$. Let its value be m_2. Then there must exist a cylinder path r_2 from P' to $X = m_1 + m_2 + 1$, such that

$$t(r_2, \omega) \leqq t_2 . \tag{5.2.13}$$

Hence consider the connected cylinder path $r_1 * l * r_2$ which links the origin to $X = m_1 + m_2 + 1$. Then

$$t(r_1 * l * r_2, \omega) \leqq t_1 + t_2 + U_1 . \tag{5.2.14}$$

Therefore

$$y_{t_2 + t_2 + U_1}(\omega) \geqq m_1 + m_2 + 1 = y_{t_1}(\omega) + y_{t_2}(P', \omega) + 1 . \tag{5.2.15}$$

Taking expected values of (5.2.15), we have

$$Y(t_1 + t_2 + U_1) \geq Y(t_1) + 1 + Ey_{t_2}(P', \omega) . \qquad (5.2.16)$$

P' is a random node, and it can only be one of a finite number of nodes $\{A_i\}_{i=1}^{k}$ since the distribution is bounded. Now for any fixed A_i the random variable $y_{t_2}(A_i, \omega)$ is independent of the random variable

$$z(A_i, \omega) = \begin{cases} 1 \text{ if } P' = A_i , \\ 0 \text{ if } P' \neq A_i , \end{cases} \qquad (5.2.17)$$

since the values of these two random variables are determined by two disjoint sets of arcs. Thus, using (5.2.8), we have

$$
\begin{aligned}
Ey_{t_2}(P', \omega) &= E \Sigma_i y_{t_2}(A_i, \omega) \, z(A_i, \omega) \\
&= \Sigma_i E\left[y_{t_2}(A_i, \omega) \, z(A_i, \omega)\right] \\
&= \Sigma_i E y_{t_2}(A_i, \omega) \, Ez(A_i, \omega) \\
&= \Sigma_i Y(t_2) \, Ez(A_i, \omega) \qquad (5.2.18) \\
&= Y(t_2) \, E \Sigma_i z(A_i, \omega) \\
&= Y(t_2) \, E\, 1 = Y(t_2) .
\end{aligned}
$$

Substitution of (5.2.18) into (5.2.16) yields (5.2.10).

Proof of Theorem 5.2.7. From (5.2.4) and (5.2.10) we see that $-1 - Y(t - U_1)$ is a measurable subadditive function of t which is bounded below by $-1 - (t - U_1)/U_0 \geq -t/U_0$. Hence $\{-1 - Y(t - U_1)\}/t$ tends to a finite limit as $t \to \infty$; and Theorem 5.2.7 follows at once. We also obtain immediately

Corollary 5.2.19. *For all $t \geq U_1$,*

$$X(t) \leq Y(t) \leq (t + U_1)\, \lambda(U) - 1 . \qquad (5.2.20)$$

Notice that the proof of Lemma 5.2.9 will not work for $X(t)$ in place of $Y(t)$, because $t_{0n}(\omega)$ is not necessarily an increasing function of n and we cannot find two disjoint sets of arcs to justify the analogue of (5.2.18).

Despite the fact that the argument of Lemma 5.2.9 fails for $X(t)$, we shall nevertheless prove in the next section that

Theorem 5.2.21. *Provided the U-distribution is bounded,*

$$\lim_{t \to \infty} X(t)/t = \lambda(U) , \qquad (5.2.22)$$

where $\lambda(U)$ is the same constant as the one in Theorem 5.2.7.

We conjecture, but cannot yet prove, that Theorems 5.2.7 and 5.2.21 remain valid for unbounded U-distributions.

5.3. *Relationship between $\lambda(U)$ and $\mu(U)$.* The inverse relationship between reach theory and first-passage theory suggests

Theorem 5.3.1. *For a bounded distribution U,*

$$\lambda(U) = 1/\mu(U) . \tag{5.3.2}$$

We shall prove Theorems 5.2.21 and 5.3.1 via a series of lemmas. First we note, as a trivial consequence of (5.2.5) and Theorem 5.2.7,

$$\limsup_{t \to \infty} X(t)/t \leq \lambda(U) . \tag{5.3.3}$$

Lemma 5.3.4. $\liminf_{t \to \infty} X(t)/t \geq 1/\mu(U).$

Proof. Define $y_i = t_{(i-1)n,\,in}(\omega)$. Define $S_m^n(\omega)$ by

$$S_m^n(\omega) = y_1 + y_2 + \cdots + y_m . \tag{5.3.5}$$

$S_m^n(\omega)$ is therefore the mth partial sum of independent, nonnegative, identically distributed random variables. Define

$$N_n(t) = \sup m \mid S_m^n(\omega) \leq t , \qquad (t > 0) . \tag{5.3.6}$$

Then, since $t_{0,\,mn}(\omega) \leq S_m^n(\omega)$, we have $x_t(\omega) \geq nN_n(t)$ for all t. Therefore, taking expectations, we get

$$X(t) \geq nEN_n(t) = nH_n(t) , \qquad (t \geq 0) , \tag{5.3.7}$$

say. Now provided n is fixed, the elementary renewal theorem (Smith, 1958) gives

$$\lim_{t \to \infty} \frac{H_n(t)}{t} = 1/Ey_t = [\tau(0, n)]^{-1} . \tag{5.3.8}$$

Hence, combining (5.3.7) and (5.3.8), we have

$$\liminf_{t \to \infty} \frac{X(t)}{t} \geq n \lim_{t \to \infty} \frac{H_n(t)}{t} = n\,[\tau(0, n)]^{-1} . \tag{5.3.9}$$

Let $n \to \infty$ in (5.3.9) to yield (5.3.4) upon use of (4.1.6).

Lemma 5.3.10. $\lambda(U)\,\mu(U) \leq 1.$

Proof. Consider the random variable $y_{s_{0m}(\omega)}(\omega)$. This will identically equal m, for all ω, because U_0, the minimum time coordinate of any arc is strictly positive. Hence for any integer m,

$$y_{s_{0m}(\omega)}(\omega) = Ey_{s_{0m}(\omega)}(\omega) = m , \qquad (\omega \in \Omega) . \tag{5.3.11}$$

Since, by Theorem 4.2.22, $s_{0m}(\omega)/m$ converges in probability to μ as $m \to \infty$ there exists $m_0(\varepsilon, \eta)$ for prescribed $\varepsilon > 0, \eta > 0$ such that

$$P\,[s_{0m}(\omega) \leq (\mu - \varepsilon)\,m] < \eta , \qquad (m \geq m_0) . \tag{5.3.12}$$

Define $\Omega_m = \{\omega : s_{0m}(\omega) \leq (\mu - \varepsilon)\,m\}$. Since $y_t(\omega)$ is nondecreasing in t for fixed ω,

$$y_{(\mu-\varepsilon)\,m}(\omega) \leq y_{s_{0m}(\omega)}(\omega) = m , \qquad (\omega \in \Omega - \Omega_m) . \tag{5.3.13}$$

Since $y_t(\omega) \leq t/U_0$ for all $\omega \in \Omega$

$$y_{(\mu-\varepsilon)\,m}(\omega) \leq (\mu - \varepsilon)\,m/U_0 , \qquad (\omega \in \Omega_m) . \tag{5.3.14}$$

Hence, considering the expected value of $y_{(\mu-\varepsilon)\, m}(\omega)$

$$Y[m(\mu-\varepsilon)] \leq P(\Omega_m)(\mu-\varepsilon)m/U_0 + mP(\Omega-\Omega_m). \quad (5.3.15)$$

Thus, by (5.3.12),

$$Y[m(\mu-\varepsilon)] \leq m(\mu-\varepsilon)\eta/U_0 + m(1-\eta), \qquad (m \geq m_0), \quad (5.3.16)$$

or in other words

$$\frac{Y[m(\mu-\varepsilon)]}{m} \leq \frac{(\mu-\varepsilon)\eta}{U_0} + 1 - \eta. \quad (5.3.17)$$

Thus, letting $m \to \infty$ in (5.3.17), we obtain

$$(\mu-\varepsilon)\lambda = \lim_{m\to\infty} \frac{Y[m(\mu-\varepsilon)]}{m} \leq 1 + \eta\mu U_0^{-1} - \eta\varepsilon U_0^{-1} - \eta. \quad (5.3.18)$$

Since ε, η may be taken arbitrarily small and $U_0 > 0$

$$\mu\lambda \leq 1. \quad (5.3.19)$$

This proves Lemma 5.3.10. From (5.3.10), (5.3.4), (5.2.5), and (5.2.7)

$$\lambda \leq 1/\mu \leq \liminf_{t\to\infty} X(t)/t \leq \limsup_{t\to\infty} X(t)/t \leq \lim_{t\to\infty} Y(t)/t = \lambda, \quad (5.3.20)$$

which establishes Theorems 5.2.21 and 5.3.1.

5.4. The convergence of $x_t(\omega)$ as $t \to \infty$. We have shown (Theorem 4.1.11) that $m^{-1}t_{0m}(\omega)$ converges in probability for an arbitrary distribution U, and (Theorem 4.1.13) converges in mean square when the distribution U is bounded. Correspondingly, for the reach functions $x_t(\omega), y_t(\omega)$ we have

Theorem 5.4.1. *As the parameter $t \to \infty$, $x_t(\omega)/t$ and $y_t(\omega)/t$ both converge in mean square to the constant $\lambda(U) = [\mu(U)]^{-1}$.*

The proof of this theorem follows quite easily from

Theorem 5.4.2. *As $t \to \infty$, $\mathrm{var}[x_t(\omega)]$, and $\mathrm{var}[y_t(\omega)]$ are both $o(t^2)$.*

Comparing Theorem 5.4.2 with the corresponding result in renewal theory (SMITH 1958),

$$\mathrm{var}\, N_t \sim Kt, \qquad (K = \text{constant}), \quad (5.4.3)$$

we conjecture that Theorem 5.4.2 is a rather weak result. However, we see no method of proving as strong a result as the convergence of $t^{-1}\,\mathrm{var}\,x_t$ as $t \to \infty$.

Proof of Theorem 5.4.2. Consider the random variable, $y_{s_{0m}(\omega)}(\omega)$ which for all integers m, and all ω is identically equal to m,

$$\mathrm{var}\, y_{s_{0m}(\omega)}(\omega) = 0. \quad (5.4.4)$$

As before, for prescribed $\varepsilon > 0, \eta > 0$, define

$$\Omega_m = \{\omega; s_{0m}(\omega) \leq (\mu-\varepsilon)m\}. \quad (5.4.5)$$

Then by Theorem 4.2.22 there exists $m_0 = m_0(\varepsilon, \eta)$ for which

$$P(\Omega_m) \leq \eta, \qquad (m \geq m_0). \quad (5.4.6)$$

Thus since $y_t(\omega)$ is nondecreasing in t, for fixed ω, and $y_t(\omega) \leq t/U_0$ for all ω.

$$[y_{m(\mu-\varepsilon)}(\omega)]^2 \leq m^2, \qquad\qquad (\omega \in \Omega - \Omega_m) \qquad (5.4.7)$$

$$[y_{m(\mu-\varepsilon)}(\omega)]^2 \leq m^2(\mu-\varepsilon)^2/U_0^2, \qquad (\omega \in \Omega_m). \qquad (5.4.8)$$

Hence, by (5.4.7) and (5.4.8)

$$E[y_{m(\mu-\varepsilon)}(\omega)]^2 \leq m^2(1-\eta) + \eta m^2(\mu-\varepsilon)^2/U_0^2, \qquad (m \geq m_0). \qquad (5.4.9)$$

Hence,

$$\frac{\operatorname{var} y_{(m\mu-\varepsilon)}(\omega)}{m^2} \leq 1 - \eta + \frac{\eta(\mu-\varepsilon)^2}{U_0^2} - \frac{\left\{Y[m(\mu-\varepsilon)]\right\}^2}{m^2}. \qquad (5.4.10)$$

Taking the limit as $m \to \infty$ in (5.4.10) we obtain, since $\lambda\mu = 1$,

$$\limsup_{m \to \infty} \frac{\operatorname{var} y_{m(\mu-\varepsilon)}(\omega)}{m^2} \leq \eta\left[\frac{(\mu-\varepsilon)^2}{U_0^2} - 1\right] + \frac{\varepsilon}{\mu^2}(2\mu - \varepsilon); \qquad (5.4.11)$$

and since ε, η may be taken arbitrarily small we have proved Theorem 5.4.2 for $y_t(\omega)$. For x_t, we use $x_t(\omega) \leq y_t(\omega)$ to give

$$\begin{aligned}
\operatorname{var} x_t = Ex_t^2 - (Ex_t)^2 &\leq Ey_t^2 - (Ex_t)^2 \\
&\leq \operatorname{var} y_t + (Ey_t)^2 - (Ex_t)^2 \\
&\leq \operatorname{var} y_t + 2t(Ey_t - Ex_t)/U_0 \\
&= o(t^2) \text{ as } t \to \infty.
\end{aligned} \qquad (5.4.12)$$

Proof of Theorem 5.4.1.

$$\begin{aligned}
E\left[\frac{y_t(\omega)}{t} - \lambda\right]^2 &= E\left[\frac{y_t(\omega)}{t}\right]^2 - \frac{2\lambda Y(t)}{t} + \lambda^2 \\
&= \frac{\operatorname{var} y_t(\omega)}{t^2} + \frac{[Y(t)]^2}{t^2} - \frac{2\lambda Y(t)}{t} + \lambda^2.
\end{aligned} \qquad (5.4.13)$$

Taking the limit as $t \to \infty$ we have by Theorem 5.4.2 that

$$\lim_{t \to \infty} E\left[\frac{y_t(\omega)}{t} - \lambda\right]^2 = 0. \qquad (5.4.14)$$

This proves the result for $y_t(\omega)$, and similarly for $x_t(\omega)$.

5.5. *Further conjectures concerning* $x_t(\omega)$. We close this section by mentioning some further conjectures about $x_t(\omega)$ which will be derived in the main from corresponding results in renewal theory. Conjectures will be stated only in terms of $x_t(\omega)$. However, they possibly apply equally well to $y_t(\omega)$.

A result of considerable intuitive appeal is

$$\lim_{t \to \infty} X(t + a) - X(t) = \lambda a. \qquad (5.5.1)$$

This result is derived from the famous Blackwell renewal theorem. However, the intricacy of Blackwell's proof is such as to intimidate attacks on the even more difficult problem (5.5.1).

One conjecture we make, with more hope than the above, is that the results obtained in Sections 5.2, 5.3, and 5.4 hold for unbounded U distributions. Replacement of the lower bound U_0 by zero should not present too great a difficulty. However, we see no way of replacing U_1 by $+\infty$ and obtaining even the fundamental result (5.2.10).

Similarly, if we define a reach function in terms of the absolute first-passage time we again see no way of proving (5.2.10). This is because the introduction of absolute first-passage times destroys a great deal of the independence at present at our disposal. Nevertheless, we strongly suspect the above results to hold for such a reach function.

Many of the recent results obtained in renewal theory stem from the fact that the renewal function $H(t)$ satisfies the fundamental integral equation,

$$H(t) - F(t) + \int_0^t H(t-z)\,dF(z) \tag{5.5.2}$$

where $F(z)$ is the cumulative distribution function of X.

Corresponding to this result we have attempted to obtain a tractable integral inequality for $x_t(\omega)$; but so far, without success.

5.6. *Proof that the process $x(m, t; \omega)$ is measurable on $\Omega \times R$ for fixed m.* It is sufficient to prove the results for $x_s(\omega) = x(0, s; \omega)$, because by invariance under lateral shift, the distribution of $x(m, t; \omega)$ is independent of m.

$\Omega \times R$ is the product of the phase-space Ω and $R = (0, \infty)$. The measure on $\Omega \times R$ is the product measure induced by P, the probability measure on Ω, and Lebesgue measure on R.

$x_s(\omega)$ is an integer function on $\Omega \times R$. Hence it is sufficient to show that for any integer c

$$A(c) = \left\{ (\omega, s) : x_s(\omega) \leq c \right\} \tag{5.6.1}$$

is a measurable subset of $\Omega \times R$.

Now by definition of x

$$\left\{ (\omega, x) : x_s(\omega) \leq c \right\} \tag{5.6.2}$$

$$= \left[\bigcup_{m=1}^{c} \left\{ (\omega, s) : t_m(\omega) \leq s \right\} \right] \cap \left[\bigcap_{m=c}^{\infty} \left\{ (\omega, s) : t_m(\omega) > s \right\} \right].$$

But

$$\bigcap_{m=c+1}^{\infty} \left\{ (\omega, s) : t_m(\omega) > s \right\} = \left\{ (\omega, s) : \inf_{r \in K} t(r, \omega) > s \right\} \tag{5.6.3}$$

where K is a countable set of paths on the lattice. Now

$$\left\{ (\omega : s) : \inf_{r \in K} t(r, \omega) > s \right\} = \left\{ (\omega, s) : \inf_{r \in K} [t(r, \omega) - s] > 0 \right\}. \tag{5.6.4}$$

And for any r, s, $t(r, \omega)$ is a sum of random variables and is measurable on $\Omega \times R$. Hence $\inf_{r \in K} [t(r, \omega) - s]$ is measurable on $\Omega \times R$ and hence $\bigcap\limits_{m=c+1}^{\infty}\left\{(\omega, s): t_m(\omega) > s\right\}$ is measurable on $\Omega \times R$. In similar fashion $\left\{(\omega : s): t_m(\omega) \leqq s\right\}$ is measurable on $\Omega \times R$ and thus $A(c)$ is measurable on $\Omega \times R$ for all integers c. This proves the required result. A similar argument holds for $y_t(\omega)$ in place of $x_t(\omega)$.

6. The time constant regarded as a functional of the underling distribution

The central position of the time constant $\mu(U)$ is evident from preceding sections. Here we derive some results on the functional dependence of $\mu(U)$ on U for the square lattice. The techniques used in obtaining these results should extend to other lattices straightforwardly.

6.1. *Estimation of $\mu(U)$*. In (4.1.7) we stated the obvious inequality

$$0 \leqq \mu(U) \leqq \bar{u} \tag{6.1.1}$$

and showed that the bounds herein were attainable. However, Theorem 4.1.9 shows that attainment of the bounds is exceptional. This section looks for a better upper bound for $\mu(U)$. A simple but quite efficient algorithm for travel between the origin and $X = m$ has the following simple rule. At each point (m_1, n_1) at which we arrive, we take one of three continuations, choosing that one of the three continuations with the least time coordinate:

$$\left.\begin{aligned}
r_1 &\equiv \left\{(m_1, n_1) \rightarrow (m_1, n_1 + 1) \rightarrow (m_1 + 1, n_1 + 1)\right\}; \\
r_2 &\equiv \left\{(m_1, n_1) \rightarrow (m_1 + 1, n_1)\right\}; \\
r_3 &\equiv \left\{(m_1, n_1) \rightarrow (m_1, n_1 - 1) \rightarrow (m_1 + 1, n_1 - 1)\right\}.
\end{aligned}\right\} \tag{6.1.2}$$

This algorithm yields travel between the origin and $X = m$ in at most $2m$ steps. Clearly, in the notation of Section 4.4,

$$\mu(U) \leqq \Psi_1(0, 1) \leqq \Psi(0, 1) \leqq E \min\left\{u_1 + u_2, u_3, u_4 + u_5\right\} \tag{6.1.3}$$

where u_i $(i = 1, \ldots, 5)$ are independent observations from $U(u)$.

In two simple cases, this gives

$$\mu(U) \leqq .425, \tag{6.1.4}$$

when U is the rectangular distribution on $(0, 1)$,

$$\mu(U) \leqq .629, \tag{6.1.5}$$

when U is the exponential distribution $U(u) = 1 - e^{-u}$. These represent improvements of 15 per cent and 37 per cent respectively on (6.1.1). The following method sharpens these upper bounds even further by calculating $\Psi_1(0, 1)$ in (6.1.3). Let U be the underlying distribution with cumulative distribution function $U(u)$. Define $Q(u) = 1 - U(u)$. Let v be defined as the cylinder $(0 \leqq X < 1)$ first-passage time from the origin to the line $X = 1$, over paths which lie strictly in the upper path plane. Each such path is a single upward step followed by either a step to the right or another path of the same type, whichever is shorter. Hence v is the convolution of u with the minimum of u and another such v. Thus, defining $G(x) = P[v > x]$, we have

$$1 - G(x) = \int_0^x [1 - Q(x-y)\, G(x-y)]\, dU(y)\, , \qquad (6.1.6)$$

which simplifies to

$$G(x) = Q(x) - \int_0^x Q(x-y)\, G(x-y)\, dQ(y)\, , \qquad (6.1.7)$$

Now $\Psi_1(0, 1) = \min\{u_1, v_1, v_2\}$ where v_1, v_2 are the cylinder $(0 \leq X < 1)$ first-passage times from the origin to $X = 1$ over paths lying *strictly* in the upper and lower half plane respectively and u_1 is the time coordinate of the straight line arc from the origin to $(1, 0)$. Clearly v_1, v_2 are the independent random variables with the distribution of v above, and are also independent of u_1. So

$$\Psi_1(0, 1) = \int_0^\infty Q(x)\, [G(x)]^2\, dx\, , \qquad (6.1.8)$$

where G is the solution of (6.1.7).

Example 6.1.9. *Exponential distribution* $Q(x) = e^{-x}$. Here (6.1.7) becomes

$$G(x) = e^{-x} + e^{-x} \int_0^x G(y)\, dy\, . \qquad (6.1.10)$$

Thus $G(x)$ satisfies the differential equation

$$\frac{d}{dx}\, [e^x\, G(x)] = G(x)\, , \qquad (6.1.11)$$

which has the solution

$$G(x) = \exp(1 - x - e^{-x})\, . \qquad (6.1.12)$$

On substitution of (6.1.12) into (6.1.8) we find

$$\Psi_1(0, 1) = .59726\, . \qquad (6.1.13)$$

Although this is an improvement on (6.1.6) in this particular case, (6.1.3) is usually more tractable since it does not involve the solution of the integral equation (6.1.7)

Example 6.1.14. *Bernouilli distribution with* $p = \frac{1}{2}$. Let the time coordinates of the arcs of the lattice be 0 or 1 each with probability $\frac{1}{2}$. We obtain

$$\mu(U) \leq .2813\, , \qquad \text{via (6.1.3)}\, , \qquad (6.1.15)$$

$$\mu(U) \leq .167\, , \qquad \text{via (6.1.8)}\, . \qquad (6.1.16)$$

An interesting conjecture (which we think should not be too hard to prove, though we have not yet tried seriously) is prompted by the results obtained in ordinary percolation theory:

Conjecture 6.1.17. *When U is the Bernouilli distribution* $P(u = 0)$ $= 1 - P(u = 1) = p$ *and* $p \geq \frac{1}{2}$ *(the critical percolation probability for the square lattice), then* $\mu(U) = 0$.

A Monte Carlo estimation of $\mu\,(U)$ for these and other distributions is under way and we hope to publish some results shortly.

Although we have thus obtained some upper bounds for $\mu\,(U)$, the lower bound problem is completely unsolved.

6.2. *Monotonicity of $\mu\,(U)$*.

Theorem 6.2.1. *If two underlying cumulative distribution functions U_1 and U_2 satisfy $U_1\,(u) \leqq U_2\,(u)$ for all u, then $\mu\,(U_1) \geqq \mu\,(U_2)$.*

Proof. For any cumulative distribution function $U\,(u)$, we define the usual inverse function

$$U^{-1}\,(\xi) = \inf_{U\,(u)\,>\,\xi} u\,, \qquad (0 \leqq \xi < 1)\,, \tag{6.2.2}$$

which has the properties that

$$U\,[U^{-1}\,(u)] \geq u\,, \tag{6.2.3}$$

and

$$U_1^{-1}\,(\xi) \geqq U_2^{-1}\,(\xi) \text{ if } U_1\,(u) \leqq U_2\,(u) \text{ for all } u\,. \tag{6.2.4}$$

Also, if ξ is a random variable uniformly distributed on $(0, 1)$, then $U^{-1}\,(\xi)$ is a random variable distributed with cumulative distribution function $U\,(u)$.

Thus, if ω is any point in the phase space $(\Omega_0,\,B_0,\,P_0)$ induced by the uniform rectangular distribution on $(0, 1)$, and if $U^{-1}\,(\omega)$ is the sample point which assigns a time coordinate $U^{-1}\,(u_i)$ to the arc l_i whenever ω assigns u_i to l_i, then $U^{-1}\,(\omega)$ is a typical point of the phase space induced by the underlying distribution $U\,(u)$.

Consequently, if $U_1\,(u) \leqq U_2\,(u)$ for all u,

$$t\,[r;\,U_1^{-1}\,(\omega)] \geqq t\,[r;\,U_2^{-1}\,(\omega)] \tag{6.2.5}$$

for all $\omega \in \Omega_0$ and for any path r on the lattice. Hence

$$t_{0n}\,[U_1^{-1}\,(\omega)] \geqq t_{0n}\,[U_2^{-1}\,(\omega)]\,, \qquad (\omega \in \Omega_0)\,. \tag{6.2.6}$$

Take expectations of (6.2.6), divide by n, and let $n \to \infty$. This yields the required result.

Example 6.2.7. If $U_1\,(u) = 0$ or 1 according as $u < 0.45$ or $u \geq 0.45$, and if $U_2\,(u)$ is the uniform rectangular distribution on $[0, 1]$, then by (6.4.1)

$$\mu\,(U_2) \leqq 0.425 < 0.45 = \mu\,(U_1) \tag{6.2.8}$$

although $\overline{u}_2 = 0.5 > 0.45 = \overline{u}_1$. This counterexample shows that $\mu\,(U)$ is not in general a monotone function of the mean of U.

6.3. *The effect of elementary operations on U.* In practical examples one may need to study some sort of homogeneous transformation of ω. For instance, multiplying the time coordinate for each arc by a constant, multiplies $\mu\,(U)$ by the same constant. Again let $\omega \oplus k$ denote the time state of the lattice obtained by increasing the time coordinate of each arc by a constant k. Then if $N(r)$ is the number of arcs in the path r

$$t(r; \omega \oplus k) = t(r; \omega) + kN(r) . \tag{6.3.1}$$

Therefore

$$t_{0n}(\omega \oplus k) \leq t_{0n}(\omega) + kN_n(\omega) , \tag{6.3.2}$$

where $N_n(\omega)$ is the number of steps in the route of $t_{0n}(\omega)$. Taking expectations of (6.3.2), we have

$$\tau_k(0, n) = Et_{0n}(\omega \oplus k) \leq \tau(0, n) + kEN_n(\omega) . \tag{6.3.3}$$

This holds for $k \geq 0$. It also holds for negative values of k, provided that the distribution $U \oplus k$ remains that of a nonnegative random variable. Assuming this proviso is satisfied, we get from (6.3.3)

$$\tfrac{1}{2}\left[\tau_k(0, n) + \tau_{-k}(0, n)\right] \leq \tau(0, n) . \tag{6.3.4}$$

Hence

$$\tfrac{1}{2}\mu(U \oplus k) + \tfrac{1}{2}\mu(U \oplus - k) \leq \mu(U) \tag{6.3.5}$$

whenever $U \oplus \pm k$ are distributions of nonnegative random variables. This shows that $\mu(U \oplus k)$ is a convex functional of k within its region of definition. Also (6.3.3) shows that

$$\tau_k(0, n) \leq \tau(0, n) + kn , \qquad (k \leq 0) , \tag{6.3.6}$$

because $N_n(\omega) \geq n$. (Note that k is nonpositive in this relation.) Consequently, on dividing by n and letting $n \to \infty$, we have

$$\mu(U \oplus k) \leq \mu(U) + k , \qquad (k \leq 0) . \tag{6.3.7}$$

Thus, $\mu(U \oplus k)$ is, within its region of definition, a nondecreasing function of k which (for almost all k) has a derivative (with respect to k) not less than 1.

6.4. *Continuity of the functional $\mu(U)$.* With an appropriate metric, namely

$$d(U_1, U_2) = \sup_{\xi} \left| U_1^{-1}(\xi) - U_2^{-1}(\xi) \right| \tag{6.4.1}$$

on the space of all distribution functions $\{U\}$, $\mu(U)$ is a continuous functional of U provided that U is the distribution function of a random variable which is bounded away from zero. This follows easily enough by combining Theorem 6.2.1 with the properties of $\mu(U \oplus k)$ discussed in Section 6.3. The situation does not appear to be so simple when $U(0) > 0$.

6.5. *Convexity and concavity of $\mu(U)$.* The functional $\mu(U)$ is a convex functional of U^{-1} in the sense that

$$U^{-1}(\xi) = pU_1^{-1}(\xi) + qU_2^{-1}(\xi) , \qquad (p, q > 0, p + q = 1) \tag{6.5.1}$$

implies

$$\mu(U) \geq p\mu(U_1) + q\mu(U_2) . \tag{6.5.2}$$

To prove (6.5.2) we note that for any given path r

$$t\,[r;\,U^{-1}(\xi)] = t\,[r;\,pU_1^{-1}(\xi) + qU_2^{-1}(\xi)]$$

$$= pt\,[r;\,U_1^{-1}(\xi)] + qt\,[r;\,U_2^{-1}(\xi)]\,. \qquad (6.5.3)$$

$$\geq p\,\inf_{r\in R} t\,[r;\,U_1^{-1}(\xi)] + q\,\inf_{r\in R} t\,[r;\,U_2^{-1}(\xi)]$$

for any class of paths R; and hence

$$\inf_{r\in R} t\,[r;\,U^{-1}(\xi)] \geq p\,\inf_{r\in R} t\,[r;\,U_1^{-1}(\xi)] + q\,\inf_{r\in R} t\,[r;\,U_2^{-1}(\xi)]\,. \qquad (6.5.4)$$

On taking expectations, choosing R to be the paths from the origin to $(n, 0)$, dividing by n and letting $n \to \infty$, we get (6.5.2). Notice however that (6.5.4) is more general than (6.5.2) inasmuch as it applies to an arbitrary linear graph g with a countable set of arcs. We believe that $\mu\,(U)$ is a concave functional of U:

Conjecture 6.5.5. For $p, q > 0$ and $p + q = 1$,

$$\mu\,(pU_1 + qU_2) \leq p\mu\,(U_1) + q\mu\,(U_2)\,. \qquad (6.5.6)$$

7. Subadditive processes with a superadditive component

7.1. *Flow in a porous pipe.* For many physical applications of first-passage theory, we need only consider a subset of paths joining the points under consideration. For example, when considering the first-passage time between the origin and $(n, 0)$ on the square lattice, we could, with some heuristic justification, restrict ourselves to paths with fewer than n^2 arcs, the error thereby introduced being negligible for large n.

First-passage times over such restricted sets of paths are easier to study. This is exemplified below where we study the problem of first-passage times between the origin and $(n, 0)$ over paths which lie inside a strip (or *pipe*) of fixed width $2\,k$, bounded by the ordinates $Y = \pm\,k$, where k is a constant positive integer.

This problem is of interest because:

a) It has a certain physical significance of its own right, as a model for the maximum flow rate of fluid along a porous pipe, width $2\,k$, when the radii of the pores are chance variables, and the rate of flow depends on the radius of the pore. Fluid is, of course, an abstract term: it might be a detonation front, for instance.

b) It exhibits some new techniques in dealing with first-passage times on the lattice which might be useful in more complex situations.

c) It is a good example of a subadditive stochastic process which has strong *superadditive* properties.

Define $p_{mn}^k\,(\omega)$ to be the first-passage time between $(m, 0)$ and $(n, 0)$ over paths which lie strictly inside the rectangle bounded by the lines $Y = \pm\,k$, $X = m$, $X = n$. Then for k fixed, $p_{mn}^k\,(\omega)$ is a 2-parameter stochastic process on (Ω, B, P). By the inclusion lemma it is obvious that

$$p_{mn}^k\,(\omega) \geq p_{mn}^{k+1}\,(\omega) \geq t_{mn}\,(\omega), \qquad (\omega \in \Omega, k \geq 0)\,. \qquad (7.1.1)$$

The expected value $P_k(m, n)$ of $p_{mn}^k(\omega)$ exists and satisfies

$$\bar{u}(n-m) \geqq P_k(m, n) \geqq P_{k+1}(m, n) \geqq \tau(m, n), \qquad (k \geqq 0). \qquad (7.1.2)$$

Also it is not difficult to see that when k is fixed

$$p_{mn}^k(\omega) + p_{nq}^k(\omega) \geqq p_{mq}^k(\omega), \qquad (m \leqq n \leqq q); \qquad (7.1.3)$$

and in particular $\{p_{mn}^k(\omega)\}$ is, for fixed k, an independent nonnegative subadditive stochastic process.

Hence by Theorem 3.3.3 there exists a time constant $\mu_k(U)$ such that for k fixed

$$P_k(0, n)/n \geqq \mu_k(U) = \lim_{n \to \infty} P_k(0, n)/n, \qquad (7.1.4)$$

and by (7.1.2)

$$\bar{u} \geqq \mu_k(U) \geqq \mu_{k+1}(U) \geqq \mu(U), \qquad (k \geqq 0). \qquad (7.1.5)$$

Also since $p_{0n}^k(\omega)$ is nonincreasing in k for fixed n and ω, and

$$\lim_{n \to \infty} p_{0n}^k(\omega) = t_{0n}(\omega), \qquad (7.1.6)$$

the Monotone Convergence Theorem yields

$$\lim_{k \to \infty} P_k(m, n) = \tau(m, n), \qquad (\text{fixed } m, n); \qquad (7.1.7)$$

and it is not difficult to show

$$\lim_{k \to \infty} \mu_k(U) = \mu(U). \qquad (7.1.8)$$

For $m < n < q$ let r_1, r_2, r_3 be the routes of $p_{mq}^k(\omega)$, $p_{mn}^k(\omega)$, $p_{nq}^k(\omega)$, respectively; and let A be the point where r_1 first intersects the line $X = n - 1$, and let B be the point where r_1 last intersects $X = n + 1$. Replace the segment of r_1 from A to B by a path from A direct to $(n - 1, 0)$, thence direct to $(n + 1, 0)$, and thence direct to B. This detour (composed of three straight segments) has at most $2k + 2$ steps in it; and its expected transit time is at most $2(k + 1)\bar{u}$. The new path takes no shorter time to traverse than $r_2 * r_3$, by the definitions of r_2 and r_3. Hence taking expectations, we have for fixed k

$$P_k(m, q) + 2k\bar{u} + 2\bar{u} \geqq P_k(m, n) + P_k(n, q). \qquad (7.1.9)$$

Write (7.1.9) as

$$[P_k(m, q) - 2(k + 1)\bar{u}] \geqq [P_k(m, n) - 2(k + 1)\bar{u}] +$$
$$+ [P_k(n, q) - 2(k + 1)\bar{u}]. \qquad (7.1.10)$$

Thus, $[P_k(0, n) - 2(k + 1)\bar{u}]$ is a *superadditive* function of n for fixed k and hence (HILLE, Chapter 6) there exists a constant γ such that

$$[P_k(0, n) - 2(k + 1)\bar{u}]/n \leqq \lim_{n \to \infty} [P_k(0, n) - 2(k + 1)\bar{u}]/n = \mu_k(U) \qquad (7.1.11)$$

the last step following from (7.1.4). Consequently for fixed k,

$$0 \leq n\mu_k(U) \leq P_k(0, n) \leq n\mu_k(U) + 2(k+1)\bar{u}. \qquad (7.1.12)$$

7.2. *Convergence of $p_{0n}^k(\omega)/n$ as $n \to \infty$.*

Theorem 7.2.1. *As $n \to \infty$ the random variable $p_{0n}^k(\omega)/n$ converges with probability 1 to the time constant $\mu_k(U)$.*

Proof of Theorem 7.2.1. Whilst proving (7.1.9) we showed [in the argument preceding (7.1.9)] that

$$p_{mn}^k(\omega) + p_{nq}^k(\omega) \leq p_{mq}^k(\omega) + f(\omega), \qquad (m \leq n \leq q), \qquad (7.2.2)$$

where $f(\omega)$ was a random variable on (Ω, B, P) such that

$$0 \leq Ef(\omega) \leq 2(k+1)\bar{u}. \qquad (7.2.3)$$

Let q_i be the smallest connected sets of arcs connecting $(in-1, -k)$ to $(in-1, +k)$, connecting $(in-1, 0)$ to $(in+1, 0)$, and connecting $(in+1, -k)$ to $(in+1, +k)$. q_i therefore has the shape of the letter H.

Since each q_i $(i = 1, \ldots, j-1)$ intersects the route of $p_{0, jn}^k(\omega)$ and also passes through the point $(in, 0)$ by the connection lemma 2.3.2

$$p_{0, jn}^k(\omega) + \sum_{i=1}^{j-1} t(q_i, \omega) \geq \sum_{i=1}^{j-1} p_{in, (i+1)n}^k(\omega). \qquad (7.2.4)$$

Now the sequence $\{t(q_i, \omega)\}_{i=1}^{j-1}$ is a sequence of independent identically distributed random variables with finite mean $\bar{u}(4k+2)$. Hence

$$P\left[\lim_{j \to \infty} j^{-1} \sum_{i=1}^{j-1} \{t(q_i, \omega) - \bar{u}(4k+2)\} = 0\right] = 1 \qquad (7.2.5)$$

and since the right side of (7.2.4) is a sum of independent random variables

$$P\left[\lim_{j \to \infty} j^{-1} \sum_{i=0}^{j-1} p_{in, (i+1)n}^k(\omega) = P_k(0, n)\right] = 1. \qquad (7.2.6)$$

Hence by (7.2.4), (7.2.5), (7.2.6) we have

$$P\left[\liminf_{j \to \infty} j^{-1} p_{0, jn}^k(\omega) \geq P_k(0, n)\right] = 1. \qquad (7.2.7)$$

Thus (7.1.12) yields

$$P\left[\liminf_{j \to \infty} n^{-1} j^{-1} p_{0, jn}^k(\omega) \geq \mu_k(u)\right] = 1. \qquad (7.2.8)$$

Let now $m = jn + r$ where $0 \leq r < n$. By subadditivity

$$p_{0, m}^k(\omega) + p_{m, (j+1)n}^k(\omega) \geq p_{0, (j+1)n}^k(\omega). \qquad (7.2.9)$$

Also since $p_{m, (j+1)n}^k(\omega)$ has the distribution of $p_{0, n-r}^k(\omega)$, (3.5.30) gives

$$P\left[\lim_{j \to \infty} j^{-1} p_{0, n-r}^k(\omega) = 0\right] = 1. \qquad (7.2.10)$$

Consequently

$$P\left[\liminf_{m \to \infty} m^{-1} p_{0m}^k(\omega) \geq \mu_k(u)\right] = 1. \qquad (7.2.11)$$

However, $p_{0m}^k(\omega)$ is a smotherable subadditive process; so

$$P\left[\limsup_{m \to \infty} m^{-1} p_{0m}^k(\omega) = \mu_k(U)\right] = 1 \; ; \qquad (7.2.12)$$

and Theorem 7.2.3 is thereby proved.

The above subadditive process is an example of a type which might well be called *almost superadditive*.

8. Geometrical properties of first-passage routes

8.1. *Existence of first-passage routes.* We now consider two funda-mental problems associated with the study of first-passage theory on a graph g. Does there exist an algorithm for determining the (or a) route of any first-passage time? Indeed, does this optimal route exist? These problems are the main problems associated with PERT networks (Sec-tion 1). In Section 6 we gave two algorithms for relatively economic though not generally optimal travel between a point and a line on the square lattice. However, we make no claim that these are especially good algorithms: they are just the best we have so far found.

In Section 7 we considered the effect of limitations on the set of paths over which first-passage times are taken. Before limitations such as are imposed in Section 7 are justified, however, we need to find out more about the nature of the route of these first-passage times. This is the main purpose of Section 8.

Before we can logically discuss properties of a route of a first-passage time it is obviously necessary to consider whether or not this route exists. When we are dealing with a first-passage time over a finite set of paths on *any* graph g, a route of this first-passage time must exist. The route is a random connected path on the graph g. To show that the route of $t_R(\omega)$ exists when R is an infinite set of paths is however quite difficult. Here we shall let our graph be the square lattice and consider the existence of the routes of the first-passage time $t_{0n}(\omega)$, $a_{0n}(\omega)$ and $b_{0n}(\omega)$.

First notice that, when the time coordinates u_i of the arcs l_i of the lattice satisfy the bounding restriction

$$0 < U_0 \leqq u_i \leqq U_1 < \infty \,, \qquad (U_0, \, U_1 = \text{constants}), \qquad (8.1.1)$$

the first-passage times satisfy

$$nU_0 \leqq a_{0n}(\omega), \, b_{0n}(\omega), \, t_{0n}(\omega), \, s_{0n}(\omega) \leqq nU_1 \,. \qquad (8.1.2)$$

We need only consider the first-passage times over paths with $\leqq m$ arcs, where

$$mU_0 \leqq nU_1 \,. \qquad (8.1.3)$$

Since we are now considering first-passage times over only a finite set of paths, the route of each first-passage time exists for all ω.

If, however, the distribution U is not bounded as in (8.1.1) we can only partly solve the problem.

Theorem 8.1.4. *Routes of $t_{0n}(\omega)$, $s_{0n}(\omega)$ exist with probability* 1.

It is easy to show by a counter-example that this theorem is best-possible: that is, routes may fail to exist with probability 0.

Conjecture 8.1.5. *Routes of $a_{0n}(\omega)$, $b_{0n}(\omega)$ also exist with probability* 1 *even when U is an unbounded distribution.*

Intuitively Conjecture 8.1.5 is appealing. However, it will need a new idea to prove it as there is no way of extending the proof of Theorem 8.1.4 to cover Conjecture 8.1.5 as well.

Conjecture 8.1.5 is yet another example of the simplifying effect of bounding the distribution U.

Proof of Theorem 8.1.4. For fixed x_0, consider $t_{0x_0}(\omega)$ and $s_{0x_0}(\omega)$. We use the technique and notation in the proof of Theorem 4.2.10. If a pair of sets of arcs, one of them B_y and the other the mirror image of B_y in the X-axis, are both barriers, the route must exist and lie between these barriers because only finitely many paths do not cross either barrier and because the time coordinate of every path crossing a barrier is at least equal to that along the X-axis. Hence, a route must exist unless B_y and its reflection are not barriers for $y = 1$, $1 + (x_0 + 1)$, $\ldots$, $1 + (k-1)(x_0 + 1)$; and the probability of this is at most $(1 - \pi_0)^{2k}$. But k is arbitrary; and this probability tends to zero as $k \to \infty$. The argument holds for each prescribed x_0. This completes the proof.

8.2. *The number of arcs in the routes of first-passage time.* The number of arcs in the route of the first-passage times $t_{0n}(\omega)$, $s_{0n}(\omega)$ is a very important quantity. Knowledge of its behavior would yield better inequalities in Section 6 on the functional dependence of $\mu(U)$ on U. We shall consider only the number of arcs in the routes of $s_{0m}(\omega)$ and $t_{0m}(\omega)$. The extension to absolute routes (assuming they exist) is not hard. From now on we consider only the subset Ω' of Ω, $P(\Omega') = 1$, for which the routes of $t_{0m}(\omega)$, $s_{0m}(\omega)$ exist.

Let $N_n(\omega)$ and $N_n^*(\omega)$ be the number of arcs in the routes of $t_{0n}(\omega)$ and $s_{0n}(\omega)$, respectively. Let $U(u)$ be any distribution, which is bounded away from zero: that is,

$$U(u) = 0 \text{ for } u \leqq U_0 > 0, \tag{8.2.1}$$

where U_0 is a constant. Then with the notation of Section 6.3 and the result (6.3.5) in mind, we see that $\mu(U \oplus k)$ is a nondecreasing convex function of k for all $k > - U_0$. Hence

$$\varrho(k) = \partial\mu(U \oplus k)/\partial k \tag{8.2.2}$$

exists for almost all $k > - U_0$.

Theorem 8.2.3. *Suppose the underlying distribution U satisfies (8.2.1). Then as $n \to \infty$, both $N_n(\omega)/n$ and $N_n^*(\omega)/n$ converge in probability to $\varrho(0)$, provided $\varrho(0)$ exists.*

Proof. We cite the proof for $N_n(\omega)/n$. The corresponding proof for $N^*(\omega)/n$ is exactly similar, except that $s_{0n}(\omega)$ replaces $t_{0n}(\omega)$. We have, by (6.3.2),

$$t_{0n}(\omega \oplus k) \leq t_{0n}(\omega) + kN_n(\omega) , \qquad (k > -U_0) . \qquad (8.2.4)$$

Prescribe $\varepsilon > 0$, and assume that $\varrho(0)$ exists. Choose $k = k(\varepsilon) > 0$ so that

$$\mu(U \oplus k) - \mu(U) \geq [\varrho(0) - \varepsilon] k . \qquad (8.2.5)$$

This is possible because $\mu(U \oplus k)$ is convex. From (8.2.4) and (8.2.5) we have

$$P[N_n(\omega)/n \leq \varrho(0) - 3\,\varepsilon] \leq$$

$$\leq P\left[\frac{t_{0n}(\omega \oplus k) - t_{0n}(\omega)}{nk} \leq \frac{\mu(U \oplus k) - \mu(U)}{k} - 2\,\varepsilon\right] \qquad (8.2.6)$$

$$= P\left\{\left[\frac{t_{0n}(\omega)}{n} - \mu(U) - \varepsilon k\right] + \left[\mu(U \oplus k) - \frac{t_{0n}(\omega \oplus k)}{n} - \varepsilon k\right] \geq 0\right\}$$

$$\leq P\left[\frac{t_{0n}(\omega)}{n} \geq \mu(U) + \varepsilon k\right] + P\left[\frac{t_{0n}(\omega \oplus k)}{n} \leq \mu(U \oplus k) - \varepsilon k\right]$$

and the right side tends to zero as $n \to \infty$ by virtue of Theorem 4.1.11. Thus

$$\lim_{n \to \infty} P[N_n(\omega)/n \leq \varrho(0) - 3\,\varepsilon] = 0 . \qquad (8.2.7)$$

Next choose k so that

$$\mu(U \oplus k) - \mu(U) \geq [\varrho(0) + \varepsilon] k , \qquad (-U_0 < k < 0) . \qquad (8.2.8)$$

Then

$$P[N_n(\omega)/n \geq \varrho(0) + 3\,\varepsilon] \leq$$

$$\leq P\left[\frac{t_{0n}(\omega \oplus k) - t_{0n}(\omega)}{kn} \geq \frac{\mu(U \oplus k) - \mu(U)}{k} + 2\,\varepsilon\right] \qquad (8.2.9)$$

$$= P\left\{\left[\frac{t_{0n}(\omega)}{n} - \mu(U) + \varepsilon k\right] + \left[\mu(U \oplus k) - \frac{t_{0n}(\omega \oplus k)}{n} + \varepsilon k\right] \geq 0\right\}$$

and this tends to zero as in (8.2.6). [It should be noted that k is negative throughout (8.2.8) and (8.2.9).] Consequently

$$\lim_{n \to \infty} P[N_n(\omega)/n \geq \varrho(0) + 3\,\varepsilon] = 0 ; \qquad (8.2.10)$$

and (8.2.7) and (8.2.10) complete the proof, since ε is arbitrary.

A more difficult, but interesting mathematical problem is the following: "What is the probability $p_m(r)$ that the route of $t_{0m}(\omega)$ has exactly r arcs?" Since any path to $(m, 0)$ from the origin has $m + 2k$ steps where k is an integer, it is obvious that

$$\begin{aligned}
p_m(r) &= 0 , \qquad (r < m) , \\
p_m(r) &= 0 , \qquad (r - m = \text{odd integer}) .
\end{aligned} \qquad (8.2.11)$$

We suspect that no general answer can be made to this question, the result varying considerably from distribution to distribution. However, if the problem could be solved for the uniform rectangular distribution on [0, 1] for example, the techniques used would probably be of considerable interest.

One result available, which although not solving the problems posed above may be of help in future work is the following:

Define $t_{mn}^{k}(\omega)$ to be the cylinder first-passage time between $(m, 0)$ and $(n, 0)$ over paths which have at most $k(n-m)$ arcs. Then it is easily shown that for fixed $k\{t_{mn}^{k}(\omega)\}$ is an independent (and therefore self-smothering) subadditive stochastic process; and by Theorem 3.5.38 there exists a constant $\nu_k(U)$ such that

$$t_{0n}^{k}(\omega)/n \to \nu_k(U) \text{ in probability as } n \to \infty. \tag{8.2.12}$$

Also the connection lemma 2.3.2 yields

$$t_{0n}^{k_1}(\omega) + t_{n,2n}^{k_2}(\omega) \geq t_{0,2n}^{(k_1+k_2)/2}(\omega). \tag{8.2.13}$$

Thus, taking expected values of (8.2.13), dividing by n, and letting $n \to \infty$, we get that $\nu_k(U)$ is a convex function of k for fixed U, because

$$[\nu_{k_1}(U) + \nu_{k_2}(U)]/2 \geq \nu_{(k_1+k_2)/2}(U). \tag{8.2.14}$$

8.3. *The height problem.* More important mathematically than either of the problems discussed in previous sections is the *height problem.* If the route of $s_{0n}(\omega)$ terminates at the point $[n, h_n(\omega)]$, we define $\theta(n) = E\,|\,h_n(\omega)\,|$; and the height problem is to discuss the behavior of $\theta(n)$ as $n \to \infty$. No progress has yet been made, but the importance of the problem springs from the following theorems.

Theorem 8.3.1. *The expected values $\tau(0, n)$, $\Psi(0, n)$ satisfy*

$$\tau(0, m+n) \geq \Psi(0, n) + \Psi(0, m). \tag{8.3.2}$$

Proof. Let r be the route of $t_{0,m+n}(\omega)$ and let r meet the line $X = n$ at a point P. Let r_1, r_2 be the portions of r which run from $(0, 0)$ to P and from $(m+n, 0)$ to P, respectively. Then

$$t_{0,m+n}(\omega) = t(r, \omega)$$
$$= t(r_1, \omega) + t(r_2, \omega). \tag{8.3.3}$$

Now inspection instantly shows that $t(r_1, \omega) \geq s_{0n}(\omega)$ while $t(r_2, \omega) \geq s_{m+n,n}(\omega)$. By symmetry and stationarity $s_{m+n,n}(\omega)$ has the distribution of $s_{0,m}(\omega)$ and hence taking expected values of (8.3.3) we get (8.3.2). Similarly

Theorem 8.3.4. *For m, n, any integers ≥ 0, the corresponding absolute times satisfy*

$$\alpha(0, m+n) \geq \beta(0, n) + \beta(0, m). \tag{8.3.5}$$

Theorem 8.3.6. *If* $\theta\,(n)$ *is the height function, then*

$$\tau\,(0,\,n) \leqq \Psi\,(0,\,n) + \bar{u}\theta\,(n)\ . \tag{8.3.7}$$

Proof. Annex the straight path $[n,\,h_n\,(\omega)] \to (n,\,0)$ to the end of the route of $s_{0n}\,(\omega)$. The expected transit time along this combined path from the origin to $(n,\,0)$ is the right side of (8.3.7).

Combining Theorems 8.3.1 and 8.3.6, we get

$$\tau\,(0,\,m+n) \geqq \Psi\,(0,\,m) + \Psi\,(0,\,n) \geqq \tag{8.3.8}$$
$$\geqq \tau\,(0,\,m) - \bar{u}\theta\,(m) + \tau\,(0,\,n) - \bar{u}\theta\,(n)\ ;$$

and hence the function

$$\tau^*\,(n) = \tau\,(0,\,n) - \bar{u}\theta\,(n) \tag{8.3.9}$$

satisfies

$$\tau^*\,(m) + \tau^*\,(n) \leqq \tau^*\,(m+n) + \bar{u}\theta\,(m+n)\ . \tag{8.3.10}$$

Thus $\tau^*\,(m)$ is a generalized superadditive function [HAMMERSLEY (1962)]; and accordingly, we could deduce the rate of convergence of $\tau\,(0,\,n)/n$ from a knowledge of the behavior of $\theta\,(n)$ as $n \to \infty$.

Another result which would follow from a slightly modified version of the height problem is Conjecture 4.4.5. For let the route of $b_{0n}\,(\omega)$ meet $X = n$ at $P = [n,\,h_n^b\,(\omega)]$. Then by dropping a perpendicular from P to $(n,\,0)$ we have by a simple combination of the connection and inclusion lemmas, that

$$\beta\,(0,\,n) + \bar{u}E\,|\,h_n^b\,(\omega)\,| \geqq \alpha\,(0,\,n)\ . \tag{8.3.11}$$

Since $\lim_{n\to\infty} \alpha\,(0,\,n)/n = \mu$, Conjecture (4.4.5) would be a consequence of the conjecture

$$\lim_{n\to\infty} E\,|\,h_n^b\,(\omega)\,|\,/n = 0\ . \tag{8.3.12}$$

The highways and byways problem. Let $r\,(X,\,Y)$ denote the route from the origin to the point $(X,\,Y)$. An arc l is called *byway arc* or a *highway arc* according as it belongs to the routes of $r\,(X,\,Y)$ for finitely many or infinitely many $(X,\,Y)$. Let $f\,(R)$ denote the number of highway arcs which intersect the circumference of the circle $X^2 + Y^2 = R^2$. Does $f\,(R) \to \infty$ as $R \to \infty$; and, if so, how fast?

References

BIGELOW, C. G.: Bibliography on project planning and control by network analysis 1959—61. Op. Res. **10**, 728 (1962).

DOOB, J. L.: Stochastic Processes. New York: Wiley 1952.

ERDÖS, P.: Remark on my paper "On a theorem of Hsu and Robbins" Ann. Math. Statist. **21**, 138 (1950).

FELLER, W.: An introduction to probability theory and its applications. New York: Wiley 1957.

FRISCH, H. L., and J. M. HAMMERSLEY: Percolation processes and related topics. J. Soc. Indust. Appl. Math. **11**, 894 (1963).

FULKERSON, D. R.: Expected critical path lengths in PERT networks. Op. Res. **10**, 808 (1962).

HAMMERSLEY, J. M.: Generalization of the fundamental theorem on subadditive functions. Proc. Cambridge Phil. Soc. **58**, 235 (1962).

HILLE, E.: Functional Analysis and Semigroups. Amer. Math. Soc. Colloq. Publ. **1957**, 31.

KOCHEN, M., C. ABRAHAM, and E. WONG: Adaptive man-machine concept-processing. Air Force Cambridge Research Laboratories Report No. 397 (1962).

MALCOLM, D. G., J. H. ROSEBOOM, E. E. CLARK, and W. FAZAR: Application of a technique for research and development program evaluation. Op. Res. **7**, 646 (1959).

POLLACK, M.: Solutions of the kth best route through a network — a review. To appear in J. Math. Anal. and Appl.

—, and W. WIEBENSON: Solutions of the shortest route problem — a review. Op. Res. **8**, 224 (1960).

SMITH, W. L.: Renewal theory and its ramifications. J. Roy. Statist. Soc. B. **20**, 243 (1958).

Direct Product Branching Processes
and Related Induced Markoff Chains
I. Calculations of Rates of Approach to Homozygosity*

By SAMUEL KARLIN and JAMES MC GREGOR

Department of Mathematics Stanford University, Stanford, California

In this paper we introduce a class of finite state Markoff chains of special structure which includes many cases of interest in applications. For these chains we will determine a full set of eigenvalues (Sections 3 and 4) and provide their probabilistic interpretations (Section 5) which in the present context is rather striking. An intrinsic characterization of these processes and some associated limit theorems will be elaborated in a separate publication (see also [4]).

The developments of this paper were inspired by certain genetics models and it is instructive to review this background first.

Certain idealized genetics models were proposed by S. WRIGHT and R. FISHER to investigate the fluctuation of gene frequency under the influence of mutation, migration, selection and genetic drift. For the sake of completeness we review the essential features of these models. We begin by formulating the simplest model. The model describes a haploid population of two types which under the circumstance of random mating can be interpreted as the fluctuations of a gamete population for a diploid structure involving two allelomorphs (= types). For some discussion of the biological justification and interpretation we refer the reader to WRIGHT [8], see also [1], [7].

Consider a fixed population of N elements which are either of type a or A. The next generation is formed by N independent binomial trials as follows: If the parent population consists of j a-types and $N - j$ A-types then each trial results in a or A with probabilities

$$p_j = \frac{j}{N} \qquad q_j = 1 - \frac{j}{N}$$

Repeated samplings are made with replacement.

By this procedure we generate a Markov chain $\{X_n\}$ where X_n is the number of a-genes in the nth generation in a population size of N elements. The state space consists of $N + 1$ values $\{0, 1, 2, \ldots, N\}$. The transition matrix is computed according to the binomial distribution as

$$Pr\{X_{n+1} = k \mid X_n = j\} = P_{jk} = \binom{N}{k} p_j^k q_j^{N-k}. \tag{1}$$

---

* Prepared under Auspices of National Institutes of Health GM 10452-01 A 1.

Notice that states 0 and N are permanent absorbing (or sometimes referred to as states of fixation). One of the standard questions of interest is to determine the probability, under the condition $X_0 = i$, that the population will attain fixation consisting only of a-types $(A$-types). It is also pertinent to determine the rate of fixation. It is in this respect that the knowledge of the eigenvalues are important.

FELLER [1] observed that the transition matrix (1) transforms polynomials into polynomials and exploiting this property he was able to determine the eigenvalues of (1). It is easy to prove and quite well-known that the largest eigenvalue less than 1 gives the rate of approach to homozygosity (fixation). The other eigenvalues are valuable in analyzing for the corresponding multi-type version the rate at which a certain number of types disappear from the population. (Results of this kind and extensions will be developed in Section 5 below.)

The eigenvectors of the Wright model are essentially unknown even for the simple situation of (1). Their knowledge in addition to providing a representation for P_{ij}^n would be useful for determining various probabilistic quantities of interest. In this paper one of the objectives is to develop the method of Feller considerably further. We will determine the eigenvalues for several other Markov processes appropriate in describing population growth and gene frequency fluctuations. In a separate paper we will present several extensions of these methods to treating various biological stochastic models describing general genetic mating systems (e.g., positive assortative mating, combinations of random mating and assortativeness, geographical spread, bisexual models, etc.).

We return now to review briefly the variations of the Wright model taking account of mutation, migration and selection forces. To simplify the discussion we will introduce these factors one at a time.

The following variant of Wright's model takes account of mutation pressures. We assume that prior to the formation of each new generation, each type has the possibility to mutate, that is, to change into a type of the other kind. Specifically, we assume that the mutation

$$a \to A \text{ occurs with probability } \alpha_1,$$
$$A \to a \text{ occurs with probability } \alpha_2.$$

Again we assume that the composition of the next generation results from N independent binomials trials. The relevant values of p_j and q_j when the parent population consists of the j a-types are determined according to the formulas.

$$p_j = \frac{j}{N}(1 - \alpha_1) + \left(1 - \frac{j}{N}\right)\alpha_2$$

$$q_j = \frac{j}{N}\alpha_1 + \left(1 - \frac{j}{N}\right)(1 - \alpha_2) \tag{2}$$

The transition probability matrix of the associated MC is calculated from (1) using the values of p_j and q_j given in (2).

If $\alpha_1 \alpha_2 > 0$ then fixation will not occur in any state. Instead the distribution function of X_n will approach, as $n \to \infty$, a steady state distribution of a random variable ξ where $Pr\{\xi = k\} = \pi_k$ ($k = 0, 1, 2, \ldots, N$), $\left(\sum_{k=0}^{N} \pi_k = 1, \pi_k > 0 \right)$.

We next formulate the model where a selective advantage favors say, the a-type over the A-type. A selective advantage is assured by assuming that the relative number of offspring of a and A types per individual have expectations proportional to $1 + \sigma$ and 1 respectively where σ is positive. Accordingly we replace for p_j and q_j

$$p_j = \frac{(1+\sigma)\, j}{N + \sigma j}, \qquad q_j = 1 - p_j \tag{3}$$

and build the next generation by binomial sampling as before. The denominator in p_j is, of course, a normalizing constant. If the parent population consisted of j a-types, the expected population size of a-types and A-types respectively in the next generation is

$$N \frac{(1+\sigma)\, j}{N + \sigma j}, \qquad N \frac{(N - j)}{N + \sigma j}$$

The expected ratio of a-types to A-types at the $n + 1$th generation is obviously $[(1 + \sigma)/1] \cdot [j/(N - j)]$. In this model the states 0 and N are again absorbing.

A factor of migration can be introduced into the model as follows: We take

$$p_j = \frac{\alpha j + \gamma}{\alpha N + \gamma + \delta}, \qquad q_j = \frac{\alpha (N - j) + \delta}{\alpha N + \gamma + \delta} \tag{4}$$

where $\alpha > 0$, $\gamma > 0$ and $\delta > 0$. The interpretation of (4) is as follows. The probability of producing an a-type is proportional to the current frequency of a-types plus a constant factor γ which represents an immigration rate into the system from the outside; a similar connotation applies to q_j. The denominator in (4) is the normalizing constant so that $p_j + q_j = 1$.

The transition matrix for the process again is constructed in accordance with the rules of binomial sampling.

It is very important to emphasize at this point that the selection model leads to probabilities p_j which are bona fide rational functions of j while in the mutation or migration models the function p_j is a linear function of j. This property appears to be crucial in the analysis of the mutation and migration models.

114   Samuel Karlin and James McGregor

I. Branching processes and frequency models

There is an intimate connection of the theory of branching processes and the frequency models presented above. We will now develop this relationship. There is some similarity of the considerations below and the point of view of MORAN [7].

A branching process has the following structure. An individual at the end of its lifetime produces a random number ξ of offsprings with probability distribution

$$Pr\{\xi = k\} = a_k \qquad k = 0, 1, 2, \ldots \qquad (5)$$

where, as usual, $a_k \geq 0$ and $\sum_{k=0}^{\infty} a_k = 1$. We assume that all children (offspring) act independently of each other and at the end of their lifetimes (for simplicity, the lifespan of each individual is assumed henceforth to be the same), have progeny in accordance with the probability distribution (5); and so on. The process of random variables X_n which counts the number of individuals in the nth generation is a Markov chain referred to as a branching process. Let $f(s)$ denote the generating function of ξ, i.e.,

$$f(s) = \sum_{k=0}^{\infty} a_k s^k \qquad (6)$$

and define

$$f_n(s) = f[f_{n-1}(s)] \qquad n = 1, 2, 3 \ldots \qquad (7)$$

where $f_0(s) = s$.

It is clear that $[f_n(s)]^i$ is the generating function of X_n provided $X_0 = i$.

Now consider two populations of types a and A respectively, each independently multiplying according to branching processes. The two dimensional process unfolds as a sequence of pairs of random variables $Z_n = \{X_n, Y_n\}$ where X_n (Y_n) denotes the number of a-types (A-types) in the nth generation. In the above formulation, the components X_n and Y_n generate independent branching processes so that Z_n is the direct product process. Let the generating functions of the number of offspring of a-types and A-types per individual be $f(s)$ and $g(s)$ respectively. Suppose initially we have a population of i a-types and j A-types. After one generation the joint probability distribution of the populations of a and A types has a generating function $H(s, t) = f^i(s) g^j(t)$. The joint probability that the first generation contains k a-types and a total population of M individuals of either type is the coefficient of $s^k t^M$ in the generating function $f^i(st) g^j(t)$. Symbolically

$$Pr\{X_1 = k, X_1 + Y_1 = M \mid X_0 = i, Y_0 = j\} \qquad (8)$$
$$= \text{coefficient of } s^k t^M \text{ in } f^i(st) g^j(t).$$

Moreover, the probability that the offspring generation includes k

a-types conditioned that the total progeny is M is computed by the formula

$$Pr\left\{X_1 = k \mid X_0 = i,\, Y_0 = j,\, X_1 + Y_1 = M\right\} \tag{9}$$
$$= \frac{\text{coefficient of } s^k\, t^M \text{ in } f^i\, (st)\, g^j\, (t)}{\text{coefficient of } t^M \text{ in } f^i\, (t)\, g^j\, (t)}$$

If we specialize so that

$$f(s) = e^{\lambda\,(s-1)}, \qquad g(s) = e^{\mu\,(s-1)} \qquad (\lambda > 0,\, \mu > 0)$$

then (9) becomes

$$\binom{M}{k}\left(\frac{i\lambda}{i\lambda + j\mu}\right)^k \left(\frac{j\mu}{i\lambda + j\mu}\right)^{M-k}. \tag{10}$$

A branching process for which $f\,(s) = e^{\lambda\,(s-1)}$ is called a Poisson branching process of parameter λ.

In the particular case where $\lambda = \mu$ (10) reduces to (1) provided $M = i + j = N$. That is when $M = N = i + j$, we obtain the transition probability matrix (1) by taking the direct product of two Poisson branching processes of the same parameter, one for a-types and the other for A-types and calculate the probability distribution for the number of a-types under the condition that the initial population consisted of N individuals (i of a-type and $N - i$ of A-type) and the total number of offsprings was precisely N individuals. When $\lambda \neq \mu$ and $M = N = i + j$ then (10) is precisely the transition probability matrix of Wright's selection model where $\lambda/\mu = 1 + \sigma$.

In the context of branching processes we can introduce the effects of migration in the following manner. We assume that in addition to the breeding of offspring there is immigration into the system independent of the population size. Let $h\,(s)$ be the probability generating function for the number of a-type immigration and $k\,(t)$ the probability generating function of immigration of A-type. The joint generating function for the population of a-type and of A-type in the next generation is $f^i\,(s)\,h\,(s)\,g^j\,(t)\,k\,(t)$ where i and j are the number of a-types and A-types respectively of the parent generation. The generating function of the total progeny population is $f^i\,(s)\,h\,(s)\,g^j\,(s)\,k\,(s)$. Finally, analogously to (9) we see that the expression

$$Pr\left\{X_1 = k \mid X_0 = i,\, Y_0 = j,\, X_1 + Y_1 = M\right\} = P\,(k \mid i, j, M) \tag{11}$$
$$= \frac{\text{coefficient of } s^k\, t^M \text{ in } f^i\,(st)\,h\,(st)\,g^j\,(t)\,k\,(t)}{\text{coefficient of } t^M \text{ in } f^i\,(t)\,h\,(t)\,g^j\,(t)\,k\,(t)}$$

gives the probability that the next generation is composed of k a-types conditioned that the total progeny is M.

Mutation may be introduced in two ways depending upon when this transformation occurs. We will assume that each a-type individual may mutate into an A-type with probability α_1 ($0 \leq \alpha_1 \leq 1$) and each A-type can mutate into an a-type with probability α_2 ($0 \leq \alpha_2 \leq 1$). With respect to mutation pressures, individuals act independently. To form the next generation we may postulate that mutation occurs first followed by branching multiplication or in the other order. They lead to different Markov chains.

It is easiest to describe the mechanism of the mutation model directly in terms of the corresponding probability generating functions. We begin with the case where mutation follows reproduction. Let $f(s)$ $[g(t)]$ represent, as before, the generating function of the number of progeny of one individual of type a (A). Suppose on reproduction each offspring of an a-type (A-type) can produce individuals of both kinds with generating function $A(s, t)$ $[B(s, t)]$. In other words, we postulate two stages of reproduction, the first corresponding to the usual multiplication process where the children are replicas of the parent while the second phase corresponds usually to a transformation of one type into the other type. The final generating function culminating both states of reproduction depicting the offspring population stemming from one individual of type a is $f[A(s, t)]$ and that of a type A individual is $g[B(s, t)]$. The formulation of a multiplicative process with several stages of reproduction and growth should now be obvious to the reader.

The mutation mechanism is obtained by specializing as follows. Take

$$A(s, t) = (1 - \alpha_1) s + \alpha_1 t \text{ and } B(s, t) = \alpha_2 s + (1 - \alpha_2) t. \qquad (12)$$

Its interpretation is clear. The second reproduction stage is a conversion process. With probability $1 - \alpha_1$ the offspring of an a-parent remains of type a and with probability α_1 mutates and becomes an individual of type A. A similar interpretation is ascribed to the generating function $B(s, t) = \alpha_2 s + (1 - \alpha_2) t$.

In summary, the generating function of the offspring population resulting from reproduction of one individual of type a affected afterwards by mutation pressures is

$$f(s, t) = f[(1 - \alpha_1) s + \alpha_1 t]. \qquad (13)$$

In a similar way, we see that the progeny population due to one individual of type A taking account of mutation pressures is

$$g(s, t) = g[(1 - \alpha_2) t + \alpha_2 s]. \qquad (14)$$

On the other hand if mutation occurs before reproduction the generating functions become

$$f^*(s, t) = (1 - \alpha_1) f(s) + \alpha_1 g(t), \qquad g^*(s, t) = (1 - \alpha_2) g(t) + \alpha_2 f(s).$$

The preceding discussion suggests the following general construction. We postulate that each individual of type a can produce offspring of both types. We denote its generating function as $f(s, t)$. Similarly, we assume that an individual of type A may produce individuals of both types and let $g(s, t)$ designate the generating function of the progeny. Let $h(s, t)$

represent the generating function of the number of a and A types immigrating into the system during each period. Let $(X_n, Y_n) = Z_n$, $n = 0, 1, \ldots$, denote the resulting two-dimensional branching process. The generating function of Z_1 for the initial condition $X_0 = i$, $Y_0 = j$ is

$$[f(s, t)]^i [g(s, t)]^j h(s, t) . \tag{15}$$

The transition probability matrix obtained by conditioning that the population size has a fixed size is calculated in the usual way. We get

$$Pr\{X_1 = k \mid X_0 = i, Y_0 = j; X_1 + Y_1 = M\} = P(k \mid i, j; M) \tag{16}$$
$$= \frac{\text{coefficient of } s^k t^M \text{ in } f^i(st, t) g^j(st, t) h(st, t)}{\text{coefficient of } t^M \text{ in } f^i(t, t) g^j(t, t) h(t, t)} .$$

For the special choice $f(s, t) = \exp\{\lambda [(1 - \alpha_1) s + \alpha_1 t] - 1]\}$, $g(s, t) = \exp\{\mu [(1 - \alpha_2) t + \alpha_2 s - 1]\}$ [see (13) and (14)] and $h(s, t) \equiv 1$ the expression (16) provided $i + j = N = M$ reduces to Wright's transition probability matrix for the model involving mutation and selection. If we take $h(s, t) = \exp[a(s - 1) + b(t - 1)]$ keeping $f(s, t)$ and $g(s, t)$ unaltered when $M = N = i + j$ then (16) becomes

$$P_{ik}(N) = \binom{N}{k} [(1 - \alpha_1) \lambda i + \mu(N - i) \alpha_2 + a]^k$$
$$\times \frac{[\alpha_1 \lambda i + (1 - \alpha_2) \mu(N - i) + b]^{N-k}}{[\lambda i + \mu(N - i) + a + b]^N} . \tag{17}$$

(Here N is a fixed parameter representing the constant population size.)

The formula (16) when $i + j = N = M$ by various specifications of $f(s, t)$, $g(s, t)$ and $h(s, t)$ provides a variety of interesting transition probability matrices P_{ik} associated with Markov chains on the state space $(0, 1, 2, \ldots, N)$.

Returning to the simplest direct product branching process we record other examples of (16) which arise in different biological situations.

A. If $f(s) = g(s) = (1 - p + ps)^\varkappa$, $\qquad i + j = M = N$
then
(13) $P(r \mid i, j; M) = \text{Prob}\{r \ a\text{-types and } M - r \ A\text{-types} \mid \text{total of } M \text{ offsprings}\}$

$$= \frac{\binom{i\varkappa}{r}\binom{j\varkappa}{M-r}}{\left[\begin{matrix}(i+j)\varkappa \\ M\end{matrix}\right]} \qquad r = 0, 1, \ldots, M \tag{18}$$

where the initial population composition consisted of i a-types and j A-types. Notice that this formula is independent of p.

B. Suppose the a and A types reproduce according to negative binomial distributions with parameter p. Specifically

$$f(s) = \frac{(1-p)^\alpha}{(1-ps)^\alpha}, \qquad g(t) = \frac{(1-p)^\beta}{(1-pt)^\beta}$$

118   SAMUEL KARLIN and JAMES MCGREGOR

then (9) becomes

$$P(r \mid i, j; M) = \frac{\binom{i\alpha + r - 1}{r}\binom{j\beta + M - r - 1}{M - r}}{\binom{i\alpha + j\beta + M - 1}{M}}.$$

Example A with $\varkappa$ equal to 2 occurs as a model proposed by KIMURA describing polysomic inheritance. The state variable of this process is the number of mutant subunits. Specifically each chromosone consists of N subunits and suppose a mutation has occurred in one of them. The subunits duplicate to produce $2N$ which divide at random into two daughter chromosones of N subunits. A single line of descent is observed. The state in each generation designates the number of mutant subunits contained in the cell. The transition probabilities are given by

$$P_{nm} = \frac{\binom{2n}{m}\binom{2N - 2n}{N - m}}{\binom{2N}{N}}$$

and this is the same as (18) for $n = i$, $m = r$, $i + j = N = M$ and $\varkappa = 2^*$.

The generating function $f(s) = (1 - p)^\gamma / (1 - ps)^\gamma$ of example B may be given the following interpretation. Consider a heterogeneous population of different subspecies. Each separate individual is multiplying according to a Poisson branching process of parameter λ, where the value of λ depends on the subspecies. From a viewpoint of the large population we can regard λ as a random variable with density function $u(\lambda)\,d\lambda = \Gamma^{-1}(\gamma)\,b^\gamma\,\lambda^{\gamma-1}\,e^{-b\lambda}\,d\lambda$ $(b > 0, \lambda > 0)$. The generating function of the total offspring population is an average of the generating function associated with each species. Thus

$$f(s) = b^\gamma\,[\Gamma(\gamma)]^{-1} \int_0^\infty e^{\lambda(s-1)}\,\lambda^{\gamma-1}\,e^{-b\lambda}\,d\lambda = \frac{(1-p)^\gamma}{(1-ps)^\gamma}$$

where $p = (1 + b)^{-1}$.

2. Multi-type models

The formulation of the preceding theory for the case of any number of types is direct. We consider a multi-type branching process with p types of individuals labeled respectively, $A_1, A_2, \ldots, A_p$. Suppose each individual of type A_k in one generation yields progeny of all types whose generating function is given by

$$f_k(s_1, s_2, \ldots, s_p) \qquad k = 1, 2, \ldots, p. \tag{19}$$

Individuals are assumed to act independently. Let $[X_1(n), X_2(n), \ldots, X_p(n)]$ denote the associated branching process where $X_k(n)$ represents the number of A_k-type at the start of the nth generation. The probability generating function of the progeny in one generation is

---

* This example was first formulated by KIMURA [5].

$$f_1^{i_1}(s_1, \ldots, s_p) f_2^{i_2}(s_1, \ldots, s_p) \ldots f_p^{i_p}(s_1, \ldots, s_p) h(s_1, \ldots, s_p) \quad (20)$$

where $X_k(0) = i_k$ $(k = 1, 2, \ldots, p)$ and $h(s_1, \ldots, s_p)$ denotes the generating function of the various types immigrating into the system.

As in the case of two types, the Markov chain arising by fixing the population size has an interpretation as a frequency model, described as follows. The state space consists of all p-tuples of non-negative integers $\bar{k} = (k_1, k_2, \ldots, k_p)$ obeying the constraint $\sum\limits_{\nu=1}^{p} k_\nu = N$. The transition probability matrix is constructed from the branching process as follows: Let $\bar{k} = (k_1, k_2, \ldots, k_p)$, $\bar{l} = (l_1, l_2, \ldots, l_p)$ then

$$P_{\bar{k}, \bar{l}} = Pr\{X_1(1) = l_1, X_2(1) = l_2, \ldots, X_p(1) = l_p \,|\, X_\nu(0) = k_\nu \,(\nu = 1, \ldots, p),$$

$$\sum_{\nu=1}^{p} X_\nu(0) = N, \ \sum_{\nu=1}^{p} X_\nu(1) = N\} \quad (21)$$

$$= \frac{\text{coefficient } s_1^{l_1} s_2^{l_2} \ldots s_p^{l_p} \text{ in } f_1^{k_1}(\bar{s}) f_2^{k_2}(\bar{s}) \ldots f_p^{k_p}(\bar{s}) h(\bar{s})}{\text{coefficient } t^N \text{ in } f_1^{k_1}(\bar{t}) f_2^{k_2}(\bar{t}) \ldots f_p^{k_p}(\bar{t}) h(\bar{t})}$$

where $\bar{s} = (s_1, s_2, \ldots, s_p)$ and $\bar{t} = (t, t, \ldots, t)$.

In the special case

$$f_i(\bar{s}) = \exp\left\{ \lambda_i \left(\sum_{\nu=1}^{p} \alpha_{i\nu} s_\nu - 1 \right) \right\} \qquad i = 1, 2, \ldots, p \quad (22)$$

$$h(\bar{s}) = \exp\left\{ \sum_{\nu=1}^{p} c_\nu (s_\nu - 1) \right\},$$

$$\alpha_{i\nu} \geq 0, \ \sum_{\nu=1}^{p} \alpha_{i\nu} = 1, \ \lambda_i > 0 \ (\nu, i = 1, \ldots, p), \ c_\nu > 0, \ (\nu = 1, \ldots, p) \quad (23)$$

the transition probability matrix (21) reduces to

$$P_{\bar{k}, \bar{l}} = \frac{\binom{N}{l_1, l_2, \ldots, l_p} \prod\limits_{\nu=1}^{p} \left[\sum\limits_{i=1}^{p} k_i \lambda_i \alpha_{i\nu} + c_\nu \right]^{l_\nu}}{\left[\sum\limits_{i, \nu} k_i \lambda_i \alpha_{i\nu} + c_\nu \right]^{N}} \quad (24)$$

where

$$\binom{N}{l_1, l_2, \ldots, l_p} = \frac{N!}{l_1! \, l_2! \ldots l_p!}.$$

The parameters occurring in (24) are to be interpreted as follows:

$\alpha_{i\nu}$ represents the chance that an A_i-type individual after birth will mutate into an A_ν-type individual, λ_i represents the relative selection (= fitness) coefficient of an A_i-type, c_i represents the average rate at which A_i-type individuals are immigrating into the population.

The probability matrix (24) is plainly the multi-type version of Wright's gene frequency Markov chain stochastic model allowing for mutation, migration and selection.

We close this section by recording two other important Markov chains of the conditioned process (21). These correspond to the multitype versions of examples A and B.

120   SAMUEL KARLIN and JAMES MCGREGOR

A'. Let

$$f_r(\bar{s}) = (1 - p + ps_r)^\gamma \quad (r = 1, 2, \ldots, p; \gamma = \text{a positive integer} \geq 2)$$

$h(\bar{s}) = 1$. Then (21) becomes

$$P_{\bar{k}, \bar{i}} = \frac{\binom{\gamma k_1}{l_1}\binom{\gamma k_2}{l_2} \ldots \binom{\gamma k_p}{l_p}}{\binom{\gamma N}{N}}, \quad \sum_{\nu=1}^{p} k_\nu = \sum_{\nu=1}^{p} l_\nu = N . \tag{25}$$

B'. Let

$$f_r(\bar{s}) = \frac{(1-p)^\alpha}{(1-ps)^\alpha} \quad (r = 1, 2, \ldots, p; \alpha = \text{a positive integer})$$

$h(\bar{s}) = 1$. Then (21) becomes

$$\frac{\binom{\alpha k_1 + l_1 - 1}{l_1}\binom{\alpha k_2 + l_2 - 1}{l_2} \ldots \binom{\alpha k_p + l_p - 1}{l_p}}{\binom{\alpha N + N - 1}{N}} \quad \sum_{\nu=1}^{p} k_\nu = \sum_{\nu=1}^{p} l_\nu = N .$$

Polyploid selfing

A natural multi-type Markov chain of the form (21) suggested by certain non-random mating systems is the following.

We consider a population in which every individual has p homologous chromosomes, each of which, at a given locus, may be either of two types A or a. If an individual has ν chromosomes of type A and $p - \nu$ of type a, the individual is of type $A^\nu a^{p-\nu}$. To form progeny the p chromosomes duplicate, forming $2p$ in all, and from these p chromosomes are selected at random to build a new individual. The probability that a progeny from a parent of type $A^\mu a^{p-\mu}$ is of type $A^\nu a^{p-\nu}$ is

$$\alpha_{\mu, \nu} = \frac{\binom{2\mu}{\nu}\binom{2p - 2\mu}{p - \nu}}{\binom{2p}{p}} .$$

We assume the generating function for the number of progeny of any individual is $f(s)$.

The induced Markov chain frequency model describing the fluctuations of this population is governed by the transition probability matrix

$$P_{\bar{i}, \bar{k}} = \frac{\text{coefficient } t_0^{k_0} t_1^{k_1} \ldots t_p^{k_p} \text{ in } \prod_{n=0}^{p} f^{i_\mu}\left(\sum_{\nu=1}^{p} \alpha_{\mu\nu} t_\nu\right)}{\text{coefficient of } t^N \text{ in } f^N(t)} \tag{26}$$

where $\alpha_{\mu\nu}$ is defined above. Here, $\bar{i} = (i_0, i_1, \ldots, i_p)$, $\bar{k} = (k_0, k_1, \ldots, k_p)$, $i_\nu \geq 0$, $k_\nu \geq 0$, $\Sigma i_\nu = \Sigma k_\nu = N$ where i_ν denotes the number of individuals in the parent population of type $A^\nu a^{p-\nu}$ and k_ν is the number of individuals of the same type in the offspring population.

3. Eigenvalues of Markov chains of frequency models

In this section we study the nature of the eigenvalues and eigenvectors of the induced Markov chains.

To ease the discussion and underscore the essential ideas we start with the simplest class of examples. Consider the model of two types (labeled A and a). Let the probability generating function of the offspring population be $f(s)$, the same for A and a individuals.

The associated Markov chain has a transition probability matrix $P = \| P_{ik} \|$ where

$$P_{ik} = \frac{\text{coefficient } t^k s^N \text{ of } f^i(ts) f^{N-i}(s)}{\text{coefficient } s^N \text{ of } f^N(s)} \qquad i, k = 0, 1, \ldots, N . \tag{27}$$

Theorem 1. *Let $f(s)$ be a probability generating function of a nonnegative integer valued random variable, i.e.,*

$$f(s) = \sum_{r=0}^{\infty} c_r s^r \qquad c_r \geq 0 , \ \sum_{r=0}^{\infty} c_r = 1 .$$

The eigenvalues of the matrix (27) *are*

$$\lambda_0 = 1, \lambda_1 = 1, \lambda_r = \frac{\text{coefficient } s^{N-r} \text{ of } f^{N-r}(s) \ [f'(s)]^r}{\text{coefficient } s^N \text{ of } f^N(s)} \qquad r = 2, 3, \ldots, N . \tag{28}$$

If $c_0 \cdot c_1 \cdot c_2 > 0$ then

$$1 > \lambda_2 > \lambda_3 > \ldots > \lambda_N > 0 . \tag{29}$$

[Actually much weaker assumptions suffice to guatantee the validity of (29)*.]*

Moreover, the right eigenvector α_r (apart from a constant factor) corresponding to λ_r $(r = 2, 3, \ldots, N)$ is associated with a polynomial $Q_r(z)$ of degree r such that

$$\alpha_r = [Q_r(0), Q_r(1), \ldots, Q_r(N)] \qquad r = 2, \ldots, N . \tag{30}$$

Two linearly independent right eigenvectors associated with $\lambda_0 = \lambda_1 = 1$ are

$$\alpha_0 = \left(1, 1 - \frac{1}{N}, 1 - \frac{2}{N}, \ldots, 0\right) \quad and \quad \alpha_1 = \left(0, \frac{1}{N}, \frac{2}{N}, \ldots, \frac{N}{N}\right) . \tag{31}$$

[They are displayed in this manner to facilitate their probabilistic interpretation (see Section 5). Of course, any linear combination of α_0 and α_1 is also an eigenvector for $\lambda_0 = 1$.]

The left eigenvectors β_r (apart from a constant factor) are of the form

$$\beta_r = [\tbinom{N}{0}(-1)^0 R_{N-r}(0), \tbinom{N}{1}(-1)^1 R_{N-r}(1), \ldots, \tbinom{N}{i}(-1)^i R_{N-r}(i) ,$$
$$\ldots, \tbinom{N}{N}(-1)^N R_{N-r}(N)] \qquad r = 2, \ldots, N \tag{32}$$

where $R_l(z)$ is a polynomial of degree l. Also

$$\beta_0 = (0, 0, \ldots, 0, 1) , \qquad \beta_1 = (1, 0, 0, \ldots, 0) . \tag{33}$$

The eigenvectors $\{\alpha_n\}_{n=0}^{N}$ and $\{\beta_m\}_{m=0}^{N}$ constitute a biorthogonal system.

In order to ease the exposition of the proof of the theorem, we divide it into several steps.

Lemma 1. *If $c_0 c_1 c_2 > 0$ then the values* (28) *are distinct as indicated in* (29).

Proof. Notice that for any power series $g(s)$ and any integer M,

$$M \cdot \text{coefficient of } s^M \text{ in } g(s) = \text{coefficient of } s^{M-1} \text{ in } g'(s).$$

Therefore, it follows that

coefficient s^{N-r} in $f^{N-r}(s)\,[f'(s)]^r$

$$
\begin{aligned}
&= \frac{1}{(N-r)}\left[\text{coefficient } s^{N-r-1} \text{ in } \frac{d}{ds}\left\{f^{N-r}(s)\,[f'(s)]^r\right\}\right] \\
&= \text{coefficient } s^{N-r-1} \text{ in } f^{N-r-1}(s)\,[f'(s)]^{r+1} + \\
&\quad + \frac{r}{N-r}\,[\text{coefficient } s^{N-r-1} \text{ in } f''(s)\,f^{N-r}(s)\,[f'(s)]^{r-1}] \\
&\qquad\qquad\qquad\qquad\qquad r = 1, 2, 3, \ldots, N-1.
\end{aligned}
\tag{34}
$$

For $r = 0$, coefficient of s^N in $[f(s)]^N = $ coefficient s^{N-1} in $f^{N-1}(s)\,f'(s)$. Now $f(s)$ is a power series with non-negative coefficients. Therefore

$$\text{coefficient } s^{N-r-1} \text{ in } \left\{f''(s)\,f^{N-r}(s)\,[f'(s)]^{r-1}\right\} \text{ is non-negative.} \tag{35}$$

Owing to the condition $c_0 > 0,\ c_1 > 0,\ c_2 > 0$ we conclude that the coefficients in (35) are actually strictly positive. This fact, in conjunction with the relation (34), implies the assertion of the lemma.

Our next lemma describes an important transformation property essential for the subsequent analysis.

Lemma 2. *Let P_{ik} be defined as in* (27). *Then*

$$\sum_{k=0}^{N} P_{ik}\, k^\nu = i^\nu\, \lambda_\nu + H_{\nu-1}(i) \qquad \nu = 0, 1, 2, \ldots \tag{36}$$

where $H_{\nu-1}(x)$ is a polynomial of degree at most $\nu - 1$, $H_{-1}(x) \equiv 0$ and $\lambda\nu$ is given in (28).

Remark: An equivalent way to express the content of (36) is that the matrix transformation P carries a polynomial $U_m(x)$ of degree m [i.e., a vector whose kth component is the value $U_m(k)$] into a vector $[V_m(0), V_m(1), \ldots, V_m(N)]$ where $V_m(x)$ is also a polynomial of degree m.

Proof: We start with the generating function $\sum_{k=0}^{N} P_{ik}\, t^k$ which on account of the definition of P_{ik} becomes

$$\sum_{k=0}^{N} P_{ik}\, t^k = \frac{\text{coefficient } s^N \text{ of } f^i(ts)\, f^{N-i}(s)}{\text{coefficient } s^N \text{ of } f^N(s)} \qquad (i = 0, 1, \ldots, N).$$

Now differentiating ν times with respect to t and then setting $t = 1$ produces on the left

$$\sum_{k=0}^{N} P_{ik}\,[k(k-1)\ldots(k-\nu+1)] = \Sigma\, P_{ik}\,[k^\nu + U_{\nu-1}(k)]$$

where $U_{\nu-1}(\cdot)$ is a polynomial of degree $\leq \nu - 1$.

When $f^i(st) f^{N-i}(s)$ on the right is differentiated ν times with respect to t we obtain

$$\frac{d^\nu}{dt^\nu} \left[f^i(st) f^{N-i}(t) \right]$$

$$= \left[i(i-1) \ldots (i-\nu+1) \right] f^{i-\nu}(st) \left[f'(st) \right]^\nu f^{N-i}(t) s^\nu + W_\nu \left[i, f(s) \right]$$

where $W_\nu(i, f)$ denotes the other contributions resulting from differentiation. The first term is the outcome of continued differentiation of only the factor $f(st)$ to its power ν times. The terms that comprise W_ν occur by differentiating at least once one of the auxiliary factors like $f'(st)$ or $f''(st)$, etc. that arise from previous differentiations. This means that the result of such differentiations produce terms whose coefficients include at most $\nu - 1$ of the factors $i, i-1, \ldots, i-\nu+1$. In each such term the dependence on f is of the form

$$f^{i-l}(st) f^{N-i}(s) \left[f'(st) \right]^{\alpha_1} \left[f''(st) \right]^{\alpha_2} \ldots$$

where $k, \alpha_1, \alpha_2, \ldots$ are integers satisfying $0 \leq l, \alpha_1, \alpha_2, \ldots \leq \nu$. When we put $t = 1$ the factors $f^{i-l}(s) f^{N-i}(s)$ combine into $f^{N-l}(s)$ and the dependence on i disappears. It follows that for $t = 1$, $W_\nu(i, f)$ is a polynomial in the variable i of degree at most $\nu - 1$.

Now suppose we have proved (36) inductively for ν up to $r - 1$. In view of the result of differentiation as described above we have (put $t = 1, \nu = r$)

$$\sum_{k=0}^{N} P_{ik} k^r + \sum_{k=0}^{N} P_{ik} U_{r-1}(k) = i^r \lambda_r + V_{r-1}(i)$$

where $V_{r-1}(\cdot)$ is a polynomial of degree $r - 1$. Applying the induction hypotheses to the second term on the left we conclude the validity of (36) with $\nu = r$ as desired to be shown.

We explicitly have

$$\nu = 0; \quad \sum_{k=0}^{N} P_{ik} 1 = 1 \tag{37a}$$

$$\nu = 1; \quad \sum_{k=0}^{N} P_{ik} k = i \frac{\text{coefficient } s^{N-1} \text{ of } f^{N-1}(s) f'(s)}{\text{coefficient } s^N \text{ of } f^N(s)} = i\lambda_1 = i \tag{37b}$$

$$\nu = 2; \quad \sum_{k=0}^{N} P_{ik} k^2 = i^2 \lambda_2 + i \frac{\text{coefficient } s^{N-2} \text{ of } f^{N-2}(s) \left[f'(s) \right] f''(s)}{\text{coefficient } s^N \text{ of } f^N(s)} -$$

$$- i(\lambda_2 - \lambda_1) = i^2 \lambda_2 + i(\gamma_1 - \lambda_2 + \lambda_1) \tag{37c}$$

where γ_1 is defined in the obvious manner.

We are now ready to prove the first half of the theorem.

It is obvious from (37a) and (37b) that the vectors $\tilde{\alpha}_0 = (1, 1, \ldots, 1)$ and $\tilde{\alpha}_1 = (0, 1, 2, \ldots, N)$ are eigenvectors belonging to the eigenvalue $\lambda_0 = \lambda_1 = 1$. Plainly α_0 and α_1 are linearly combinations of $\tilde{\alpha}_0$ and $\tilde{\alpha}_1$ and these are linearly independent.

If we introduce the vectors

$$y_\nu = (0^\nu, 1^\nu, 2^\nu, \ldots, N^\nu) \qquad \nu = 0, 1, 2, \ldots, N$$

as a basis then the matrix P, in this coordinate system, is upper triangular of the form

$$A^{-1}PA = \begin{pmatrix} \lambda_0 & * & * & & * \\ 0 & \lambda_1 & * & & * \\ 0 & 0 & \lambda_2 & & \\ \vdots & & & \ddots & * \\ 0 & 0 & \cdots & & \lambda_N \end{pmatrix}$$

where A is the matrix composed of the column vectors y_ν. This is clear by virtue of relation (36). Since the eigenvalues are invariant under change of basis we conclude from the representation of P above that $\lambda_0, \lambda_1, \lambda_2, \ldots, \lambda_N$ consitute a complete set of eigenvalues of P. This proves statement (28) of the theorem.

The polynomial character of the eigenvectors as asserted in the theorem can be established from closer study of (36). We proceed in an equivalent but slightly different manner.

Consider now a polynomial $q(x)$ of the form

$$q(x) = a_0 x^r + a_1 x^{r-1} + \ldots + a_{r-1} x + a_r \qquad r \geq 2 \qquad (38)$$

where the constants $a_1, a_2, \ldots, a_r$ are to be determined and $a_0 = 1$. We associate with (38) a vector whose kth component is $q(k)$ $(k = 0, 1, \ldots, N)$ and examine the result of applying the matrix P to it. According to (36) the ith component of the image vector is

$$(Pq)_i = \sum_{\nu=0}^{r} a_\nu \left[\lambda_{r-\nu} \, i^{r-\nu} + H_{r-\nu-1}(i) \right].$$

We seek to determine $a_1, a_2, \ldots, a_r$ so that

$$\sum_{\nu=0}^{r} a_\nu \left[\lambda_{r-\nu} \, i^{r-\nu} + H_{r-\nu-1}(i) \right] = \lambda_r \, q(i) = \lambda_r \sum_{\nu=0}^{r} a_\nu \, i^{r-\nu} \qquad (39)$$

is satisfied identically in i which shows that $[q(0), q(1), \ldots, q(N)]$ qualifies as an eigenvector for λ_r. The method set forth below will apply for $r \geq 2$. The calculation of $\{a_i\}_{i=1}^{r}$ is done recursively as follows: Recall that $a_0 = 1$. Equating the coefficients of i^{r-1} in (39) imposes a condition of the form

$$a_1 \lambda_{r-1} + \text{a known term} = a_1 \lambda_r. \qquad (40)$$

By Lemma 1 we know since $r \geq 2$ that $\lambda_r < \lambda_{r-1}$ and therefore (40) can be solved for a_1. Next equating the coefficient of i^{r-2} in (39) and using the fact that a_0 and a_1 are already determined, produces an identity of the form

$$a_2 \lambda_{r-2} + \text{a known term} = a_2 \lambda_r. \qquad (41)$$

We solve (41) for a_2. Proceeding thus, we may determine all the constants so that (39) is valid for all i. We see that the ith component of the eigenvector associated with λ_r is the value at i of a polynomial $q(x)$ of degree r. The proof of assertions (28, (29) and (30) are complete. It remains to identify the left eigenvectors as described in (31) and (32).

The left eigenvectors (33) associated with $\lambda_0 = \lambda_1 = 1$ can be checked directly. Since $\lambda_2, \lambda_3, \ldots, \lambda_N$ are distinct and $\neq 1$ the standard theory of matrix analysis affirms the existence of $\beta_2, \beta_3, \ldots, \beta_N$, and $\{\beta_i\}_{i=0}^N$ are biorthogonal to $\{\alpha_j\}_{j=0}^N$. Also, the left and right eigenvectors of P for $\lambda_0 = \lambda_1 = 1$ were prescribed in a manner so that they are automatically biorthogonal.

In order to continue our discussion of the left eigenvectors, we digress briefly to review certain properties of the Krawtchouk polynomial which will be needed.

Let $K_l(x; 1/2, N) = K_l(x)$ denote the lth Krawtchouk polynomial with respect to the parameters $p = 1/2$ and N. We know that

$$\sum_{x=0}^{N} K_l\left(x; \tfrac{1}{2}, N\right) U_m(x) \binom{N}{x} = 0 \qquad l > m \tag{42}$$

for any polynomial $U_m(x)$ of degree $< l$.

Consider the vectors

$$\gamma_l = \left[\binom{N}{0} K_l(0), \binom{N}{1} K_l(1), \ldots, \binom{N}{N} K_l(N)\right], \qquad l = 0, 1, \ldots, N.$$

These are linearly independent since $K_l(\cdot)$ are an orthogonal system of polynomials. We claim that γ_l is orthogonal to α_i whenever $l > i$. This is obvious from (42) since α_i is a polynomial of degree i.

For $i \geq 2$ let $\mathcal{M}_i$ be the linear space consisting of all vectors orthogonal to each of the vectors $\alpha_0, \alpha_1, \ldots, \alpha_{i-1}$. Then $\mathcal{M}_i$ is of dimension $N - i + 1$ and is spanned by either of the linearly independent sets $\beta_i, \beta_{i+1}, \ldots, \beta_N$ or $\gamma_i, \gamma_{i+1}, \ldots, \gamma_N$. It follows that for $i \geq 2$, β_i is a linear combination of $\gamma_i, \gamma_{i+1}, \ldots, \gamma_N$, that is

$$\beta_{N-l} = b_0 \gamma_N + b_1 \gamma_{N-1} + \ldots + b_l \gamma_{N-l}, \qquad l \leq N - 2 \tag{43}$$

where the b_ν are constants. It is clear that $b_l \neq 0$.

Next we use the symmetry relations [3]

$$K_l\left(x; \tfrac{1}{2}, N\right) = K_x\left(l; \tfrac{1}{2}, N\right)$$
$$K_x\left(N - l; \tfrac{1}{2}, N\right) = (-1)^x K_x\left(l; \tfrac{1}{2}, N\right) \tag{44}$$

valid for $l, x = 0, 1, \ldots, N$. From these it follows that

$$K_l\left(x; \tfrac{1}{2}, N\right) = (-1)^x K_{N-l}\left(x; \tfrac{1}{2}, N\right)$$

and hence using (43) the xth component of β_{N-l} is

$$(-1)^x \binom{N}{x} \sum_{\nu=0}^{l} b_\nu K_\nu\left(x; \tfrac{1}{2}, N\right).$$

This is a polynomial of degree l multiplied by $(-1)^x \binom{N}{x}$. Thus the proof of Theorem 1 is complete.

We close this section with a remark and some examples.

Remark 2. Since α_r ($r = 2, \ldots, N$) is orthogonal to β_0 and β_1, this implies that $Q_r(x)$ vanishes at $x = 0$ and N. Thus

$$Q_r(x) = x(N-x)P_{r-2}(x) \qquad r = 2, 3, \ldots .$$

Actually slightly more care exploiting suitable symmetries shows that

$$Q_{2m}(x) = T_m[x(N-x)], \quad Q_{2m+1}(x) = (N-2x)\widetilde{T}_m[x(N-x)]$$

where T_m is a polynomial of degree m and similarly for $\widetilde{T}_m$.

It seems very difficult to express T_m and $\widetilde{T}_m$ in closed form. Nevertheless, the polynomials $Q_r(x)$, in principle, can be computed recursively; in fact by the very method of the proof of Theorem 1. It is also possible to devise a recursive procedure for computing $R_r(\cdot)$. We do not enter into details.

Examples. It is worthwhile to list some particular examples of Theorem 1.

A. Let $f(x) = e^{\lambda(s-1)}$ then

$$P_{ik} = \binom{N}{k}\left(\frac{i}{N}\right)^k\left(1-\frac{i}{N}\right)^{N-k}.$$

The eigenvalues are

$$\lambda_0 = \lambda_1 = 1, \quad \lambda_r = \frac{\text{coefficient } s^{N-r} \text{ in } [f'(s)]^r [f(s)]^{N-r}}{\text{coefficient } s^N \text{ in } [f(s)]^N} = \frac{N!}{(N-r)!}\frac{1}{N^r}$$

$$r = 2, \ldots, N .$$

B. Let $f(s) = (1 - p + ps)^\gamma$; $0 < p < 1, \gamma = $ a positive integer ≥ 2.

Then

$$P_{ik} = \frac{\binom{\gamma i}{k}\binom{\gamma(N-i)}{N-k}}{\binom{\gamma N}{N}} \qquad i, k = 0, 1, \ldots, N .$$

The eigenvalues are

$$\lambda_0 = \lambda_1 = 1, \quad \lambda_r = \frac{\gamma^r \binom{N\gamma - r}{N - r}}{\binom{N\gamma}{N}} \qquad r = 2, \ldots, N .$$

C. Let $f(s) = \frac{(1-p)^\alpha}{(1-ps)^\alpha} \qquad (0 < p < 1; \alpha > 0)$

$$P_{ik} = \frac{\binom{\alpha i + k - 1}{k}\binom{\alpha(N-i) + N - k - 1}{N - k}}{\binom{\alpha N + N - 1}{N}} .$$

The eigenvalues are

$$\lambda_0 = 1, \quad \lambda_r = \frac{\alpha^r \binom{N\alpha + N - 1}{N - r}}{\binom{N\alpha + N - 1}{N}} \qquad r = 1, 2, \ldots, N .$$

Asymptotically, as $N \to \infty$: in example A, $\lambda_2 = 1 - \dfrac{1}{N}$; in example B, $\lambda_2 \sim 1 - \dfrac{\gamma - 1}{\gamma} \dfrac{1}{N}$; in example C, $\lambda_2 \sim 1 - \dfrac{\alpha + 1}{\alpha} \dfrac{1}{N}$. Notice that the Poisson generating function leads to an asymptotic value (as $N \to \infty$) of λ_2 intermediate between that of the binomial family and the negative binomial family.

4. Eigenvalues of multitype mutation model

In the previous section we analyzed the character of the eigenvalues and eigenvectors of the transition matrix for the two type induced MC with no mutation. In this section we will study the induced MC of p ($p \geq 2$) types with general mutation rates. Specifically, we will investigate the form of the eigenvalues and eigenvectors for the Markov chain with transition matrix

$$P_{\bar{i}, \bar{k}} = \frac{\text{coefficient of } s_1^{k_1} s_2^{k_2} \ldots s_p^{k_p} \text{ in } \prod_{\nu=1}^{p} f^{i_\nu} \left(\sum_{\mu=1}^{p} \alpha_{\nu\mu} s_\mu \right)}{\text{coefficient of } s^N \text{ in } f^N(s)} \tag{46}$$

where $\bar{i} = (i_1, i_2, \ldots, i_p)$, $\bar{k} = (k_1, k_2, \ldots, k_p)$, i_ν and k_ν are integers

$$\sum_{\nu=1}^{p} i_\nu = \sum_{\nu=1}^{p} k_\nu = N \qquad \left. \begin{matrix} i_\nu \geq 0 \\ k_\nu \geq 0 \end{matrix} \right\} \text{ all } \nu .$$

(We will designate by Δ_p the set of all such p-tuples $\bar{i}$.) $P_{\bar{i}, \bar{k}}$ is the transition probability function of the conditioned frequency process of a p type branching process where the νth type produces progeny with probability generating function $f(s)$, after which each A_ν offspring mutates into an A_j type with probability $\alpha_{\nu j}$ ($j = 1, 2, \ldots, p$). The conditioning involves keeping the population size constant.

By its very meaning the matrix $\Gamma = \| \alpha_{\nu\mu} \|_{\nu, \mu=1}^{p}$ is a Markov chain matrix. We will *assume* Γ is diagonalizable, that is, for its p eigenvalues

$$\gamma_1 = 1, \gamma_2, \ldots, \gamma_p \qquad (\,|\gamma_i| \leq 1 \text{ for } i = 2, \ldots, p)$$

there exists p linearly independent eigenvectors. Let $u^{(1)}, u^{(2)}, \ldots, u^{(p)}$ denote a complete set of right eigenvectors where we may take $u^{(1)} = (1, 1, 1, \ldots, 1)$ since the sum of the elements of any row in the Γ matrix equals 1.

The matrix (46) can also be represented in the following form:

$$P_{\bar{i}, \bar{k}} = \frac{\text{coefficient of } s^N t_1^{k_1} t_2^{k_2} \ldots t_p^{k_p} \text{ in } \prod_{\nu=1}^{p} f^{i_\nu} \left(s \sum_{\mu=1}^{p} \alpha_{\nu\mu} t_\mu \right)}{\text{coefficient } s^N \text{ in } f^N(s)} . \tag{47}$$

We will now form the probability generating function of $P_{\bar{i}, \bar{k}}$. We obtain

$$G(t_1, t_2, \ldots, t_p) = \sum_k P_{\bar{i}, \bar{k}}\, t_1^{k_1}\, t_2^{k_2} \ldots t_p^{k_p}$$

$$= \frac{\text{coefficient of } s^N \text{ in } \prod_{\nu=1}^{p} f^{i_\nu}\left(s \sum_{\mu=1}^{p} \alpha_{\nu\mu}\, t_\mu\right)}{\text{coefficient of } s^N \text{ in } f^N(s)}. \tag{48}$$

Before passing to the general problem of characterizing the eigenvalues of $P = \| P_{\bar{i}, \bar{k}} \|$ we will consider two special subcases which will help to clarify the general method employed for determining all the eigenvalues of the P matrix.

Case 1.

Differentiating (48) once with respect to t_μ and then setting $t_1 = t_2 = \ldots = t_p = 1$ yields the identity

$$\frac{d}{dt_\mu} G(t_1, \ldots, t_p)\Big|_{t_1 = \cdots = t_p} = \sum_k k_\mu\, P_{\bar{i}, \bar{k}}$$

$$= \text{coefficient of } s^N \text{ in } \left\{\frac{si_1\, \alpha_{1\mu}\, f^{N-1}(s)\, f'(s) + si_2\, \alpha_{2\mu}\, f^{N-1}(s)\, f'(s)}{\text{coefficient } s^N \text{ in } f^N(s)}\right.$$

$$\left. + \cdots + \frac{si_p\, \alpha_{p\mu}\, f^{N-1}(s)\, f'(s)}{1}\right\} \tag{49}$$

$$= \frac{\left(\sum_{\nu=1}^{p} i_\nu\, \alpha_{\mu\nu}\right) \text{coefficient of } s^{N-1} \text{ in } f^{N-1}(s)\, f'(s)}{\text{coefficient of } s^N \text{ in } f^N(s)}.$$

This equation can be written compactly in the form

$$\sum_k k_\mu\, P_{\bar{i}, \bar{k}} = \lambda_1 \sum_{\nu=1}^{p} i_\nu\, \alpha_{\nu\mu} \qquad (\mu = 1, 2, \ldots, p) \tag{50}$$

valid for all $\bar{i}$ where λ_1 is defined in (28).

Next, multiply both sides of (50) by $u_\mu^{(q)}$ and sum on μ. We obtain

$$\sum_\mu u_\mu^{(q)} \sum_{\bar{k}} P_{\bar{i}, \bar{k}}\, k_\mu = \lambda_1 \sum_\mu u_\mu^{(q)} \sum_{\nu=1}^{p} i_\nu\, \alpha_{\nu\mu}$$

or rearranging the order of summation

$$\sum_{\bar{k}} P_{\bar{i}, \bar{k}} \left(\sum_{\mu=1}^{p} u_\mu^{(q)}\, k_\mu\right) = \lambda_1 \sum_{\nu=1}^{p} i_\nu \sum_{\mu=1}^{p} a_{\nu\mu}\, u_\mu^{(q)}. \tag{51}$$

But we know that $\sum_{\mu=1}^{p} \alpha_{\nu\mu}\, u_\mu^{(q)} = \gamma_q\, u_\nu^{(q)}$ $(\nu, q = 1, 2, \ldots, p)$ since $u^{(q)}$ is an eigenvector of $\| \alpha_{\nu\mu} \|$ for the eigenvalue γ_q. At this point it is convenient to introduce the quantities $\sum_{\mu=1}^{p} u_\mu^{(q)}\, k \ - L_q(\bar{k})$. (*Note*: $L_q(\bar{k})$ is a linear function of $\bar{k}$ and particularly $L_1(\bar{k}) = \sum_{\mu=1}^{p} k_\mu = N$.)

Then (51) becomes

$$\sum_{\bar{k}} P_{\bar{i}, \bar{k}}\, L_q(\bar{k}) = \lambda_1\, \gamma_q\, L_q(\bar{i}); \qquad (q = 1, 2, \ldots, p; \text{ all } \bar{i}) \tag{52}$$

which shows that $\lambda_1 \gamma_q$ is an eigenvalue of the P matrix, and that $L_q(\bar{\imath})$ is a corresponding eigenvector.

Note that for $q = 1$, $L_1(\bar{\imath}) \equiv N$ (a constant independent of $\bar{\imath}$). We will treat $L_1(\bar{\imath})$ to some extent differently from the other $L_q(\bar{\imath})$, $q = 2, \ldots, p$ which are bona fide linear functions of $\bar{\imath}$. The eigenvectors $L_q(\bar{\imath})$, associated with the eigenvalues $\lambda_1 \gamma_q$ $(q = 1, 2, \ldots, p)$ are linearly independent since the eigenvectors $u^{(q)}$ of $\| \alpha_{\nu\mu} \|$ are linearly independent.

Since $L_1(\bar{\imath}), L_2(\bar{\imath}), \ldots, L_q(\bar{\imath})$ are linearly independent, it is evident that any linear function $\mathscr{L}(\bar{\imath})$ of $\bar{\imath}$ can be represented as a linear combination of $\{L_q(\bar{\imath})\}_{q=1}^{p}$, i.e., there exists constants $b_1, b_2, \ldots, b_p$ such that

$$\mathscr{L}(\bar{\imath}) = \sum_{\nu=1}^{p} b_\nu L_\nu(\bar{\imath}) \qquad \text{for all } \bar{\imath}.$$

In view of (52) and this remark we see that the matrix P maps linear functions into linear functions, i.e., if $\mathscr{L}(\bar{\imath})$ is linear, then

$$\Sigma\, P_{\bar{\imath},\bar{\jmath}}\, \mathscr{L}(\bar{\jmath}) = \mathscr{L}^*(\bar{\imath}) \tag{53}$$

is also linear.

Case 2. This is similar to Case 1 except that the algebra is a little more complicated and in certain respects incorporates the general argument.

Differentiating equation (48) twice with respect to any two t_μ's (they may be the same) and then setting $t_1 = t_2 = \ldots = t_p = 1$ we obtain for the left side of (48)

$$\frac{d^2}{dt_m\, dt_n} G(t_1, \ldots, t_p) = \sum P_{\bar{\imath},\bar{k}}\, [k_m k_n + \mathscr{L}_{m,n}(\bar{k})] \tag{54}$$

$$m, n = 1, 2, \ldots, p$$

where

$$\mathscr{L}_{m,n}(\bar{k}) = \begin{cases} -k_m & \text{if } m = n \\ 0 & \text{otherwise} \end{cases}.$$

The right side of (48) becomes

$$= \frac{\text{coefficient of } s^N \text{ in } \left\{ s^2 \left(\sum\limits_{\nu=1}^{p} i_\nu \alpha_{\nu m} \right) \left(\sum\limits_{\nu=1}^{p} i_\nu \alpha_{\nu n} \right) f^{N-2}(s)\, [f'(s)]^2 \right\}}{\text{coefficient of } s^N \text{ in } f^N(s)} \tag{55}$$

$$+ \text{[linear polynomial in } (i_1, i_2, \ldots, i_p)\text{]}.$$

We now multiply equation (54) by $\mu_m^{(q)} \mu_n^{(q')}$ $(q, q' = 1, 2, \ldots, p)$ and sum over all m, n.

The left side of (55) becomes

$$\sum_m \sum_n u_m^{(q)} u_n^{(q')} \sum_{\bar{k}} P_{\bar{\imath},\bar{k}}\, [j_m j_n + \mathscr{L}_{m,n}(\bar{k})]$$

and interchanging the order of summation we get

$$\sum_{\bar{k}} P_{\bar{\imath},\bar{k}} \left\{ [\sum_m u_m^{(q)} k_m]\, [\sum_n u_n^{(q')} k_n] + \widetilde{\mathscr{L}}(\bar{k}) \right\}$$

where $\widetilde{\mathscr{L}}(\bar{k})$ is also a linear function of $\bar{k}$. Using our previous notation we can write the last expression in the form

$$\sum_{\bar{k}} P_{\bar{i},\bar{k}} \left[L_q(\bar{k}) L_{q'}(\bar{k}) + \widetilde{\mathscr{L}}(\bar{k}) \right] .$$

Executing the analogous operations on the expression of (55), we obtain

$$= \lambda_2 \left[\sum_{\nu=1}^{p} i_\nu \sum_{m=1}^{p} \alpha_{\nu m} u_m^{(q)} \right] \left[\sum_{\nu=1}^{p} i_\nu \sum_{n=1}^{p} \alpha_{\nu n} u_n^{(q')} \right] + \sum_{m=1}^{p} \sum_{n=1}^{p} u_m^q u_n^{(q')} \Lambda_{m,n}(\bar{i}) \quad (56)$$

where $\Lambda_{m,n}(\bar{i})$ for each m and n is a linear function of $\bar{i}$. Once again we use the fact that $\sum_{m=1}^{p} \alpha_{\nu m} u_m^{(q)} = \gamma_q u_\nu^q$ by definition of the eigenvectors of $\Gamma = \| \alpha_{\nu\mu} \|$. This simplifies the expression of (56) to

$$= \lambda_2 \gamma_q \gamma_{q'} \left(\sum_{\nu=1}^{p} u_\nu^q i_\nu \right) \left(\sum_{\nu=1}^{p} u_\nu^q i_\nu \right) + \sum_{n=1}^{p} \sum_{m=1}^{p} u_m^q u_n^{(q')} \cdot \Lambda_{mn}(\bar{i})$$

$$= \lambda_2 \gamma_q \gamma_{q'} L_q(\bar{i}) L_{q'}(\bar{i}) + L^*(\bar{i})$$

where

$$L^*(\bar{i}) = \sum_{m=1}^{p} \sum_{n=1}^{p} u_n^{(q)} u_n^{(q')} \Lambda_{m,n}(\bar{i})$$

is a linear function of $(i_1, i_2, \ldots, i_p)$.

The upshot of these calculations is the formula

$$\sum_{\bar{k}} P_{\bar{i},\bar{k}} \left[L_q(\bar{k}) L_{q'}(\bar{k}) + \widetilde{\mathscr{L}}(\bar{k}) \right] = \lambda_2 \gamma_q \gamma_{q'} L_q(\bar{i}) L_{q'}(\bar{i}) + L^*(\bar{i}) .$$

Now transposing $\sum_{\bar{k}} P_{\bar{i},\bar{k}} L(\bar{k})$ to the right side we obtain

$$\sum_{\bar{k}} P_{\bar{i},\bar{k}} L_q(\bar{k}) L_{q'}(\bar{k}) = \lambda_2 \gamma_q \gamma_{q'} L_q(\bar{i}) L_{q'}(\bar{i}) + \mathscr{L}(\bar{i}) \quad (57)$$

where $\mathscr{L}(\bar{i}) = L^*(\bar{i}) + \sum_{\bar{k}} P_{\bar{i},\bar{k}} \widetilde{\mathscr{L}}(\bar{k})$ is a linear function of $\bar{i}$ [by (53)].

For simplicity of the discussion we assume momentarily that $\lambda_1 \gamma_l \neq \lambda_2 \gamma_q \gamma_{q'}$ for all l, q, q' ($2 \leq l, q, q \leq p$); we exclude $q' = 1$, $q = 1$ so that the expressions $L_q(\bar{k}) L_{q'}(\bar{k})$ are all quadratic functions of $\bar{k}$.

Equation (57) almost exhibits the eigenvalue relation except for the term $\mathscr{L}(\bar{i})$.

We claim that $\lambda_2 \gamma_q \gamma_{q'}$ is an eigenvalue of the matrix $P = \| P_{\bar{i},\bar{k}} \|$.

In fact, consider as a candidate an eigenvector of the form

$$a_{qq'}(i_1, i_2, \ldots, i_p) = L_q(\bar{i}) L_{q'}(\bar{i}) + K(\bar{i})$$

where $K(\bar{i})$ is linear in $\bar{i}$ which is to be determined. This is a non-null vector since the quadratic part can never be cancelled by the linear part. Now we write the representation

$$K(\bar{i}) = \sum_{l=1}^{p} b_l L_l(\bar{i}) ,$$

which is possible as explained in (53).

The following construction of the eigenvector resembles the method used to construct the eigenvector for the transition matrix associated with the two type model with *no* mutation. In the above expressions for the eigenvector, the b_k's are variables to be determined. In order that $a_{qq'}(\bar{i}) = a_{qq'}(i_1, \ldots, i_p)$ be an eigenvector it must satisfy the identity

$$\sum_{\bar{k}} P_{\bar{i},\bar{k}}\left\{L_q(\bar{k}) L_{q'}(\bar{k}) + \sum_{l=1}^{p} b_l L_l(\bar{k})\right\}$$

$$= \lambda_2 \gamma_q \gamma_{q'} \left[L_q(\bar{i}) L_{q'}(\bar{i}) + \sum_{l=1}^{p} b_l L_l(\bar{i})\right] \qquad (58)$$

for all $\bar{i}$. By (51) and (57) we know that the left side of the above equation is equal to

$$\lambda_2 \gamma_q \gamma_{q'} L_q(\bar{i}) L_{q'}(\bar{i}) + \mathcal{L}(\bar{i}) + \lambda_1 \sum_{l=1}^{p} b_l \gamma_l L_l(\bar{i}).$$

Equating this expression to the right side of (58) and cancelling common terms we obtain the relations

$$\sum_{l=1}^{p} c_l L_l(\bar{i}) + \lambda_1 \sum_{l=1}^{p} b_l \gamma_l L_l(\bar{i}) = \lambda_2 \gamma_q \gamma_{q'} \sum_{l=1}^{p} b_l L_l(\bar{i}) \qquad \text{all } \bar{i}$$

where $\mathcal{L}(\bar{i}) = \Sigma c_l L_l(\bar{i})$ is a specific known linear function of $\bar{i}$.

Since the $L_l(\bar{i})$'s $(l = 1, 2, \ldots, p)$ are linearly independent it follows that

$$\lambda_1 b_l \gamma_l + c_l - \lambda_2 \gamma_q \gamma_{q'} b_l = 0 \qquad l = 1, 2, \ldots, p$$

and solving for b_l we obtain

$$b_l = \frac{c_l}{\lambda_2 \gamma_q \gamma_{q'} - \lambda_1 \gamma_l}, \qquad l = 1, 2, \ldots, p.$$

The eigenvector can now be written explicitly as

$$a_{qq}(\bar{i}) = L_q(\bar{i}) L_{q'}(\bar{i}) + \sum_{l=1}^{p} \frac{c_l}{\lambda_1 \gamma_l - \lambda_2 \gamma_q \gamma_{q'}} L_l(\bar{i}) \qquad q, q' = 2, 3, \ldots, p.$$

Thus we have proved that with each value $\lambda_2 \gamma_q \gamma_{q'}$ there is an associated eigenvector $a_{qq'}(\bar{i})$. There are $\binom{p}{2}$ linearly independent homogeneous quadratic polynomials in the $p-1$ variables $L_q(\bar{i})$, $q = 2, 3, \ldots, p$, for example the set $L_q(\bar{i}) L_{q'}(\bar{i})$, $2 \le q \le q' \le p$. Hence the eigenvectors $a_{qq'}(\bar{i})$, $2 \le q < q' \le p$ are linearly independent.

General Case. We will now sketch the general case. Differentiating equation (48), r_i times in t_i $(i = 1, \ldots, p)$ and then setting $t_1 = t_2 = \cdots = t_p = 1$ yields the identity

$$\sum_{\bar{k}} P_{\bar{i},\bar{k}} (k_1)_{r_1} (k_2)_{r_2}, \ldots, (k_p)_{r_p}$$

$$= \lambda_R \left(\sum_{l=1}^{p} i_l \alpha_{l1}\right)^{r_1} \left(\sum_{l=1}^{p} i_l \alpha_{l2}\right)^{r_2} \cdots \left(\sum_{l=1}^{p} i_l \alpha_{lp}\right)^{r_p} \qquad (59)$$

$$+ \text{ polynomial in } (i_1, i_2, \ldots, i_p) \text{ of degree } < R$$

132   SAMUEL KARLIN and JAMES McGREGOR

where $R = \sum_{i=1}^{p} r_i$,

$$\lambda_R = \frac{\text{coefficient of } s^{N-R} \text{ in } f^{N-R}(s) \; [f'(s)]^R}{\text{coefficient of } s^N \text{ in } f^N(s)},$$

and

$$(k_1)_{r_1} = (k_1)(k_1 - 1) \ldots (k_1 - r_1 + 1), \text{ etc.}$$

The first term on the right side of (59) is a homogeneous polynomial of degree R in $(i_1, i_2, \ldots, i_p)$. This term is obtained from differentiating the right side of (48) and noting that as soon as a $f''(s)$ term appears in the differentiation the degree of its final coefficient in $\bar{i}$ has to be less than R since in order to obtain $f''(s)$ we have to differentiate $f'(s)$ which is *not* taken to an i_νth power $(\nu = 1, 2, \ldots, p)$. With the aid of the formula (59) we can now proceed to determine the eigenvalues of the transition probability matrix (46).

The first eigenvalue is 1 and its eigenvector has all equal components. This is so because the sum of the row elements of a probability transition matrix equals one. When $R = 1$, we have case 1 and when $R = 2$, we have case 2 considered previously. In case 1 we obtained $\binom{p-1}{0} + \binom{p-1}{1}$ linearly independent eigenvectors. The $\binom{p-1}{0}$ corresponds to the eigenvalue equal to 1. The $\binom{p-1}{1} = (p-1)$ term is for the eigenvalues $\lambda_1 \gamma_q$, $q = 2$, $3, \ldots, p$. In case 2 there were $\binom{p}{2}$ linearly independent eigenvectors corresponding to the eigenvalues of the form $\lambda_2 \gamma_q \gamma_{q'}, 2 \leq q \leq q' \leq p$.

The procedure used in cases 1 and 2 is now repeated inductively and we thereby construct $\binom{r+p-2}{r}$ linearly independent eigenvectors which are polynomials of degree r in the variables

$$\sum_{\nu=1}^{p} u_\nu^{(1)} i_\nu = N, \; L_2(\bar{i}) = \sum_{\nu=1}^{p} u_\nu^{(2)} i_\nu, \; \ldots, \; L_q(\bar{i}) = \sum_{\nu=1}^{p} u_\nu^{(q)} i_\nu,$$

$$\ldots, \; L_p(\bar{i}) = \sum_{\nu=1}^{p} u_\nu^{(p)} i_\nu$$

for the eigenvalues

$$\lambda_r \gamma_{q_1} \gamma_{q_2}, \ldots, \gamma_{q_r}, \qquad 2 \leq q_1 \leq q_2 \leq \ldots \leq q_r \leq p. \tag{60}$$

The explicit eigenvectors possess the form

$$a_{q_1, q_2, \ldots, q_r}(i_1, i_2, \ldots, i_p) \tag{61}$$

$$= \prod_{k=1}^{r} L_{q_k}(\bar{i}) + \text{polynomial in } (i_1, \ldots, i_p) \text{ of degree} < r.$$

In the course of the proof we use the fact that a polynomial in $\bar{i}$ of degree $< r$ can be represented as a linear combination of all eigenvectors associated with the eigenvalues $\lambda_k \gamma_{q_1} \ldots \gamma_{q_k}, k \leq r - 1$.

The number of eigenvalues of the form $\lambda_r \gamma_{q_1} \gamma_{q_2}, \ldots, \gamma_{q_r}, q = 2, \ldots, p$ as demonstrated above is clearly equal to the number of homogeneous polynomials of degree r in the $p - 1$ variables $\{L_2(\bar{i}), \ldots, L_p(\bar{i})\}$. This is well known to be $\binom{r+p-2}{r}$.

The eigenvectors enumerated above (they are all linearly independent) total

$$\binom{p-2}{0} + \binom{p-1}{1} + \binom{p}{2} + \cdots + \binom{r+p-2}{r} + \cdots + \binom{N+p-2}{N} = \binom{N+p-1}{N}$$

which is precisely the number of states in the Markov chain of the matrix (46). We summarize the conclusions obtained in the following theorem.

Theorem 3. *Let* $\Gamma = \|\, \alpha_{\nu\mu}\,\|^p_{\nu,\,\mu=1}$ *denote a stochastic matrix, i.e.,* $\alpha_{\nu\mu} \geq 0,\ \sum\limits_{\mu} \alpha_{\nu\mu} = 1$ *of a Markov chain. Let* $\gamma_1 = 1, \gamma_2, \ldots, \gamma_p$ *denote the eigenvalues and assume that*

$$\lambda_r\, \gamma_{q_1}\, \gamma_{q_2} \cdots \gamma_{q_r} \neq \lambda_{r+1}\, \gamma_{q'_1}\, \gamma_{q'_2} \cdots \gamma_{q'_{r+1}} \tag{62}$$

for all choices of γ_{q_k} *and* $\gamma_{q'_l}$ $(2 \leq q_k, q'_l \leq p),\ r \geq 1,$ *where*

$$\lambda_0 = 1, \qquad \lambda_r = \frac{\text{coefficient of } s^{N-r} \text{ in } f^{N-r}(s)\, [f'(s)]^r}{\text{coefficient } s^N \text{ in } f^N(s)}, \qquad r = 1, 2, \ldots, N.$$

Suppose that $u^{(1)} = (1, 1, \ldots, 1), u^{(2)}, \ldots, u^{(p)}$ *constitutes a complete set of eigenvectors of* Γ, *i.e.,* Γ *is diagonalizable.*

Consider the Markov chain of p *types whose transition probability matrix is given by (46). For* $r \geq 1$ *there exists* $\binom{r+p-2}{r}$ *linearly independent eigenvectors which are polynomials of degree* r *in the variables* $(i_1, i_2, \ldots, i_p)$ *through the functions,*

$$L_q(\vec{i}) = \sum_{\nu=1}^{p} u^{(q)}_\nu\, i_\nu, \qquad q = 2, \ldots, p.$$

The corresponding eigenvalues are displayed in (60). The right eigenvectors are of the form (61). These eigenvectors plus the constant vector span a space of dimension $\binom{N+p-1}{N}$ *which is the order of the matrix* $\|\, P_{\vec{i},\,\vec{k}}\,\|$.

The conditions of Theorem 3 are more stringent than required. In fact, by virtue of relations (59) we can introduce a basis in the manner of Theorem 1 so that P defined in (46) achieves a triangular form whose diagonal elements are the eigenvalues

$$\lambda_0 = 1, \lambda_r\, \gamma_{q_1}\, \gamma_{q_2} \cdots \gamma_{q_r},$$
$$r = 1, 2, \ldots, N;\ 2 \leq q_1 \leq q_2 \leq \cdots \leq q_r \leq p.$$

This argument does not use assumption (62). By a standard perturbation procedure we obtain

Theorem 4. *Let* $\Gamma = \|\, \alpha_{\nu\mu}\,\|_{\nu,\,\mu=1}$ *denote a stochastic matrix of a Markov chain. Let* $\gamma_1 = 1, \gamma_2, \ldots, \gamma_p$ *denote its eigenvalues. Then the eigenvalues of the Markov chain matrix (46) are listed in (60).*

Under the general stipulations of this theorem, the existence of a full set of eigenvectors is not guaranteed. In fact, in the general case there may actually occur elementary divisors.

We present some examples of Theorem 3.

Example 1. Let $\Gamma =$ the identity matrix. In other words there are no mutation pressures. In this case (46) is the direct generalization to p types of the two type model. Thus there are p types labeled A_1, A_2, ..., A_p. Each independently multiplies by the laws of a branching process characterized by the probability generating function $f(s)$. The induced MC has the transition probability matrix

$$P_{\bar{i},\bar{k}} = \frac{\text{coefficient of } t_1^{k_1} t_2^{k_2} \ldots t_p^{k_p} \text{ in } \prod_{\nu=1}^{p} f^{l_\nu}(t_\nu)}{\text{coefficient of } t^N \text{ in } f^N(t)} \,. \tag{63}$$

The state space consists of the integral points of the simplex

$$\Delta_p = \left\{ \bar{i} = (i_1, i_2, \ldots, i_p) \mid i_\nu \text{ integers} \geq 0 , \quad \sum_{\nu=1}^{p} i_\nu = N \right\}.$$

In this case $\gamma_1 = \gamma_2 = \cdots = \gamma_p = 1$, so that $\lambda_r \gamma_{q_1} \gamma_{q_2} \ldots \gamma_{q_r} = \lambda_r$ for all choices $2 \leq q_1, q_2, \ldots, q_r \leq p$. Let λ_r be defined as in (64). Notice that condition (62) is trivially satisfied since $\lambda_r > \lambda_{r+1}$ for $r \geq 1$. Theorem 3 asserts that λ_r is an eigenvalue of multiplicity $\binom{r+p-2}{r}$, $r = 0, 1, 2, \ldots, N$, i.e., there exists $\binom{r+p-2}{r}$ linearly independent eigenvectors associated with the eigenvalue λ_r. The right eigenvectors associated with λ_r are polynomials of degree r in the variables $L_q(\bar{i}) = \sum_{\nu=1}^{p} u_\nu^q i_\nu = i_q$ $(q = 2, 3, \ldots, p)$ since we can take $u_\nu^q = \delta_{qv}$. More explicitly, to each

$$x_1^{r_1} x_2^{r_2} \ldots x_{p-1}^{r_{p-1}} \qquad r_i \text{ integers}, \qquad \sum_{i=1}^{p-1} r_i = r$$

there exists a polynomial of degree r of the form

$$W_{r_1,\ldots,r_{p-1}}(u_1, u_2, \ldots, u_{p-1}) = u_1^{r_1} u_2^{r_2} \ldots u_{p-1}^{r_{p-1}} + T_{r_1,\ldots,r_{p-1}}(u_2 \ldots u_{p-1})$$

where T is a polynomial of degree $\leq r - 1$ such that the vector whose $\bar{i}$ component is

$$\alpha_{r_1,\ldots,r_{p-1}}(\bar{i}) = W_{r_1,\ldots,r_{p-1}}(i_1, \ldots, i_{p-1})$$

is a right eigenvector for λ_r. The left eigenvectors for λ_r are indexed in the same manner, $\beta_{r_1, r_2, \ldots, r_{p-1}}$. The eigenvectors

$$\left\{ \beta_{r_1, r_2, \ldots, r_{p-1}} \right\} \qquad \text{and} \qquad \left\{ \alpha_{r_1, r_2, \ldots, r_{p-1}} \right\}$$

are specified to be biorthogonal.

It is convenient for later reference to state the preceding details as a theorem

Theorem 5. *The transition probability matrix* (63) *on the state space* Δ_p *possesses the eigenvalues*

$$\lambda_0 = 1, \qquad \lambda_r = \frac{\text{coefficient } t^{N-r} \text{ in } f^{N-r}(t) \, [f'(t)]^r}{\text{coefficient } t^N \text{ in } f^N(t)} \qquad r = 1, 2, \ldots, N \tag{64a},$$

the eigenvalue λ_r *occurs with multiplicity* $\binom{p+r-2}{r}$. *The matrix* (63) *is*

diagonalizable if $\lambda_2 > \lambda_3 > \ldots > \lambda_N$ *(e.g., if* $c_0 c_1 c_2 > 0$ *where* $f(s)$
$= \sum\limits_{m=0}^{\infty} c_m s^m$ *) .*

We append some further properties of the eigenvectors needed in connection with the probabilistic interpretations of these quantities (section 5). For ease of exposition we divide the discussion into a series of steps.

1. In accordance with the theorem we have a total of p linearly independent right eigenvectors associated with the eigenvalue $\lambda_0 = \lambda_1 = 1$. A convenient expression of these eigenvectors is

$$\alpha_q = \delta_q (i_1, i_2, \ldots, i_p) = \frac{i_q}{N} \qquad (q = 1, 2, \ldots, p) . \qquad (64)$$

Adding these eigenvectors

$$\sum_{q=1}^{p} \delta_q (i_1, i_2, \ldots, i_p) \equiv 1 \qquad \text{independent of } (i_1, \ldots, i_p)$$

which is the constant eigenvector usually assigned to the eigenvalue λ_0.

It is also useful to point out that the left eigenvectors for $\lambda_0 = \lambda_1 = 1$ possess a very simple form. We list these eigenvectors. They are

$$\beta_q = \epsilon_q (i_1, i_2, \ldots, i_{p+1}) = \begin{cases} 1 \text{ if } i_q = N \\ 0 \text{ otherwise} \end{cases} (q = 1, \ldots, p + 1) . \quad (65)$$

Notice that ϵ_q has a single nonzero component located at one of the appropriate vertices of Δ_p. The system of eigenvectors (64) and (65) are mutually biorthogonal as a direct computation shows. Of course, ϵ_q is automatically orthogonal to all $\alpha_{r_1, r_2, \ldots, r_{p-1}}$ for $r = \sum\limits_{i=1}^{p} r_i > 1$ since $\lambda_r < 1 \ (r \geq 1)$.

2. Consider a lower dimensional face of the simplex Δ_p determined by prescribing the components

$$i_{n_1} = i_{n_2} = \cdots = i_{n_k} = 0 \qquad (0 \leq n_1 < n_2 < \cdots < n_k \leq p)$$

and allowing the other components to vary arbitrarily. We denote this face by the symbol $\Delta_p (n_1, n_2, \ldots, n_k)$. It is clear that this describes the collection of states in Δ_p where the types $A_{n_1}, A_{n_2}, \ldots, A_{n_k}$ have disappeared from the population. It is clear that the transition probability matrix (63) reduces to the corresponding version involving the remaining types.

The matrix $\widetilde{P}$ restricted to the components of indices in $\Delta_p (n_1, n_2, \ldots, n_k)$ is clearly a copy of the matrix (63) entailing $p - k$ variables. The eigenvalues of $\widetilde{P}$ are again $\lambda_0, \lambda_1, \ldots, \lambda_N$ which plainly does not depend on p; only their relative multiplicity depends on p. Let

$$\left\{ \alpha_j^{n_1, \ldots, n_k} \right\} \qquad \text{and} \qquad \left\{ \beta_j^{n_1, \ldots, n_k} \right\} \qquad (66)$$

denote a complete biorthogonal set of right and left eigenvectors for the reduced matrix $\tilde{P}$. The number of these eigenvectors is $\binom{N+p-k-1}{N}$. The components of each vector in (66) are indexed by the points of the simplex Δ_p $(n_1, n_2, \ldots, n_k)$.

We extend each eigenvector $\beta_j^{n_1, \cdots, n_{v_0}}$ defined on Δ_p $(n_1, n_2, \ldots, n_k)$ to a vector $\tilde{\beta}_j^{n_1, \cdots, n_{v_0}}$ defined on Δ_p by specifying the value of the component of $\beta_j^{n_1, \cdots, n_{v_0}}$ corresponding to a point of $\Delta_p - \Delta_p$ $(n_1, n_2, \ldots, n_k)$ [the part of Δ_p outside of Δ_p $(n_1, n_2, \ldots, n_k)$] equal to zero. It is then straightforward to verify that $\tilde{\beta}_j^{n_1, \cdots, n_k}$ are linearly independent left eigenvectors of the matrix P.

The following converse is also of relevance. Let β $(\bar{i})$ denote a left eigenvector of P whose only non-zero components occur for the indices of some subface F of Δ_p. Then necessarily β $(\bar{i})$ with $\bar{i}$ restricted to F is a left eigenvector of the matrix $\tilde{P}$ obtained from P by deleting the rows and columns corresponding to points outside F.

These considerations are valid for each face of Δ_p and will play a fundamental role in the anslysis below.

3. Associated with each vertex of Δ_p is a left eigenvector of P whose only nonzero component corresponds to that vertex. We normalize these vectors so that the nonzero component has value 1. These eigenvectors all belong to the eigenvalue $\lambda_0 = \lambda_1 = 1$. They number precisely p and are manifestly linearly independent. The multiplicity of $\lambda_0 = \lambda_1 = 1$ is p so that they span the eigenmanifold of the eigenvalue 1.

4. For each edge E of Δ_p there exists apart from those enumerated under paragraph 3, $N - 2$ additional left eigenvectors, one for each eigenvalue $\lambda_2, \lambda_3, \ldots, \lambda_N$ with the property that they vanish for all components of Δ_p not contained in the edge E. The number of edges is $\binom{p}{2}$. Therefore, the number of eigenvectors belonging to λ_r $(r = 2, 3, \ldots, p)$ which vanish for all components except for those whose indices are contained in edges is $\binom{p}{2}$.

Theorem 5 tells us that the multiplicity of the eigenvalue λ_2 is $\binom{p}{2}$ and therefore a complete set of left eigenvectors for λ_2 are all of the form that the only nonzero components occur for indices of edges of Δ_p.

5. Next consider the subfaces of dimensions 2 correspond to a triplet of types. For each prescribed face F of this kind, we can construct in accordance with Theorem 5, $\binom{r+1}{r}$ eigenvectors for λ_r $(r = 2, 3, \ldots, N)$ whose only nonzero components correspond to indices of the face. In order not to duplicate any of the eigenvectors listed under paragraphs 3 or 4, we need to count only those eigenvectors which have at least one nonzero component for an index value interior to F. The number of left eigenvectors of this kind is

$$\binom{r+1}{r} - A_1(r)$$

where $A_1(r)$ is the number of left eigenvectors for λ_r whose nonzero components are restricted to an edge of F. There is one in each edge and there are 3 edges for each two dimensional simplex. Hence, $A_1(r) = 3$ and therefore $\binom{r+1}{r} - 3 = \binom{r-2}{1} =$ number of left eigenvectors in each two dimensional face F of Δ_p all of whose components vanish outside F and for which at least one nonzero component occurs for an index value interior to F.

If we total the number of independent left eigenvectors for λ_2 corresponding to all two dimensional faces of Δ_p, we have

$$\binom{p}{2}\binom{r-2}{0} + \binom{p}{3}\binom{r-2}{1}. \tag{67}$$

In fact, $\binom{p}{2}$ is the number of edges in Δ_p,

$\binom{r-2}{0} = 1$ is the number of eigenvectors for λ_2 in any given edge,

$\binom{p}{3}$ is the number of 2-faces of Δ_p,

$\binom{r-2}{1}$ is the number of eigenvectors for λ_2 in any given face with nonzero components for some index associated with an interior point of that face.

With these interpretations the validation of (67) is clear.

6. Let G be a three-dimensional face of Δ_p. The number of left eigenvectors for λ_r, $r \geq 3$ with its nonzero components restricted to G is $\binom{r+2}{r}$. Of these $\binom{r-2}{0}\binom{4}{2}$ possess nonzero components only in edges of G and $\binom{r-2}{1}\binom{4}{3}$ possess nonzero components genuinely confined to the 2-faces of G. Therefore, the number of left eigenvectors restricted to G with at least one nonzero component interior to G is

$$\binom{r+2}{r} - \binom{r-2}{0}\binom{4}{2} - \binom{r-2}{1}\binom{4}{3} = \binom{r-2}{2}.$$

To sum up the total number of left eigenvectors for λ_r ($r \geq 4$) of the form that the nonzero components are confined to 3-faces (i.e., three dimensional faces) of Δ_p is

$$\binom{r-2}{0}\binom{p}{2} + \binom{r-2}{1}\binom{p}{3} + \binom{r-2}{2}\binom{p}{4} \leq \binom{p+r-2}{r}$$

with strict inequality unless $r = 4$. In general, we have the identity

$$\sum_{k=0}^{r-2}\binom{p}{k+2}\binom{r-2}{k} = \binom{p+r-2}{4} \qquad r \geq 2 \tag{68}$$

from which the previous inequality follows.

7. The general pattern is now clear. We proceed by induction. Suppose we have proved that for any specified k-face ($k = 1, 2, \ldots, l$) the number of linearly independent left eigenvectors for λ_r whose nonzero components are restricted to this face with at least one nonzero value for an index interior to this face is $\binom{r-2}{k-1}$.

Now consider a fixed $l + 1$-dimensional-face H. The number of eigenvectors for λ_r confined to the boundary of H is (using the induction hypothesis

$$\binom{r-2}{0}\binom{l+2}{2} + \binom{r-2}{1}\binom{l+2}{3} + \cdots + \binom{r-2}{l-1}\binom{l+2}{l+1}. \tag{69}$$

The total number of eigenvectors for λ_r restricted to this face is $\binom{r+l}{r}$ (Theorem 5). Appealing to (68), we find that the difference of $\binom{r+l}{r}$ and (69) is $\binom{r-2}{l}$. This is the number of eigenvectors confined to H with some nonzero component for an index value interior to H. This advances the induction to the case of an $l + 1$ face of Δ_p.

Now the number of eigenvectors for λ_r $(r \geq 2)$ whose nonzero components are restricted to any $l + 1$-face is

$$\sum_{k=0}^{l} \binom{p}{k+2}\binom{r-2}{k}. \tag{70}$$

Comparing with (68), we see that this quantity $\leq \binom{r+p-2}{r}$ with equality if and only if $r \leq 2 + l$.

We sum up the preceding discussion in the following theorem.

Theorem 6. *Consider the transition probability matrix* (63) *on the state space* Δ_p. *All the left eigenvectors for* λ_r *($r = 2, \ldots, p$) have all their nonzero components associated with the indices of faces of* Δ_p *of dimension* $\leq r - 1$.

For any prescribed k-face F *($2 \leq k < p$) of* Δ_p *the number of left eigenvectors for* λ_r *($2 \leq r \leq N$) whose nonzero components are confined to the indices of* F *but to no subface of* F *is* $\binom{r-2}{k-1}$.

The total number of eigenvectors for λ_r *($2 \leq r \leq N$) confined to l-faces is*

$$\sum_{k=0}^{l-1} \binom{r-2}{k}\binom{p}{k+2}.$$

In particular, all left eigenvectors for λ_2 possess their nonzero components only for indices of edges of Δ_p. Similarly, for λ_3 a complete set of left eigenvectors are all of the form that the only nonzero components occur for indices confined to 2-faces of Δ_p, etc.

Example 2. The eigenvalues of the polyploid selfing model defined in (26) are easy to describe. In this case the mutation matrix is

$$\alpha_{\mu\nu} = \frac{\binom{2\mu}{\nu}\binom{2p-2\mu}{p-\nu}}{\binom{2p}{p}} \qquad \mu, \nu = 0, 1, \ldots, p.$$

Its eigenvalues are given in example B at the close of section 3. They are

$$\gamma_0 = \gamma_1 = 1, \qquad \gamma_r = \frac{2^r \binom{2N-r}{N-r}}{\binom{2N}{N}}, \qquad r = 2, 3, \ldots, N.$$

The matrix $\Gamma = \| \alpha_{\mu\nu} \|$ is diagonalizable as attested to by Theorem 1. By Theorem 4 the eigenvalues of the matrix (26) are

$$\lambda_r \, \gamma_{q_1} \gamma_{q_2} \cdots \gamma_{r_q} \qquad 1 \le q_1 \le q_2 \le \cdots \le q_r \le p \qquad r = 1, 2, \ldots, N$$

where

$$\lambda_r = \frac{\text{coefficient of } t^{N-r} \text{ in } f^{N-r}(t) \, [f'(t)]^r}{\text{coefficient of } t^N \text{ in } f^N(t)}.$$

5. Probabilistic Interpretations of the Eigenvalues

With the aid of Theorems 5 and 6, we are prepared to give some probabilistic interpretations for the eigenvalues

$$\lambda_0 = 1, \qquad \lambda_r = \frac{\text{coefficient of } s^{N-r} \text{ in } f^{N-r}(s) \, [f'(s)]^r}{\text{coefficient of } s^N \text{ in } f^N(s)} \tag{71}$$

and their associated eigenvectors of the multi-type transition matrix (63). We postulate throughout what follows that $\lambda_r > \lambda_{r+1}$ $(r > 1)$. This is guaranteed specifically when, for example, $c_0 c_1 c_2 > 0$ where $f(s) = \sum_{m=0}^{\infty} c_m s^m$. The reader should recall that λ_r occurs with multiplicity $\binom{r+p-2}{r}$ in the multiple type model of p types (Theorem 5).

We begin our discussion with the two type model $(p = 2)$. In this case all the eigenvalues, apart from $\lambda_0 = \lambda_1 = 1$ have simple multiplicity. By Theorem 1 we know that the Markov matrix (27) is diagonalizable. In terms of the biorthogonal set of eigenvectors displayed we have the representation

$$P_{ij}^t = \sum_{r=0}^{N} \lambda_r^t \, \alpha_r(i) \, \beta_r(j) \tag{72}$$

where $\alpha_r(i)$ is the ith component of the rth right eigenvector and $\beta_r(j)$ is the jth component of the rth left eigenvector as listed in Theorem 1. (Here t is a non-negative integer and α_r and β_r, are biorthogonal.) It is useful to separate the two terms of (72) for $\lambda_0 = \lambda_1 = 1$ so it has the form

$$P_{ij}^t = \alpha_0(i) \, \beta_0(j) + \alpha_1(i) \, \beta_1(j) + \sum_{r=2}^{N} \lambda_r^t \, \alpha_r(i) \, \beta_r(j) \tag{73}$$

where α_i and β_j are defined in (30) — (34). The sum in (73) goes to zero at the rate λ_2^t $(\lambda_2 < 1)$. Moreover inspection of the explicit expression of $\beta_0(j)$ and $\beta_1(j)$ [cf. (33)] reveals that P_{ij}^t is precisely the sum term when $0 < j < N$. It follows that

$$\lim_{t \to \infty} P_{ij}^t = \alpha_0(i) \, \beta_0(j) + \alpha_1(i) \, \beta_1(j)$$

$$= \begin{cases} 0 & j \neq 0, N \\ \dfrac{N-i}{N} & j = 0 \\ \dfrac{i}{N} & j = N \end{cases} \tag{74}$$

and the rate of convergence is geometric of order λ_2, i.e., λ_2 is the "rate of approach to homozygosity". Equivalently, the probability that the system is not in a homozygous state (0 or N) behaves like λ_2^t as $t \to \infty$. Furthermore, since $\lambda_3 < \lambda_2$ we see that

$$\lim_{t \to \infty} \frac{P_{ij}^t}{\lambda_2^t} = \alpha_2(i)\,\beta_2(j), \qquad i, j \neq 0, N\ . \tag{75}$$

It is easy to prove that $\alpha_2(i) \neq 0$ for $0 < i < N$. Otherwise $\alpha_2(i) \equiv 0$ for $0 < i < N$. But always $\alpha_2(0) = \alpha_2(N) = 0$ since α_2 is orthogonal to β_0 and β_1. Then $\alpha_2(i) \equiv 0$ for $0 \le i \le N$ and this is impossible since α_2 is an eigenvector of λ_2. By selecting the normalization of α_2 so that $\alpha_2(i_0) > 0$ we conclude on the basis of (75) that $\beta_2(j) \ge 0$, $0 < i < N$. We claim that $\beta_2(j)$ is not identically zero for $0 < j < N$. The proof runs similar to that used to prove $\alpha_2 \not\equiv 0$.

It follows that $\sum\limits_{j=1}^{N-1} \beta_2(j) > 0$. Actually we can prove that $\beta_2(j) > 0$, $0 < j < N$.

Indeed, the transient states $T = \{1, 2, \ldots, N-1\}$ all communicate, i.e., starting from any state of T, it is possible (with positive probability) to reach any other state of T. This implies that the rate at which P_{ij}^t tends to zero $(t \to \infty)$ for $i, j \in T$ is independent of the choice of i and j in T. We have already proved that for some $i = i_0$ and $j = j_0$, $\alpha_2(i_0) > 0$, $\beta_2(j_0) > 0$. But

$$P_{ij}^t \sim \lambda_2^t\,\alpha_2(i)\,\beta_2(j) \qquad i, j \in T\ .$$

Therefore $\alpha_2(i)\,\beta_2(j) > 0$ for all $i, j \in T$ which shows that $\alpha_2(i)$ keeps a strict constant sign for $i \in T$ and the same holds for the vector β_2.

The expression (75) can now be interpreted to the effect that the limiting probability of being in state j, given $j \neq 0, N$ is

$$\frac{\beta_2(j)}{\sum\limits_{j=1}^{N-1} \beta_2(j)}\ . \tag{76}$$

For ease of exposition we develop interpretations of the eigenvalues for the three type model whose transition probability matrix is (63) with $p = 3$. This analysis embodies the arguments of the general case. We denote the simplex of the state space by the symbol Δ_3 and its edges as E_1, E_2 and E_3, i.e., E_k consists of all $\bar{i} \in \Delta_3$ for which $i_k = 0$ ($k = 1, 2, 3$). The multiplicity of λ_r is $r + 1$ (Theorem 5). We list the corresponding right and left eigenvectors in the form

$$\alpha_{rk} = \alpha_{rk}(i_1, i_2, i_3); \qquad \beta_{rk} = \beta_{rk}(i_1, i_2, i_3), \quad k = 0, 1, \ldots, r \tag{77}$$

where (i_1, i_2, i_3) index the components of the vectors; here always $i_1, i_2, i_3 \ge 0$ and $i_1 + i_2 + i_3 = N$. The vectors (77) are specified in such a way that they are biorthognal.

The transition matrix possesses the representation

$$P^t_{\bar{i}, \bar{j}} = \sum_{r=0}^{N} \lambda^t_r \sum_{k=0}^{r} \alpha_{rk}(\bar{i}) \beta_{rk}(\bar{j}) \qquad \bar{i} = (i_1, i_2, i_3), \bar{j} = (j_1, j_2, j_3) \ . \tag{78}$$

We rewrite (78) by separating the terms involving the first two eigenvalues. Thus

$$\begin{aligned}
P^t_{\bar{i}, \bar{j}} = {} & \alpha_{00}(\bar{i}) \beta_{00}(\bar{j}) + \alpha_{10}(\bar{i}) \beta_{10}(\bar{j}) + \alpha_{11}(\bar{i}) \beta_{11}(\bar{j}) + \\
& + \lambda^t_2 \left[\alpha_{20}(\bar{i}) \beta_{20}(\bar{j}) + \alpha_{21}(\bar{i}) \beta_{21}(\bar{j}) + \alpha_{22}(\bar{i}) \beta_{22}(\bar{j}) \right] + \\
& + \sum_{r=3}^{N} \lambda^t_r \left[\sum_{k=0}^{r} \alpha_{rk}(\bar{i}) \beta_{rk}(\bar{j}) \right] .
\end{aligned} \tag{79}$$

We pointed out in our previous discussion that all the nonzero components of $\beta_{00}(\bar{j})$, $\beta_{10}(\bar{j})$ and $\beta_{11}(\bar{j})$ are confined to the vertices of Δ_3. Moreover, it was shown that the nonzero components of $\beta_{20}(\bar{j})$, $\beta_{21}(\bar{j})$ and $\beta_{22}(\bar{j})$ are confined to the indices in the edges of Δ_3. Actually $\beta_{20}(\bar{j})$ agrees with the vector $\beta_2(\bar{j})$ defined on one of edges, say E_1, and extended to be zero on the rest of the simplex. Similarly $\beta_{21}(\bar{j})$ is the vector $\beta_2(\bar{j})$ on one of the other edges, say E_2, and extended equal to zero otherwise and $\beta_{22}(\bar{j})$ is also the vector $\beta_2(\bar{j})$ on the third edge E_3, and equal to zero elsewhere. Finally, exactly one of the vectors amongst $\beta_{30}, \beta_{31}, \beta_{32}, \beta_{33}$ possesses nonzero components interior to Δ_3, the other vectors are confined to the indices in the edges of Δ_3. Actually, as noted above we can identify

$$\beta^{\overline{(i)}}_{30} = \beta_3(\bar{i}) \text{ on } E_1 \text{ and } 0 \text{ elsewhere,}$$

$$\beta^{\overline{(i)}}_{31} = \beta_3(\bar{i}) \text{ on } E_2 \text{ and } 0 \text{ elsewhere,}$$

$$\beta^{\overline{(i)}}_{32} = \beta_3(\bar{i}) \text{ on } E_2 \text{ and } 0 \text{ elsewhere.}$$

We claim that $\beta_{33}(\bar{j})$ cannot vanish identically zero in $\Delta^0_3 =$ the interior of Δ_3. Indeed each one dimensional eigenvector $\beta_r(\cdot)$ $(r = 2, \ldots, N)$ determines on each edge E_1 an eigenvector extended to the rest of Δ_3 equal to 0. This gives $N - 1$ linear independent eigenvectors whose only nonzero values are confined to the indices of E_1. Similarly we get $N - 1$ other vectors whose nonzero values are confined to E_2 and $N - 1$ corresponding to E_3. Finally, we have β_{00}, β_{10} and β_{11} with nonzero values only at the vertices of E_3. In total, this gives $3N$ linearly independent eigenvectors with nonzero components associated only with $\bar{j}$ in the boundary of Δ_3. We call the set of these eigenvectors V. There is precisely $3N$ states in the boundary of Δ_3. Therefore, the listed eigenvectors span the linear space of all vectors having only nonzero coordinates for indices in the boundary of Δ_3.

Now $\beta_{33}(\bar{j})$ which is not one of β_r since it belongs to the eigenvalue λ_3 and we already listed $\{\beta_{30}, \beta_{31}, \beta_{32}\}$ those of the form β_3 properly extended. If $\beta_{33}(\bar{j}) \equiv 0$ for $\bar{j} \in \Delta^0_3$ then $\beta_{33}(\bar{j})$ is linearly dependent on the eigenvectors of V and this is impossible since β_{33} is independent of $\beta_{30}, \beta_{31},$

and β_{32} by construction and certainly independent of the other vectors in V since the others in V are associated with different eigenvalues $(\neq \lambda_3)$.

We proved above that $\beta_{33}(\bar{j})$ is not identically zero for $\bar{j} \in \Delta_3^o$. It is also true that $\alpha_{33}(\bar{i})$ is $\neq 0$ for $\bar{i} \in \Delta_3^o$. Indeed, if we suppose the contrary that $\alpha_{33}(\bar{i}) \equiv 0$ for $\bar{i} \in \Delta_3^o$ then since α_{33} is orthogonal to every vector in V it follows that $\alpha_{33} \equiv 0$ for all $i \in \Delta_3$ which contradicts the definition of α_{33}. [This argument further shows that $\alpha_{33}(\bar{i})$ necessarily vanishes on the boundary of Δ_3.]

We observe next for $\bar{i}, \bar{j} \in \Delta_3^o$ that

$$P_{\bar{i},\bar{j}}^t \sim \lambda_3^t \, \beta_{33}(\bar{j}) \, \alpha_{33}(\bar{i}) \,, \qquad t \to \infty \,. \tag{80}$$

This shows that $\beta_{33}(\bar{j}) \, \alpha_{33}(\bar{i}) \geq 0$ for all $\bar{i}$ and $\bar{j} \in \Delta_3^o$ that is $\beta_{33}(\bar{j})$ and $\alpha_{33}(\bar{i})$ have a fixed sign (indeed the same sign) on the interior of Δ_3^o; we can choose the multiplicative constant ± 1 defining the vectors such that

$$\beta_{33}(\bar{j}) \geq 0 \qquad \bar{j} \in \Delta_3^o \neq 0 \,,$$
$$\alpha_{33}(\bar{i}) \geq 0 \qquad \bar{i} \in \Delta_3^o \neq 0 \,.$$

Actually $\alpha_{33}(\bar{i}) \, \beta_{33}(\bar{j}) > 0$ for all $\bar{i}, \bar{j} \in \Delta_3^o$. In fact, we know that all states interior to Δ_3 communicate. Moreover, we already proved that $\alpha_{33}(\bar{i}_0) \, \beta_{33}(\bar{j}_0) > 0$ for some $\bar{i}_0, \bar{j}_0 \in \Delta_3^o$. Since all states in Δ_3^o communicate, it follows that $P_{\bar{i},\bar{j}}^t$ tends to zero at the rate λ_3^t independent of the choice of $\bar{i}$ and $\bar{j}$ in Δ_3^o. Therefore $\alpha_{33}(\bar{i}) \, \beta_{33}(\bar{j}) > 0$ for all $\bar{i}$ and $\bar{j}$ and Δ_3^o.

In possession of the properties of the eigenvectors $\alpha_{00}, \ldots, \alpha_{33}$ and $\beta_{00}, \ldots, \beta_{33}$ we can now assert

a) The rate at which absorption ($=$ fixation $=$ homozyosity) occurs into the vertices is λ_2 since for $\bar{j} \neq$ vertex of Δ_3, the expression of $P_{\bar{i},\bar{j}}^t$ in (78) reduces to the sum from $r = 2$ on.

The conditional distribution for t large given that fixation has not occurred but that the first type has been lost from the population is proportional to $\beta_{22}(\bar{j})$, where $\beta_{22}(\bar{j})$ is the eigenvector whose nonzero components are contained in the edge representing only the second and third types. The eigenvectors $\beta_{21}(\bar{j})$ and $\beta_{22}(\bar{j})$ possess analogous interpretations. The proof of this fact is the same as the discussion of (76).

b) The rate at which absorption into the edges occurs (i.e., the rate at which one of the types, without specifying which, is lost from the population) is λ_3. In fact for $\bar{i}, \bar{j} \in \Delta_3^o$ (interior of Δ_3)

$$P_{\bar{i},\bar{j}}^t = \sum_{r=3}^{N} \lambda_r^t \left[\sum_{k=0}^{k} \alpha_{rk}(\bar{i}) \, \beta_{rk}(\bar{j}) \right]$$

The dominant term is λ_3^t since $\alpha_{33}(\bar{i}) \, \beta_{33}(\bar{j})$ is different from zero in Δ_3^o while $\beta_{30}(\bar{j}), \beta_{31}(\bar{j}), \beta_{32}(\bar{j})$ have all their nonzero components for indices confined to the faces of Δ_3.

The conditional distribution of the state variable $\bar{j}$ for t large given that all types are present is asymptotically equal to

$$\frac{\beta_{33}(\bar{j})}{\sum_{\bar{j}\in\varDelta_3^o}\beta_{33}(\bar{j})}\qquad \bar{j}\in\varDelta_3^o\,.$$

The meaning of the right eigenvectors are easily discerned.

$c\,\alpha_{00}(\bar{i}) =$ the probability starting from $\bar{i}$ of being absorbed into the vertex $(N,0,0)$,

$c\,\alpha_{10}(\bar{i}) =$ the probability starting from $\bar{i}$ of being absorbed into $(0,N,0)$,

$c\,\alpha_{11}(\bar{i}) =$ the probability starting from $\bar{i}$ of being absorbed into $(0,0,N)$

where $c = [\alpha_{00}(\bar{i}) + \alpha_{10}(\bar{i}) + \alpha_{11}(\bar{i})]^{-1}$.

The eigenvectors $\alpha_{20}(\bar{i})$, $\alpha_{21}(\bar{i})$ and $\alpha_{22}(\bar{i})$ can be interpreted in a similar way. For this purpose we consider the limiting conditional distribution that absorption into a vertex (fixation) has not taken place yet. This clearly yields a distribution situated on the edges of $\varDelta_3$. Clearly for $\bar{j} \neq$ vertex but on an edge of $\varDelta$,

$$P_{\bar{i},\bar{j}}^t \sim \lambda_2^t\,[\alpha_{20}(\bar{i})\,\beta_{20}(\bar{j}) + \beta_{21}(\bar{j})\,\alpha_{21}(\bar{i}) + \alpha_{22}(\bar{i})\,\beta_{22}(\bar{j})]\quad t\to\infty\,.$$

Recall that

$$\beta_{20}(\bar{j}) = \begin{cases} \beta_2(\bar{j}) & \bar{j}\in E_1 \\ 0 & \text{elsewhere} \end{cases}$$

and similarly for the others. Hence

$$\sum_{\substack{\bar{j}\\ \bar{j}\notin\text{ vertex}}}\beta_{20}(\bar{j}) = \sum_{\substack{\bar{j}\\ \bar{j}\notin\text{ vertex}}}\beta_{21}(\bar{j}) = \sum_{\substack{\bar{j}\\ \bar{j}\notin\text{ vertex}}}\beta_{22}(\bar{j}) = \sum_l \beta_2(l) > 0\,.$$

Since β_{20}, β_{21} and β_{22} are non-negative and each possess their nonzero values apart from the vertices on disjoint edges we infer that $\alpha_{20}(\bar{i})$, $\alpha_{21}(\bar{i})$ and $\alpha_{22}(\bar{i})$ are non-negative for all $\bar{i} \neq$ vertex and at least one of these vectors is positive. The conditional distribution of $P_{\bar{i},\bar{j}}^t$ $(t\to\infty)$ for $\bar{j} \neq$ vertex becomes

$$\frac{\alpha_{20}(\bar{i})\,\beta_{20}(\bar{j}) + \alpha_{21}(\bar{i})\,\beta_{21}(\bar{j}) + \alpha_{22}(\bar{i})\,\beta_{22}(\bar{j})}{[\sum_{l\notin\text{ vertex}}\beta_2(l)]\cdot[\alpha_{20}(\bar{i}) + \alpha_{21}(\bar{i}) + \alpha_{22}(\bar{i})]}\tag{81}$$

Note that for each $\bar{i}$ and $\bar{j}$ only one term of the numerator is positive since the product of any two $\beta_{20}(\bar{j})\cdot\beta_{21}(\bar{j}) = 0$ for $\bar{j} \notin$ vertex. The probability that starting from $\bar{i}$ absorption into an edge will be in E_1 rather than E_2 or E_3 (vertices are automatically excluded) is obtained by summing the expression (81) over the indices $\bar{j}\in E_1, \bar{j}\notin$ vertex. This give since $\beta_{21}(\bar{j}) = \beta_{22}(\bar{j}) = 0$ for $\bar{j}\in E_1$ the identity.

$$\text{Prob}\begin{pmatrix}\text{absorption occurs in } E_1\\ \text{and not into } E_2\cup E_3\end{pmatrix}\Bigg|\, X_0 = i\end{pmatrix} = \frac{\alpha_{20}(\bar{i})}{\alpha_{20}(\bar{i}) + \alpha_{21}(\bar{i}) + \alpha_{22}(\bar{i})}\,.$$

Similarly

$$\text{Prob}\left(\begin{array}{l}\text{absorption occurs in } E_2 \\ \text{and not into } E_1 \ \cup \ E_3.\end{array}\middle|\ X_1 = i\right) = \frac{\alpha_{21}(\bar{i})}{\alpha_{20}(\bar{i}) + \alpha_{21}(\bar{i}) + \alpha_{22}(\bar{i})}\ .$$

Finally

$$\text{Prob}\left(\begin{array}{l}\text{absorption occurs in } E_3 \\ \text{and not into } E_1 \ \cup \ E_2\end{array}\middle|\ X_0 = i\right) = \frac{\alpha_{22}(\bar{i})}{\alpha_{20}(\bar{i}) + \alpha_{21}(\bar{i}) + \alpha_{22}(\bar{i})}\ .$$

This shows that $\alpha_{20}(\bar{i})$, $\alpha_{21}(\bar{i})$ and $\alpha_{22}(\bar{i})$ are positive throughout $\bar{i} \in \Delta_3^o$, $\alpha_{20}(\bar{i}) = 0$ for $\bar{i} \in E_2^o \cup E_3^o$, $\alpha_{21}(\bar{i}) = 0$ for $\bar{i} \in E_1^o \cup E_3^o$ and $\alpha_{22}(\bar{i}) = 0$ for $\bar{i} \in E_1^o \cup E_2^o$.

All these considerations extend to the general case of several types. We state the results. The proofs involve a straightforward extension of the analysis set forth above and will therefore be omitted.

Theorem 7. *Let P denote the transition probability matrix* (63) *of the induced MC of p types with no mutation. Then*

(i) *The rate of absorption (fixation in a single pure type) is λ_2, i.e., if $\bar{i}$ and $\bar{j}$ are not vertices then $P_{\bar{i},\bar{j}}^t \sim C_{\bar{i},\bar{j}}\lambda_2^t$ ($t \to \infty$) where $C_{\bar{i},\bar{j}}$ is a constant depending on $\bar{i}$ and $\bar{j}$ but not on t.*

(ii) *The rate at which the population loses all but k types ($k \le p$) without specifying which they are is λ_k. Equivalently the probability that the population at the tth generation includes at least k types $\sim C_{\bar{i}}\lambda_k^t$ ($C_{\bar{i}}$ is a constant depending on the initial state but not on t). In particular the probability that the population contains all types at the tth generation decreases to zero at the rate λ_p^t.*

In the case of the transition matrix (46) and general mutation matrix, the Markov chain is irreducible and a stationary distribution $\{\pi_{\bar{i}}\}$ exists. In this case we know of no probabilistic interpretation for the eigenvalues like that of Theorem 7. The fact that the right eigenvectors are polynomials determined recursively can be used to compute moments of $\pi_{\bar{i}}$. We illustrate the procedure in the two type model with mutation. Let

$$P_{ik} = \frac{\text{coefficient of } t^k s^{N-k} \text{ in } f^i\left[(1-\alpha_1)\,t + s\alpha_1\right] f^{N-i}\left[\alpha_2 t + (1-\alpha_2) s\right]}{\text{coefficient of } t^N \text{ in } f^N(t)}$$

$$i, k = 0, 1, \ldots, N;\ 0 < \alpha_1, \alpha_2 < 1, 0 < \alpha_1 + \alpha_2 < 1\ .$$

The eigenvalues are

$$(1 - \alpha_1 - \alpha_2)^r \lambda_r \qquad\qquad r = 0, 1, 2, \ldots$$

and λ_r are defined as in (28) (see Theorem 1). The corresponding right eigenvectors have the form $\alpha_r = [Q_r(0), Q_r(1), \ldots, Q_r(N)]$ where $Q_r(\cdot)$ is a polynomial of degree r. Now if $\pi = \{\pi_i\}$ denotes the stationary distribution then since π is the left eigenvalue associated with $\lambda_0 = 1$ we have $\sum_{k=0}^{N} \pi_k Q_r(k) = 0$, $r = 1, 2, \ldots, N$. The successive moments of $\{\pi_k\}$ up to order N are now computable recursively from these orthogonality relations.

A related discussion concerning the probabilistic interpretation of eigenvalues of stochastic models of fluctuations of gene frequency is given in KIMURA [6] using diffusion approximations.

References

[1] FELLER, W.: Diffusion processes in genetics. Proc. 2nd Berkeley Symp. Mathematical Statistics and Probability. University of California Press 1951, p. 227.

[2] FISHER, R. A.: The Genetical Theory of Natural Selection. Oxford: University Press 1930.

[3] KARLIN, S., and J. McGREGOR: The Hahn polynomials, formulas and an application. Scripta Math. 26, 33 (1961).

[4] — — Direct product branching processes and related Markoff chains, Proc. Nat. Acad. Sci., 51 (1964) 598.

[5] KIMURA, M.: Some problems of stochastic processes in genetics. Ann. Math. Stat. 38, 882 (1957).

[6] — Stochastic processes and gene frequencies. Cold Spr. Harb. Symp. Quant. Biol. 20, Population Genetics 1955, 33.

[7] MORAN, P. A. P.: Statistical Processes in Evolution Theory. Oxford: Clarendon Press 1962.

[8] WRIGHT, S.: Evolution in Mendelian populations. Genetics, 16, 97 (1931).

Automatically Controlled Sequence of Statistical Procedures

By **Tosio Kitagaw**

Kyushu University

1. Introduction and Summary

The object of this paper is to give an explanation of an automatically controlled sequence of statistical procedures (ACSSP). In Section 2 we shall start with a notion of an automatically controlled sequence of procedures (ACSP). Four examples are then given of a successive process of statistical inferences and controls, each of which is an ACSP in our terminology.

We give these examples with the idea that a definition of ACSSP must be broad enough to contain these ACSP's in its domain of definition and that an ACSSP approach must be in some sense a generalization of that of a successive process of statistical inferences and controls which the author has been investigating since 1950. We shall not give a definition of an ACSSP in Section 2 as one specification of an ACSP, because it is not a simple matter to distinguish statistical procedures from procedures in general. However we statisticians have a backlog of statistical procedures gathered from various statistical activities which can be organized into an ACSP system. Consequently it should not be too difficult to analyse the main characteristics an ACSP system should have in order for us to classify it as an ACSSP. Our argument in Section 3 as well as in Sections 4 and 5 is not based upon a strict definition of ACSSP but rather upon a premised understanding derived from our experiences with statistical activities in various fields. In fact we are seeking for a definition of ACSSP throughout the discussion developed in these sections.

In Section 3 we shall point out three characteristic features of ACSSP approaches. In Section 4 the principles of statistical analysis using large electronic computers given by Terry [*34*] are discussed in reference to the characteristic aspects of ACSSP approaches given in Section 2. Our discussion will be concerned with both their logical foundations and their statistical programming techniques. Section 5 is devoted to a general review of data analysis with particular emphasis on the connection between automatic data analysis and ACSSP approaches. In Section 6 we shall give a definition of statistical procedures and then that of ACSSP which will be based on ACSP in the notion defined in Section 2. These definitions are based upon observations made in the previous sections, and are intended to be broad enough to encompass a sufficiently large area of statistical activities, including all aspects of successive processes of statistical inferences, within the domain of the definition of an ACSSP. On the other hand we shall not aim at giving a definition which would cover all the possible statistical activities that have been or that will be experienced. To aim for a fixed setup for statistical activities is not consistent with our philosophy that human statistical activities constitute a learning process where work is done under tentative and for the most part incompletely specified patterns and models.

It is noted that the present paper is a continuation of the 1963 paper KITAGAWA [26], and many revisions are given to make several relevant notions clearer than those given there. Many citations from several papers presented to the 34th Session of the International Statistical Institute are given to illustrate an intimate connection of our ACSSP approaches with recent advancements in statistical programming of designed experiments, quality controls and surveys as well as those resulting from changing uses of official statistics. The integrated organization of many divisions of statistics into a science along the line of extensive uses of the ACSSP approach is suggested in Sections 5 and 6.

2. Automatically controlled sequence of procedures (ACSP)

The purpose of this section is to lead to an understanding of the notion of an automatically controlled sequence of procedures (ACSP) through defining each of its several constituent notions step by step. In doing so, we shall rely heavily upon several notions which have been or can be defined in their respective domain in order to make our definitions much simpler than they would otherwise be.

We shall postpone giving the definition of statistical procedures and hence that of automatically controlled sequences of statistical procedures (ACSSP) until Section 6, for the reason given in the latter part of this section.

Definition 2.1. A *sequence* is an ordered set of elements called components. A set of sequences of components $\{\alpha_{i_1}, \beta_{i_2}, \gamma_{i_3}, \ldots\}$ where a finite or infinite sequence of the indices $(i_1, i_2, i_3, \ldots)$ runs through a set I is denoted by the notation $S = \{\alpha_{i_1}, \beta_{i_2}, \gamma_{i_3}, \ldots; (i_1, i_2, i_3, \ldots) \in I\}$.

Definition 2.2. A *path* $0 \; \alpha_{i_1} \beta_{i_2} \gamma_{i_3} \ldots$ is a set of the connected segments corresponding to the sequence $\alpha_{i_1}, \beta_{i_2}, \gamma_{i_3}, \ldots$, which is contsructed in the following way:

a) There exists one and only one point 0, which is called the *bottom point* of the path.

10*

b) There corresponds to each component α one and only one point (α) lying on a *horizontal line* L_α such that to any two different values α_i and α_j there correspond two different points (α_i) and (α_j) on L_α and similarly for each component β, γ, δ, ... and their corresponding horizontal lines L_β, L_γ, L_δ, ...

c) The heights of the horizontal lines L_α, L_β, L_γ, L_δ, ... are in strictly ascending order, and the line L_α is higher than the bottom point 0.

d) A path $\overline{0\ \alpha_{i_1}\ \beta_{i_2}\ \gamma_{i_3}}$ ... is a set of segments each of which joins two adjacent points of the sequence of points (α_{i_1}), (β_{i_2}), (γ_{i_3}), (δ_{i_4}), ... lying on L_α, L_β, L_γ, L_δ, ... respectively, starting at the bottom point 0 which is connected with the point (α_{i_1}) on L_α.

In this case the tree T is defined as the set of all paths $0\ \alpha_{i_1}$, β_{i_2}, γ_{i_3}, ... when $(i_1,\ i_2,\ i_3,\ ...)$ runs through the set I and is denoted by

$$(2.1) \qquad T = \left\{ 0\ \alpha_{i_1}\ \beta_{i_2}\ \gamma_{i_3}\ ... : (i_1, i_2, i_3, ...) \in I \right\} .$$

Definition 2.3. A tree T is said to consist of a set of *automatically controlled paths* with respect to an assigned automatic computer C when there exists a set of single-valued functions $\{f_k\}$ $(k = 1, 2, 3, ...)$ satisfying the following conditions:

a) For any assigned path $0\ \alpha_{i_1}\ \beta_{i_2}\ \gamma_{i_3}$ ... the set of indices is determined in the following way:

$$(2.2) \qquad (1^\circ)\ i_1 = f_1\,(0)$$
$$(2^\circ)\ i_k = f_k\,(i_1, i_2, ..., i_{k-1};\ 0)\ (k \geq 2).$$

b) The values of the right sides of (2.2) can be automatically computed by use of the automatic computer C.

This definition relies upon the notion of computability by an assigned automatic computer, which is not clear unless we define each computer through a set of programming languages including symbolic machine language, flow diagram, representation of information, compilers, problem oriented languages as well as subroutines, interpreters and generators. We are not however much concerned with the sophisticated notion of computability by which to show the set of all possible function $\{f_k\}$ but rather with a set of functions $\{f_k\}$ which can be effectively handled by the assigned automatic computer C with an available software.

Now we have to turn to the problem of how to define statistical procedures so as to include automatic data processing. In this connection it is worthwhile to consider in general the roles and implications of data processing and data in its general form.

For the past forty years a notable characteristic of statistics has been its division into two rather independent branches, descriptive statistics and statistical inference theory, having no common subject matter and sharply distinct logical frameworks, although having some common computational aspects.

Descriptive statistics is concerned with the calculation of statistics by which to reduce an amount of data into a set of relevant numerical values, without any use of the concepts of parent population and samples drawn from it. Descriptive statistics cannot analyze data with reference to any framework, since no assumptions are made as to how the data is generated; it reduces to the mechanical application of arithmetical operations on data in order to obtain means, variances, correlation coefficients, etc., with no insight into the background from which the data have come.

On the other hand statistical inference theory in general does require a framework upon which random samples from a population is definitely prescribed and which yields some insight into the background from which data has come. This framework is indeed the basis upon which modern statistical theories can be built. However data analysis cannot confine itself to a prescribed framework for the data, because in data analysis we should discuss more or less the adequacy of any proposed framework and therefore can not start with some mathematical model so definitely defined as in current statistical inference theories.

Having these observations on data analysis in mind, how should we then define statistical procedures? From the computer point of view, however, every automatic statistical procedure is a combination of logical and arithmetical operations performed by the programming of a computer, and it does not seem useful to try to define statistical programming without first having crystalized the characteristic features of statistical approaches which distinguish them from those of mathematics. It is true that some types of logical and computational procedures are more frequently used in statistical than in mathematical analysis and for other types the situation is converse, but this fact can hardly be enough to define statistical programming in sharp distinction to mathematical programming, because these procedures can be decomposed into elementary operations which are common to both of them. We believe that the difference between statistical programming and mathematical programming comes rather from the difference between the mathematical frameworks which generate their data and from the difference between the informative patterns within which their data are discussed. Several examples in what follows come from the realm of successive processes of statistical inferences and controls. Each of them is an ACSP in the sense of this section, and furthermore each should certainly be an ACSSP when the latter have been defined adequately.

Example 2.1. *Pooling of data.* Let us assume each of two sets of observations O_{n_i}: $(X_{i1}, X_{i2}, \ldots, X_{in_i})$, $i = 1, 2$, to be a random sample drawn from the respective population Π_i, $i = 1, 2$.

The population means ξ_1 and ξ_2 are unknown to us. The distinction between ξ_1 and ξ_2 is hypothetical. In the previous papers Kitagawa [17] and [24], sometimes pooling of data is formulated as an estimation of the population mean ξ_1 after a preliminary test of significance. This approach is based upon two kinds of assumptions. In the first place each Π_i, $i = 1,2$, is assumed to be a normal population denoted respectively by $N(\xi_i, \sigma^2)$, $i = 1,2$, with a common but unknown variance σ^2. Secondly it is explicitly assumed that the object of our experiment is to estimate the population mean ξ_1. In dealing with this problem we make use of the sample means $\overline{x}_1$ and $\overline{x}_2$ and the amalgamated unbiased estimate of variance s^2 and then the statistic t. The tree associated with this sequence of statistical procedures can be enunciated in the following way (see Fig. 1):

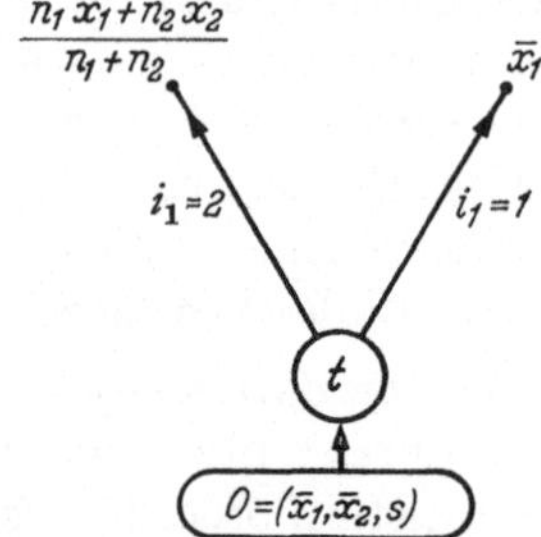

1 The bottom point is $O = (O_{n_1}, O_{n_2})$,

Two samples give a set of statistics

$$(2.3) \qquad \overline{x}_1,\ \overline{x}_2,\ s^2 = \frac{n_1\,s_1^2 + n_2\,s_2^2}{n_1 + n_2 - 2}$$

and

$$(2.4) \qquad t = (x_1 - x_2)/s\left(\frac{1}{n_1} + \frac{1}{n_2}\right).$$

2. The decision function $f_1(0)$ takes the value either 1 or 2 according to whether $|t|$ is greater than or not greater than the α — significant point of $|t|$, that is, $t_\nu(\alpha)$ with $\nu = n_1 + n_2 - 2$ degrees of freedom.

3. If $|t| \geq t_\nu(\alpha)$, we define

$$\overline{\overline{x}} = \overline{x}_1$$

4. If $|t| < t_\nu(\alpha)$, we define

$$\overline{\overline{x}} = \frac{(n_1\,\overline{x}_1 + n_2\,\overline{x}_2)}{(n_1 + n_2)}.$$

The characteristic aspects of the statistical analysis can be enunciated in the following way.

a) Our mathematical model is incompletely specified in the same sense as in Bozivitch, Bancroft and Hartley [5].

b) Under our incompletely specified model we may be eager to ascertain whether or not ξ_1 is equal to ξ_2. In actual cases we may not be satisfied with having one estimate $\bar{\bar{x}}$ after a preliminary test, although such an estimate $\bar{\bar{x}}$ may be required from the viewpoint of operational use of the estimate. For this purpose we should rather store information by keeping with us the statistics $\bar{x}_1$, $\bar{x}_2$, s and t as well as $\bar{\bar{x}}$. The result of the test of significance may be of the same importance to us in giving a pattern recognition of our hypothetical populations.

Example 2.2. *Several different processes of successive poolings of data.* These are described in a paper by KITAGAWA [25] which was devoted to the discussion of the relativistic logic of mutual specification in statistics. These processes were introduced in order to explain the idea of logic nets which seems indispensable in dealing with a sequence of tests of statistical hypotheses. Looking back carefully at their procedures, however, one can easily recognize that these processes of successive pooling of data are nothing but automatically controlled sequences of procedures which can be automatically performed by most current electronic computers.

In this connection the following remarks may be of some use in making clear the notion of ACSSP.

Remark 2.1. Each statistical test gives one and only one of two alternatives, significance (S) and non-significance (N).

Remark 2.2. A path belonging to the tree is determined by a sequence of values of decision functions each of which gives us either S or N.

Remark 2.3. From the informative use of our data, there are three aspects:

a) Pattern recognition: on the basis of the path which has been automatically determined by the data we shall be capable of establishing a stratification scheme by which to classify the sequence of the population means $\{\xi_i\}$ $(i = 1, 2, 3, \ldots)$ into possibly several strata. It is to be noted that such a stratification scheme is essentially tentative and always has the possibility of being changed into another stratification scheme by future data.

b) From the operational use of our information obtained through successive pooling of data, it may be emphasized that the final goal of our statistical procedure is to abtain an estimate of a common population mean when the population means become coincident with each other after a certain stage of the sequence. It is also important to know the operational characteristic of our ACSP when the population means will not ever coincide.

c) It is to be noted that we shall have established a storage of many statistics through this ACSP, which can be used in combination with future data.

Example 2.3. *Evolutionary Operations Program (EVOP).* A paper of the author KITAGAWA [21] gives a set of objective rules by which to determine a sequence of statistical procedures based on data obtained from a production process using a designed cycle of controlled factors in the sense of Box and his school, such as enunciated in Box-Hunter [3] and [4]. The set of these objective rules defines a successive process of statistical inferences and controls, and is also one example of an ACSP in the terminology of the present section. In this connection, the operational characteristics aspect of our ACSP is discussed in relation to our mathematical model which specifies the data to come from a certain stochastic process. It is intended that our mathematical model should be capable of amendment in the light of future data. It is to be noted that in practical application of the EVOP no automatically controlled sequence of statistical procedures may be recommended, further procedures are recommended by some authors, including Box himself, based on complete free choice by engineers and statisticians who may appeal to their intuitions.

An elimination of intuitive approaches is intrinsic in our formulation of ACSSP and its implication should be discussed in the last section of this paper.

Example 2.4. *Sampling inspection plans with automatic adjustment operations.* A classification of various sampling inspection plans used in engineering industries was given in Japan by the Committee for Specification of Sampling Inspection Plans sponsored by the Japanese Association of Standards into four main types, namely 1. standard, 2. screening, 3. adjustment and 4. continuous production types. The theory of testing hypotheses due to Neyman-Pearson [31] appealing to the power function in nonsequential sampling procedures and that of Wald [36] in sequential ones are particularly suited for a discussion of the operating characteristic curve of a sampling plan of the type 1. The single and double sampling inspection Tables due to H. F. Dodge and H. G. Romig [10] has a screening power as their main property, which characterizes the type 2. On the other hand, the last type 4 is concerned with continuous production as in the sampling inspection plans introduced by Dodge [9].

Examples of the type 3 include the sampling inspection plans called JAN-Standard 105, and MIL-Standard 105A [30]. No theoretical foundations for these Tables belonging to the type 3 have been formulated by Japanese statisticians and engineers. It has been felt quite important and even indispensable to develop a theoretical basis for these tables in order to justify dicisions as to when and how much they should be used.

However there exists one circumstance which makes any theoretical treatment difficult. This is the fact that each of these tables is associated with some technical convenience which, although very important in practice, may make some essential theoretical aspects of their Inspection Plans too complicated for a theoretical analysis. Our attitude in dealing with such a circumstance is to introduce a somewhat simplified formulation of each Inspection Table in order to make it possible to obtain its operating characteristic. This is possible when we can introduce an ACSP which can be recognized as a theoretical approximation to the original sampling inspection plan. The ACSP gives us a cybernetical formulation [37] in the sense that the Plan will be performed automatically by some adequate automatic computer in view of data obtained by samplings, since any possible action can be objectively defined by a sequence of decision rules.

Now let us outline the treatment of a sampling inspection plan of the type 3 in this fashion. It is a common feature of this type to have three kinds of sampling inspection procedures, namely, reduced (R), normal (N), and tightened (T) ones, and hence to have a matrix of transition rules among these three procedures such as

$$(2.5) \quad C = \begin{pmatrix} C_H(R, R) & C_H(R, N) & C_H(R, T) \\ C_H(N, R) & C_H(N, N) & C_H(N, T) \\ C_H(T, R) & C_H(T, N) & C_H(T, T) \end{pmatrix}$$

where $C_H(A, B)$ denotes the set of conditions under which we should make a transfer to the plan B when we are now in the state of using the plan A with the past history H, and $C_H(A, A)$ the set of conditions under which we should stay at the plan A when we are now in the state of using the plan A, with the past history H.

Now the actual descriptions of each of these nine conditions should first be carefully scrutinized to discover any factors which can not be objectively described but can only be determined by the judgement of inspectors or by that of their supervisors. From our standpoint of appealing to ACSP we must replace such elements of the transition conditions by some other objectively defined procedure. The situation may be greatly simplified in practice, because no condition can be given to make a jump from the reduced inspection plan R to the tightened T one and vice versa.

We are just giving an indication of the prosesses to be adopted along this line of approach, not entering into any detailed discussion of them. Theoretical treatment will be much simplified by starting with the case when a Markov chain approach can be used, that is, our conditions $C_H(A, A)$ are independent of the past history H.

It can easily be recognized that the operating characteristic property of our sampling inspection plan will be a certain average of three individual operating characteristic functions $L_R(p)$, $L_N(p)$ and $L_T(p)$ with the weights giving the limiting transition probabilities $w_R(p)$, $w_N(p)$ and $w_T(p)$ if these exist; that is,

$$(2.6) \quad L(p) = w_R(p) L_R(p) + w_N(p) L_N(p) + w_T(p) L_T(p),$$

which will give an insight into the true merits of the uses of a sampling inspection plan of the adjustment type.

Gradual elaboration of our approach by adopting more realistic approximations to the actual Sampling Inspection Plan will provide increasing information about its characteristic operating properties and hence guide us in choosing one among possible inspection plans of the adjustment type.

3. Characteristic aspects of ACSSP

We have explained our notion of an automatically controlled sequence of procedures (ACSP), but did not specify that of an automatically controlled sequence of statistical procedures (ACSSP) as its special case. In order to give a formal definition of a statistical procedure and consequently that of an ACSSP, we should prepare ourselves with a definite answer to the basic question of what the logical principles which characterize statistical approaches in general consist of. On the other hand it should be also remarked that such an answer (if any) cannot be adequately given without careful scruting of all current statistical activities. This implies that an adequate definition of a statistical procedure (if any) should be broad anough to contain those which come from various branches of statistics, including descriptive statistics, inference theories, and data analysis, in connection with statistical (both census and sample) surveys, designed experiments, statistical quality control, and so on. This implies that any definite answer to the fundamental question should be subject to the possibility of being altered in view of advancements in statistical activities. Taking into consideration the above questions concerning the foundation of statistics, we shall now suggest several characteristic aspects of an ACSSP approach in view of the examples given in Section 2. We intend them as a preparation for a definition of ACSSP, or at least to provide a better understanding of the roles and the functions of an ACSSP, because these examples are all concerned with statistical procedures as well as being ACSP's in themselves.

(3.1.) *Three aspects of the use of information obtained from data with reference to a tentatively specified pattern.* One of the characteristic aspects of an ACSSP is that it can usually be concerned with all three fundamental aspects of the use of information based upon experience, namely a) storage, b) pattern recognition and c) operational use, which we shall enunciate more specifically:

a) Storage of information within a tentatively specified pattern,

b) Reformation of one tentatively specified pattern into another one

c) Operational use of stored information within a tentatively specified pattern.

Here we mean by a *pattern* an objective description of data in terms of its constituent elements and their mutual connection. This description need not be in mathematical terminology. By a *tentatively specified pattern* we mean a pattern whose constituent elements and/or their mutual connections are not completely known to us but are specified through data obtained hitherto which may be altered in view of coming data. By a *mathematical model of a constituent in a pattern* we mean a constituent element of a pattern which is defined mathematically. Broadly speaking, the following four situations can occur in connection with the use of information obtained from the data.

1. *Accumulation of information within a tentatively specified pattern.* This happens when new data is not contradictory to a tentatively specified pattern based upon previous data. In this case it can and will be stored as additional information within the pattern and hence will serve to give additional information about some mathematical models of constituents and/or to their logical connections.

2. *Pattern recognition.* This will happen in various different ways. The following three cases are given as typical examples.

a) *Storage of data as separate information without leading to a new pattern.* This happens when new data is contradictory to our tentatively specified pattern but is not sufficient to lead us to another pattern. Then no reduction of the data may be performed and it is stored in its totality as information separate from the specified pattern. In such a case the data is called *fragmental information* combined with the pattern. The result is in itself some sort of pattern recognition, and in conjunction with new information obtained later, may lead to a reformation of the tentatively specified pattern into another one.

b) *Rejection of the data.* Our data may be subject to gross errors due to causes whose occurrence may be traced but with which we are not interested from the standpoint of pattern recognition. In such a situation the data can and will be rejected. From the logical standpoint, a distinction between two types of situations arising in practice is crucial. In some situations the rejection is final because our general form of pattern recognition excludes decisively some features of real phenomena from our consideration, while in other situations a rejection is not final but may be tentative. In the latter situations the possibility exists for the data to recover citizenship in our storage and hence to be stored in our information storage.

Thus in these situations it may be difficult to make a sharp conceptional distinction between the present case b) and the former case a). Nevertheless it will not involve any confusion so far as an ACSSP approach is concerned, because each component procedure and mutual connections among component procedures are objectively defined.

c) *Branching to a new pattern.* New data is contradictory to the tentatively assumed model, and it is sufficient to lead us to an introduction of a new model, which is, however, again tentative.

(3.2) *Operating characteristic consideration.* The second characteristic aspect of an ACSSP is that it admits of operating characteristic consideration, provided that we set up a mathematical formulation of each tentatively specified pattern in its totality. The possibility of performing an operating characteristic consideration on an ACSSP for each assigned mathematical formulation of the pattern is based upon the very fact that each ACSSP is an automatically controlled sequence of statistical procedures in the sense illustrated by various examples given in Section 2 and that its whole sequence can be performed automatically according to a program in a computer. In short an operating characteristic consideration becomes possible due to the fact that an ACSSP is objectively defined. An operating characteristic property of an ACSSP will be helpful to the statistician by providing him with an objective criterion with which to evaluate the merits and the demerits of an ACSSP under various possible situations analogously to the uses of the operating characteristic curve of a sampling inspection plan. However, since the pattern recognition of our objective world to which an ACSSP applies is tentative, there does exist the possibility of developing quite different operating characteristic considerations from those of the classical approaches of statistical theories. Let us elaborate this possibility by discussion of some extremely simplified examples. Broadly speaking, our new attitude is not to determine a statistical procedure for an assigned problem of statistical inference unter an assigned model, but to investigate an assigned statistical procedure under various tentatively assigned models.

Example 3.1. In current elementary estimation theory we are concerned with the problem of estimating an unknow parameter Θ of our parent population, most commonly under the assumption that its distribution function $f(x; \Theta)$ has a known functional form f. Let $(x_1, x_2, \ldots, x_n)$ be a random sample of size n. This problem amounts to finding a function $A(x_1, x_2, \ldots, x_n)$ which satisfies a prescribed criterion for preference of choice, such as unbiasedness, minimum variance, or maximum likelihood. Our first critique is concerned with the current assumption that the function form f is known to us. This assumption is frequently unrealistic, at least at the beginning of most newly planned research. Therefore it may be worthwhile to appeal to another approach in which we investigate the characteristic operating properties of some particular estimaters such as the sample mean $A_1 = \Sigma\, x_i/n$, and the sample median $A_2 = \text{Median}\{x_i\}$ under various functional forms of f. We have just now merely referred to one step in a sequence of statistical procedures, but the same method of attack can be generalized to an ACSSP. In this sense there remains a large uncultivated field of statistical approaches.

(3.3) System analysis of ACSSP. An ACSSP can be considered as a system which can be decomposed into a set of various component subsystems each of which is also an ACSSP, while it can also be considered as a component subsystem of a more complex system which is also an ACSSP. For example a regression analysis can be decomposed into a set of simpler operations, but it can also be a component statistical procedure of a more complex ACSSP. Since an ACSSP may have many component subsystems and a certain set of connection rules among them, a system analysis approach seems to us indispensable for theoretical investigation of ACSSP. The systems analysis of ACSSP will be based upon two fundamental possibilities:

a) The possibility of decomposing a whole system of an ACSSP into a set of subsystems each of which is an ACSSP whose operating characteristic property is established.

b) The possibility of obtaining the operating characteristic aspects of an ACSSP as a composition of those of its component subsystems each of which is also an ACSSP.

Systems analysis is therefore concerned with several analyses such as (i) component analysis, (ii) composition analysis, (iii) stability analysis, (iv) flexibility analysis and (v) reliability analysis.

By *component analysis* we mean the operating characteristic considerations of each component subsystem. By *composition analysis* we mean a network analysis of the system as composed of a set of subsystems each of which is a blackbox in this analysis. A *blackbox* is a system for which an internal mechanism is not known but for which an input and output relation is given. In *stability analysis* we are concerned with the stability of the operating characteristic properties of an ACSSP with respect to input data varying within a certain domain. In *flexibility analysis* we are concerned with investigations of the operating characteristic properties when some subsystems are replaced by other subsystems each of which is also an ACSSP. Our ACSSP is said to be *flexible* when it will maintain some operating characteristic features under these replacements. *Reliability analysis* is concerned with changes of pattern of our objective world in which some drastic changes and catastropies can be included. An ACSSP is said to be *reliable* under a pattern change if its main operating characteristic properties can be maintained to some extent under the pattern change.

4. Statistical programming

In the first part of this section we shall be concerned with principles of statistical analysis using large electronic computers and we shall explain how far our ACSSP approach can work along these principles.

In the latter part of this section we shall discuss some particular methodologies of data analysis developed recently by various statisticians from the standpoint of an ACSSP approach.

The following principles of statistical analysis using large electronic computers are given by Terry [*34*]:

Principle 1. After the data has been processed to the point that it can be read into the computer, it must be thoroughly screened for cogency before it is used in the analysis.

Principle 2. Even after screening, the analytic algorithms must be developed under the assumption that real discordancies are still in the data.

Principle 3. Whereas the scientist will choose his units for measurement to optimize the accuracy and precision of the experiment, the analysis should be carried out in those units which yield the deepest insight into the phenomena under study. This often results in a demand that the data shall specify the appropriate units for the analysis.

Principle 4. The scientist has a right and the statistican a responsibility to organize and present data in as many cogent forms as is necessary to understand.

Terry [*34*] explained the role of the statistical programmer in the following two sentences:

(i) "The statistical programmer does not know a priori the exact analytic path that his data must follow,"

(ii) "The statistician may very well prefer to let the data speak for itself and suggest the appropriate transformation to exclude from consideration on measurement deemed discordant, or to replace such measurements by derived measurements."

In realizing these principles and the roles of the statistical programmer, Terry [*34*] suggested the broad aspect of adequate statistical programming:

(iii) "Now, with the advent of the disc file, which has the effect of increasing the storage capability of the computer to the order of two million measurements or more, we believe that it will be possible to store in this ancillary device many different statistical strategies, computational techniques, and statistical decision rules as well as large blocks of data."

(iv) "Then, by writing a program of the analytical strategy to be employed, we could permit the data to call in the appropriate analytical techniques and rules, and thus produce a much more effective final analysis."

All these descriptions and prescriptions by Terry [*34*] are very understandable from our point of view. To begin with we shall comment on each of the four principles he proposed [*34*].

Re Principle 1. This principle can be realized by some particular set of statistical procedures in an ACSSP. Indeed an application of estimation after a preliminary test of significance was discussed by KITAGAWA [17] and [24] in order to describe some theoretical aspects of interpenetrating samples advocated by MAHALANOBIS [28]. Screening data procedures can be formulated as an ACSP in so far as they can be automatically performed by an automatic computer according to its programme. Regarding the background from which data have come, we should like to transfer our discussion to that of the following Principle 2.

Re Principle 2. This principle is concerned with pattern recognition and suggests the need for making assumptions that real discordancies may possibly exist in the data. This implies that our pattern recognition should be broad enough to admit of such a possibility and that our pattern should be tentative at each stage of our recognition. In view of our emphasis on the three uses of information given in Section 3, it can be observed that our ACSSP approach is ready to work under such pattern recognition.

Re Principle 3. This principle refers in general to an adequate choice of units of measurement and in particular to a requirement that the data shall specify the appropriate units for analysis.

In this connection it is worthwhile to cite the following vivid assertion due to TERRY [34] making clear a characteristic aspect of statistical programming in light of both his assertions (i) and (ii) cited before. He says "Here (statistical programming), the discordance of a single measurement cannot, in general, be determined independently but only as a member of an aggregate, and so, the very definition of a statistical problem poses a new kind of constraint on the use of the computer."

This assertion implies an emphasis on the notion of aggregate in dealing with statistical data and suggests a logical peculiarity of statistical procedures. According to our own terminology, this assertion due to TERRY belongs to the realm of "the relativistic logic of mutual specification in statistics" whose various aspects we have discussed in a previous paper, KITAGAWA [25]. Indeed we understand that one of the main aspects of principle 3 is essentially concerned with this characteristic logic of statistics which can be adopted in our ACSSP approach with reference to tentatively specified patterns, as we have discussed in some detail in KITAGAWA [25] by giving a mathematical formulation for materializing relativistic logic of mutual specification.

Re Principle 4. This principle amounts to an emphasis on characteristic aspects of statistical analysis such as (i) tentativeness and incompleteness of specifications of patterns and models in statistics, (ii) tentative indications rather than final conclusions in statistical analysis, and (iii) the possibility of a multitude of reasoning paths as manifested in a tree. Therefore this principle is closely connected with the uses of information which we have already explained in Section 3.

As a summary of our review of these four principles we can conclude that they are also valid for our ACSSP approach, and that the reason why we should introduce several fundamental notions such as a tree, a path within a tree, a set of decision functions, tentative pattern, and tentative model may be said to be quite understandable in view of these principles. Indeed these notions may be said to supply a set of the specified realizations to the needs for satisfying these four principles. The operating characteristic considerations and systems analysis of ACSSP approaches, which Terry [34] does not seem to emphasize, will be discussed in Section 5 in a more general scientific framework.

Let us now turn to some particular topics in statistical programming. Here two topics will be chosen for our discussion. The first topic is concerned with screening and validation procedures which are particularly important in the logic of statistical approaches. The second topic is a review of comprehensive programming systems developed recently by several statisticians.

Regarding the first topic, many experts on census and large scale sample surveys have been keenly aware of different types of errors occurring in the case of large-scale sample surveys. Deming [8] gave a detailed listing and description of the different types of errors which should be taken into consideration both in designing and analyzing sample surveys. Hansen, Hurwitz, Marks and Mauldin [14] discussed response errors which are important factors influencing accuracies of surveys. Mahalanobis [28] gave the classification of different types of error into three types, and "revealed the great importance of controlling and eliminating as far as possible the mistakes which occurred at the stage of the field survey." The interpenetrating sample procedure was introduced by him as one way of doing this. An interpenetration procedure should be recognized as a fundamental tool in statistical approaches where a recognition of pattern should be tentative and where relativistic logic of mutual specification is basically important, because the procedure is concerned with pattern recognition as well as with a control of objective realities so as to reduce them to a certain realm of patterns. It should also be remarked that some of the statistical techniques associated with interpenetrating samples can be discussed from the standpoint of a successive process of statistical inferences and controls as we have developed in Kitagawa [18] and [24]. These results yield us examples how far ACSSP approaches will be useful in the field of data acquisitions and processing. From the standpoint of ACSSP approaches, we can and we should proceed further somewhat in the following way:

a) First of all, some accessible document should be prepared by experts on census and large scale sample surveys which "lays out procedures, cautions and estimates which would enable the inexperienced to undertake a major analysis of data" (TERRY [34]).

b) Then we should translate each individual process suggested in the document into machine language of some computer, with adequate modification of some parts of the document if it be necessary, so as to make as many processes automatic as possible.

c) If this second step is achieved to some extent, then we have obtained either an ACSSP corresponding to the whole system of statistical analysis or many ACSSP's corresponding to some of the subsystems inherent in the whole system.

If the three steps a), b) and c) are accomplished, then we have at least theoretically the possibility of proceeding to operational characteristic considerations and to systems analysis in the sense enunciated in Section 3.

Let us explain our idea by one example. YATES [39] referred to the general problem of preliminary editing of data before analysis and enunciated the uses of electronic computers in the following sentence. "Once appropriately instructed a computer will perform any required tests on each item of data as it is read in, and can draw attention to anomalies, reject suspicious items, or even in some cases make the appropriate correction." In our terminology, such an instruction on a computer is nothing but an ACSP which can be used for preliminary editing of data and whose operating characteristic properties can be discussed under each assumed pattern of the objective world.

Several papers or memoranda have been written by various authors which have paid considerable attention to data analysis with particular reference to screening and validation problems. We can mention (i) preliminary assessment by REED [33], (ii) autostat by DOUGLAS and MITCHELL [11], (iii) treatment of spotty data by TUKEY [35], and (iv) analysis of residuals by ANSCOMBE-TUKEY [1]. In his 1963 report COOPER [7] presented the first documented rigorous procedure for validating and controlling the presentation of data to a computer. We believe that ACSSP approaches are now in urgent demand as theoretical foundations for automatic data processing including screening and validation tests.

Several comprehensive programming systems have been prepared by some statisticians or by some institutions. In his presidential address delivered to the British Computer Society, YATES [39] pointed out several important aspects of the use of computers in research. He said: "In research statistics the analysis must in fact proceed step by step, the exact nature of the next step being determined after examination of the results of the previous step. This presents considerable problems of organization between different statistical programs, since the results at each step must be stored (clearly on magnetic tape, if available), and indexed in such a manner that the required item can be specified as data for the program performing the next step."

These sentences show the reason why a sequence of statistical procedures is required in the statistical analysis of data and suggest the special uses of computers in statistical programming. On the other hand the need for automation of sequential statistical approaches and its merits can be seen from his report on the increase in the number of experiments analyzed from four hundred to the order of three or four thousand and with the increase in the number of variate analyses from eight hundred to eleven thousand with little increase in staff and much speedier service. Thus ACSSP approaches are realized in research statistics. Yates and Smith [40] prepared a general program for the analysis of surveys which has completely revolutionized the analysis of surveys on their computer according to Yates [39].

The MUSP prepared by statisticians in Harvard University is said to consist of a set of 19 subprograms which can be called in by a program in a special purpose control language specially designed for MUSP. The sequential operation of subroutines is directed by a control program called MUSP Control Program which "accepts as input the symbolic specification of the problem to be solved in terms of a sequence of sub-routine names and parameter values, checks the specification for obvious errors such as missing parameters, translates the specification into a machine-oriented representation and then executes the resulting set of specifications interpretively." (M. G. Kendall and Wegner [16]). This explanation of the functions and roles of MUSP shows us also that ACSSP in our sense is imbedded in the programming of MUSP, and here again indicates the need for developing the statistical theory of ACSSP.

In summing up the observations given in this section, the need for and effectiveness of ACSSP approaches can be said to be urged from the standpoint of statisticians using electronic computers for their data analysis of surveys and research statistics.

5. Data analysis as a science

Tukey [35] gave a thorough consideration of various aspects of data analysis. In Section 9 of Kitagawa [26] we gave some comments on Tukey's views on data analysis as an empirical science and on the roles of automatic data processing in particular. There are three fundamental assertions which we accept as valid:

a) "Data analysis is intrinsically an empirical science" (Tukey [35], Section 46, p. 63).

b) "In order to be a science, data analysis should have its theory. A theory of data analysis can only be given by providing with ACSSP system" (KITAGAWA [*26*], Section 9, p. 127).

c) "Its theory cannot necessarily explain the whole knowledge and information accumulated in the science according to logico-deductive arguments assuming a few basic facts." (KITAGAWA [*26*], Section 9, p. 128).

We have discussed the implications of these assertions in some details in KITAGAWA [*26*], and we are not repeating the same arguments here as those which were given there, except that we have to point out that the assertion b) is indispensable to our standpoint in evaluating the roles and functions of ACSSP approaches, although the exclusive expression "only" cannot be justified until after we have defined the notion ACSSP more definitely. In view of the assertion c), we are ready to understand that at each stage of the development of a theory of data analysis there may always be something which cannot be adequately explained by ACSSP approaches and it is essential for the development of statistics not to impose an apriori fixed pattern of recognition on our data analysis as if it would have an eternal validity for the future development of statistics. TUKEY [*35*] pointed out most adequately "the needs for collecting the result of actual experiences with specific data-analytic techniques," (TUKEY [*35*], Section 45, p. 62) and "the need for a free use of adhoc informal procedures in seeking for indication" (TUKEY [*35*], Section 46, p. 62). He said most adequately: "there will also the hallmarks of stimulating science: intellectual adventure, demanding calls upon insight, and a need to find out" how things really are "by investigation and the confrontation of insights with experience." (TUKEY [*35*], Section 45, p. 63).

We believe that this sentence of TUKEY [*35*] explains most clearly and vividly why data analysis should be an empirical science, and his assertion is quite agreeable to us. Now turning to the assertions b) and c), we should be conscious of the needs for much more sophisticated mathematical models and theories than those which have been used in current mathematical statistics in order to prepare for future developments of statistical approaches in which an ACSSP system will be the theoretical tool. The roles and functions of an ACSSP system are enunciated in terms of reliability, stability and flexibility in which there are many challenging problems not yet fully attacked. The theory of statistical approaches appealing to ACSSP systems must rely upon development of two areas, first on that of the computer, and secondly on that of statistical methologies.

In ACSSP approaches, various features of the use of previous information should be carefully formulated and various uses of information accumulated in view of data with combination of previous knowledge should be carefully scrutinized in more realistic attitude than we have formulated in current mathematical statistics. It is quite important to note that data analysis is essentially a learning process in which each procedure can be altered in view of our information obtained from data.

In particular, automatic data analysis will be closely connected with the future development of automation in production which we are expecting to occur in our technology. There is an intrinsic relation between automatic data analysis processes and automatic production processes.

Some scholars used to classify statistics into two divisions, namely, (i) descriptive statistics and (ii) statistical inference theory. This classification has been currently adopted by a majority of statisticians on theoretical as well as on practical grounds. In the first place, speaking from the theoretical point of view, the domain of application of descriptive statistics should be sharply distinguished from that of statistical inference theory, because the latter is exclusively concerned with random samples from a hypothetical population, while the former does not rely upon the notion of population and sample.

In the second place, the classification has had a real significance on practical grounds since each of the two divisions has had its individual domain of application in statistical activities.

Now we are faced with the need for handling mass data on one hand, while we are equipped with high-speed electronic computers having rich memories on the other hand.

We have explained in some detail the needs for statistical programming and the effectiveness of ACSSP approaches in dealing with mass data. Illustrative examples in Section 2 will be sufficient enough to verify how far ACSSP will be useful in some statistical approaches in connection with technological problems in engineering industries. Other similar examples can be found in various fields such as largescale sample surveys explained by YATES [39] and designed experiments on biological phenomena such as those explained by COOPER [6].

Besides these areas, there is another possibility. Changing uses of official statistics are noticed by several official statisticians such as FÜRST [12], BOWMAN-MARTIN [2] and GOTO [13]. They point out that social and economic statistics are now being designed for analytical use, and that new types of statistics may be required to satisfy the demand for a greater use of microanalytic techniques. BOWMAN-MARTIN [2] said. "In face of these needs, and mounting pressures for increased accuracy in measurements used to guide important public programs, more resources should be devoted to scientific research and experimentation in the problems of measurement — research which would improve the accuracy and increase the efficiency of the statistical system."

In short, there are indications that the tasks of official statistics are becoming more and more analytic, and that a gap between the two divisions of statistics is now becoming much narrower than it has heretofore been. Some data analysis currently used by official statisticians can be formulated by means of an ACSSP system.

Here is also a challenge for the statistician to cultivate a new area of ACSSP approaches to be applied to official statistics.

In summing up our observations in this section, ACSSP approaches can be expected to be useful in various fields of statistical analysis, and after any ACSSP formulation of the problems in these areas has been successfully established, then the terminologies of the ACSSP approaches will become common in these fields. This is one of the possible procedures by which to realize an integration of statistics as a science.

6. Automatically controlled sequence of statistical procedure (ACSSP)

In Section 2 we have already defined an ACSP, but we did not specify an ACSSP as its special case. However we have analyzed the characteristic aspects of an ACSSP and those of statistical programming in Sections 3 and 4. In combination with these enunciations and observations we should not give a definition of ACSSP which has been used without giving its definition.

Now we have to enter into a discussion of the problem of how to define a statistical procedure. Every monograph on statistics and every statistical theory yields us some direct or indirect answer to the fundamental question of what statistics is, and we can gather from the statistical literature some views regarding our present problem of how to define a statistical procedure. This is not, however, the place to attempt an extensive survey of the spectrum of various possible definitions of statistical procedure. An adequate answer to the question may be given by a clear description of some fundamental aspects of statistical approaches which have not been fully discussed in any current statistical theory and which our ACSSP approaches should take into consideration. Our answer in what follows is planned to explain the functions of ACSSP approaches with reference to each of four fundamental aspects of statistical recognitions: aggregate recognition, quantitative recognition, inductive recognition and integrated recognition.

(6.1) *Aggregate recognition.* Let us consider a probability field $(\Omega, \mathscr{F}, P)$ defined by the triple components: the space Ω, the completely additive family of its subsets $\mathscr{F}$ and the completely additive probability measure on the family P. A stochastic variable is defined as a measurable function defined for each element ω belonging to the space Ω except possibly for a set of probability measure zero.

An ACSSP approach uses a probability field in this sense or a set of probability fields as its mathematical tool for understanding our objective world in our statistical recognition as do most current statistical approaches. However we have to make several immediate reservations regarding an interpretation of this mathematical tool.

Reservation 1. The definition of measurable function does not require that each element ω belonging to the space Ω can be assigned, but it does require that a specific function value $x(\omega)$ should be given if an element ω is assigned. All that is required here is the fact that for any assigned two real numbers a and b $(a < b)$, the set $[\omega: a < x(\omega) \leqq b, \omega \in \Omega]$ belongs to the family $\mathscr{F}$ and hence has its definite probability measure, but even an effective assignability of the set is not required in its definition. We rely upon such an interpretation of the definition of measurable functions in order to make clear our understanding that statistical recognition is an *aggregate recognition* in which we are interested with distribution properties of function values but not with each individual correspondence between ω and $x(\omega)$.

Reservation 2. The definition of measurable function does not exclude a regular function whose value $x(\omega)$ can be effectively assigned for each assigned ω in Ω, and indeed it does not necessarily imply any notion of randomness as its logical consequence. In spite of its being called a stochastic variable, stochasticity or randomness can only be introduced with respect to mutual relations between at least two different measurable functions. This reservation on the definition of measurable function is made here because any confusion should be avoided in an interpretation of measurable function.

Reservation 3. In our ACSSP approaches where we are concerned with the precision of measurements which are more or less limited within some range and where an accumulation of informations of objective world is performed in view of the coming data, it is adequate and some times indispensable to deal with a set of finitely additive families of subsets in Ω instead of one fixed infinitely additive family $\mathscr{F}$ which can only be considered as an idealized limit of the former ones. Indeed it is one of the most crucial recognitions that some phenomena may assume an appearance of randomness under a certain range of precision of measurement while they may reveal a pattern of regularity under a different range of precision. Broadly speaking, randomness and non-randomness are defined as their mutual correlative property among measurable sets and measurable functions with respect to some assigned probability space. This reservation is made partly in order to be capable of discussing a choice of measurement units in statistical approaches.

(6.2) *Quantitative recognition.* One of the fundamental aspects of statistical recognition is generally understood to be its quantitative approach in some generalized sense. Our ACSSP approaches will not work outside of this general understanding, but they should emphasize the need for taking into consideration all different types of errors occurring in experiments and surveys and for providing some theoretical methology to tackle with data acquisition and data processing. In what follows we shall give a brief description of various kinds of error and a methological outline for dealing with some of them in the framework of our ACSSP approach, which will serve to explain what we mean by statistical procedure. For this present purpose the classification of errors into three types made by MAHALANOBIS [28] is especially suited for our general consideration, namely: 1 sampling fluctuations, 2 observational error and 3 gross inaccuracies, where 1 and 2 may be presumed to follow probabilistic schemes, while 3 does not. The third type of error is actually a very broad type which includes all errors belonging to neither of the first two nor to a combination of them, and naturally involves various kinds of errors: inaccuracies and falsehood in statements and recording, tricks and so on. Thus the sound way to make a step forward is not to give too broad (and hence obscure) a formulation aiming to cover all types of errors that could be imagined, but rather to choose, corresponding to each stage of theoretical and practical development, some restricted domain in order to make an adequate and effective improvement of statistical approaches. In our proposed ACSSP approaches we shall follow this principle. For this purpose we propose to introduce here the notions of state, operator and scheme, which seem to us more fundamental than statistical quantities and to be indispensable for dealing with the third type of error.

An objective world has a *state* which is assumed to be independent of our surveys and experiments. The state of the objective world is revealed to an *investigator* as a response given by a *respondent* according to a *scheme* assigned by the investigator. There may occur problems of non-response and also of possible interference between investigators and respondents. An abstract concept of an investigator, which may include for example mailed questionnaires, or telephone or interview surveys, should be more relevantly represented by the notion of *operator*. Thus we shall reach the following simplification of our terminology.

Our *statistical data* is assumed to be obtained under the following process:

i There exists a state ξ of the objective world.

ii An operator α is applied to each state ξ so that it may give us variable(s) under a certain scheme S.

iii The domain of α in which, under the scheme S, some variable(s) corresponding to a state ξ can be given, does not necessarily cover all possible states of the objective world, and is denoted by $\mathscr{D}(\alpha, S)$. When ξ

belongs to $\mathscr{D}(\alpha, S)$, the variable(s) defined for a set of state ξ, operator α and scheme S is denoted by $v(\xi; \alpha, S)$.

An abstract idea of $v(\xi; \alpha, S)$ may be so broad as to include variables such as falsehood, deliberations and even strategies, for which there could not be any objective approach, unless we restrict ourselves to a certain realm of S, α and ξ. Different kinds of falsehood, deliberations and strategies may have naturally different sources. It is impossible to suggest a priori an acquisition or a processing procedure by which we should be able to measure or to control all types of these errors. These procedures might be developed through intensive investigations of human psychological, sociological, economical and even political situations and based upon developments in the realms of these social sciences as well as upon statistical techniques. In this sense our data analysis is actually an empirical science as we have already emphasized.

In our mathematical formulation of ACSSP approaches we shall deal with variables $v(\xi; \alpha, S)$ in this sense and we shall be concerned with acquisition and processing of data on these variables. Our approach is therefore based upon a *formulation of pattern* which is much broader than some of the current formulations based upon the presumption that our data are samples randomly drawn from a population. The latter case is contained as a special one in our approach, while there do exist the cases to be discussed in our formulation where at least one of the following two conditions does not necessarily hold true.

r: Data are random samples from a population.

m: Variables $v(\xi; \alpha, S)$ are measurable functions of ξ.

In our ACSSP approach we shall consider the types of error for which at least one of the following two principles can be applied.

a) *Application of randomization.* The principal object of randomization is to introduce a probabilistic scheme so that valid statistical inferences can be made. From the logical point of view one basic stone of the design of experiments in the Fisherian school is the use of this principle, and it is to be noted that the scheme which experts on sample surveys use in discussing response errors and non-response errors in surveys appeals to this principle.

b) *Application of the principle of transformation.* This principle is basically concerned with realization of the condition (m) of measurability of variables $v(\xi; \alpha, S)$. For a certain set of ξ, α and S, the variables $v(\xi; \alpha, S)$ may not be defined, and it is not certain whether $v(\xi; \alpha, S)$'s are measurable or not. In such a situation some transformation of ξ, α and S may be found useful in achieving the condition (m). Let these transformations (including the identity transformation) be denoted by $\tau\xi$, $\sigma\alpha$ and ΘS respectively. Then it may occur that $v(\tau\xi; \sigma\alpha, \Theta S)$ is measurable. Now the choice of the set of transformations is entirely based upon our previous experience and knowledge of the objective world and on the technical details of surveys and experiments.

Let us give some illustrative examples of these transformations.

Example 6.1. In actual cases it frequently occurs that we are really concerned not with the variables $v\,(\xi;\,\alpha,\,S)$ themselves but with some differential change between them. In these changes it may be possible for us to eliminate certain unmanageable factors, thanks to the difference operations. For instance, suppose that we are concerned with the two states ξ and η, and that $v\,(\xi;\,\alpha,\,S)$ and $v\,(\eta;\,\alpha,\,S)$ are not measurable. In spite of this fact it may be possible to obtain measurable variables $v\,(\xi-\eta;\,\alpha,\,S)$ defined for every set of α and S.

Example 6.2. A transformation of investigator α into $\sigma\alpha$ can be performed by the co-operation $(\alpha,\,\beta)$ of two types of investigators α and β, where α is a proper investigator who wants to obtain necessary data from respondents, while β is an auxiliary person who is not well-trained as an investigator but who has intimate knowledge of the respondents and will serve to make respondents confident enough to answer α correctly.

Example 6.3. A transformation of scheme S into σS can be realized by adding another questionnaire T to the original one S. The questionnaire T may contain a set of questions which has a similar effect on the respondents as the auxiliary person in the previous example. Instead of starting with direct questioning about domestic economies of households, it is often more effective to speak about general topics which lead them naturally to answer the desired questions.

Example 6.4. Both before and in the course of a sequence of surveys, there sometimes arises the need for some enlightenment and "education" of respondents. This may be regarded as enlarging the domain of α for which $v\,(\xi;\,\alpha,\,S)$ are defined. This domain should correspond to $v\,(L\xi;\,\alpha,\,S)$ where L stands for enlightenment to the respondents.

We have mentioned some types of techniques used by experts conducting actual statistical surveys and experiments. Our emphasis here is to point out that it will not only be possible but also necessary to give theoretical consideration to such statistical techniques. By a suitable formulation the efficiencies of such transformations ξ, $\sigma\alpha$ and ΘS and the costs for executing them should be discussed, because after application of these two principles, randomization and transformation, our variables $v\,(\tau\xi;\,\sigma\alpha,\,\Theta S)$ are expected to be random measurable functions of the variates $\tau\xi$, $\sigma\alpha$ and ΘS and hence to be subject to current statistical analysis at least in principle. Thus we can compare various efficiencies between sets of possible combinations of these transformations.

(6.3) *Inductive recognition.* In his famous monograph *A System of Logic*, JOHN STUART MILL (1843) gave a systematic account of various methods of inductive logic and classified them into five categories: 1. the mathod of agreement, 2. the method of difference, 3. the joint method of agreement and difference, 4. the method of concomitant variation and 5. the method of residues. None of these methods can be regarded, however, as a rule effective enough to find decisively an invariant relationship between phenomena or as a rule powerful enough to prove rigorously such relationship (if any). On the contrary their main merit should be ascribed to their function of rejecting any false hypothesis from a set of candidate ones. This merit which had long been recognized was greatly clarified by introduction of the Neyman-Pearson theory of testing hypotheses into statistics. Indeed NEYMAN avoided the use of the word "inductive reasoning," on the ground that reasoning refers to deduction, and he seemed to emphasize the role of testing hypotheses in getting information from data.

We have now come to the point where we have to define statistical hypotheses. From what we have discussed in (6.1) aggregate recognition and (6.2) quantitative recognition, it can be conceived that our ACSSP approaches should be penetrated into the depth of pattern recognition regarding the background from which our data has come. It is also some sort of logical consequence that all characteristic features of the statistical approach to learning by experience gained from data should be defined with reference to the two characteristic aspects of statistical recognition, namely, aggregate recognition and quantitative recognition, which have already been explicated in the previous paragraphs. It is therefore natural to consider a hypothesis to be a statistical one in a broad sense when its rejection rule is defined in terms of notions belonging to the realm of aggregate and quantitative recognition.

In this sense a general form of such a rejection rule for hypotheses in statistics can be given by the following steps:

1. Set up a hypothesis H_0.

2. Define a quantitative variable x. To each set of data D a value of the variable x is determined and is denoted by $x(D)$.

3. Find the region of all values of the variables x, under the assumption that the hypothesis H_0 is true, and choose its subregion $R(x)$ which is called a region of rejection under the hupothesis H_0.

4. Reject the hypothesis H_0 when $x(D)$ belongs to $R(x)$.

Here we shall give a few illustrative examples of this general form.

Example 6.5. Obviously the Neyman-Pearson choice of rejection region is a particular case of our general form. The characteristic aspect of their approach is that the variable x is a statistic obtained from a random sample and therefore can be regarded as a random variable. As a consequence the set of ω for which the value $x(\omega)$ belongs to the region $R\,(x)$ cannot have a relevant subset in the sense of R. A. FISHER in the space Ω.

This is indeed the situation when probabilistic considerations provide the sole basis upon which a judgement of a given hypothesis or a prediction of future phenomena can be performed. However this is rather an idealized limiting situation, and it should not be startling to observe that not all real situations encountered in testing hypotheses can be covered by this model.

Example 6.6. Let us consider a family of sine functions $f\,(t;\varrho,\omega) = \varrho \sin\,(2\,\pi\,t + \omega)$ where ω runs through the set $0 \leqq \omega < 2\,\pi$ with unknown positive parameter ϱ. Let us define a variable $x = x\,(\omega;\varrho)$ by

$$(6.1) \qquad x\,(\omega;\varrho) = \max_{1 \leqq k \leqq n} \left| f\left(\frac{k}{n};\varrho,\omega\right) \right| .$$

Let us set up a hypothesis $H_0: \varrho = 1$. For each assigned $\varrho_0 < 1$, we can assign the set $R_{\varrho_0}\,(\varrho) = [\omega; x(\omega;1) < \varrho_0]$ and hence its probability measure. Let us reject our hypothesis H_0 when and only when our observed value of $x\,(\omega;\varrho)$ that is $x\,(D)$ does belong to $R_{\varrho_0}\,(\varrho)$. A suitable value of ϱ_0 can be given, for each assigned positive number ξ, such that the measure of $R_{\varrho_0}\,(\varrho)$ is less than ξ. It is evident that steps 1, 2, 3 and 4 are taken in the present example without appealing to any randomness. The common feature of Examples 6.5 and 6.6 is the fact that both of them are based upon aggregate and quantitative recognitions. The measure of the set $R_{\varrho_0}\,(\varrho) = [\omega; x\,(\omega,\varrho) < \varrho_0]$ becomes a function of ϱ which corresponds to the power function in the sense of NEYMAN-PEARSON. In this case the set $R_{\varrho_0}\,(\varrho)$ has its relevant subset in which the range of functions can be given. Any information on relevant subsets will be useful, for instance, in a prediction, where we can make use of them because they exist in reality.

Example 6.7. Let us draw a random sample of n observation points $\tau = \{t_k\}$, $k = 1, 2, \ldots, n$ drawn independently from the interval $0 \leqq t < 2\pi$ according to the uniform distribution. Then

$$(6.2) \qquad x_\tau\,(\omega;\varrho) = \max_{1 \leqq k \leqq n} f\,(t_k;\varrho,\omega)$$

can now be regarded as a random variable. Consequently we can introduce a probabilistic scheme of testing hypotheses along lines exactly analogous to the Neyman-Pearson formulation. In one of our previous papers [*19*] stochastically approximative analysis, or simply stochastic analysis, was discussed in which some of classical analysis is reformulated under a probabilistic scheme.

One of the critiques given by R. A. FISHER of the Neyman-Pearson theory of testing hypotheses is concerned with the error of the second kind and with the power function associated with their theory. According to R. A. Fisher, these notions do have a domain of validity, including most notably acceptance sampling inspection, but they cannot claim a universal validity over the whole domain of scientific research where we are in general not able to specify a set of alternative hypotheses in the parameter space. The assumption of such a predetermined set of possible alternative hypotheses violates the principle of complete freedom of choice in research cherished by the scientist. Such a complete free choice however, can be neither formulated objectively nor performed automatically according to any program stored in a computer. This we cannot deny at all. Nevertheless what we can provide and what we should prepare in any theoretical formulation are always approximate pictures of reality. Successive approximations, always trying to extend a class of statistical procedures and their sequences so as to be capable of giving a more adequate picture for complete free choice to the scientist in choosing a sequence of statistical procedures, should be regarded as a step toward reaching practical solution to our ultimate goal of establishing an automated statistician.

Finite dimensional euclidian parameter spaces introduced by Neyman-Pearson for locating alternative hypotheses should be also considered as an approximation to reality in this sense. A substantial contribution can be expected through a generalization of classical parameter spaces into more sophisticated ones having possibly more complicated local properties in a neighborhood of each parameter point as well as having a global specification of their geometrical configuration, although such a proposal has not been fully discussed in any literature, except some remarks given in one of our previous papers [25].

Example 6.8. Let us consider a family of normal populations with mean Θ and variance $\sigma_0^{2\Theta}$. Let us consider a null hypothesis $H_0: \Theta = 0$, $\sigma^2 = 1$. Then the set of alternative statistical hypotheses is $H_1: \Theta = \Theta_0$, $\sigma^2 = \sigma_0^{2\Theta_0}$. In this case the neighborhood of the null hypothesis is given by the locus of points specified by the equation $\sigma^2 = \sigma_0^{2\Theta}$ in the (Θ, σ^2) space.

The reason why we have to consider the set of alternative hypotheses specified by the relation $\sigma^2 = \sigma_0^{2\Theta}$ in the (Θ, σ^2)-space may be based upon our a priori knowledge about the mechanism of our population or may be learned from accumulated data. It would be a stimulating challenge to the ACSSP approach to analyze such a learning process and to formulate it as an ACSSP which will lead us with sufficiently high probability to the use of the relation in specifying alternative hypotheses.

Example 6.9. Although normal populations have been most frequently used in statistics, it is not rare that 1. a serious doubt has been experienced as to their validity, and that 2. a more general functional form for the probability density functions are given. An important class of such a functional form is given by

$$(6.3) \qquad \left\{ \Gamma\left(1 + \frac{1 + \alpha}{2}\right) 2^{1 + (1 + \alpha)/2} \, \sigma \right\}^{-1} \exp\left\{ -\frac{1}{2}\left| \frac{x - \Theta}{\sigma} \right|^{2/(1 + \alpha)} \right\},$$

with the three population parameters Θ, σ and α having respective ranges: $-\infty < \Theta < \infty$, $\sigma > 0$, $-1 < \alpha \leqq 1$.

It would be a task for the ACSSP approach to formulate a learning process in which both the phases 1 and 2 mentioned just now should be formulated as statistical procedures belonging to our ACSSP.

Summing up the discussions developed in Subsections 6.1, 6.2, and 6.3, our definition of *statistical procedure* is now given in the following manner:

1. A statistical procedure is a procedure which deals with a set of data under the following two specifications:

a) Aggregate recognitions are obtained through a set of probability fields as explained in Subsection 6.1.

b) Quantitative recognitions are obtained through applications of two principles for acquisition of data, namely, randomization and transformation as explained in Subsection 6.2.

2. Recognitions obtained from statistical procedures constitute a set of inductive recognitions based upon a set of rejection rules of statistical hypotheses in the sense explained in Subsection 6.3.

In giving this definition we are not and in fact we cannot be as strict as mathematicians are in their definition of mathematical objects because statistics is an empirical science whose totality cannot be developed on the basis of logicodeductive reasonings. It may be possible to give an elaborate definition of statistical procedure in a more strict manner than we have just given, and indeed we believe it worth-while to do so after further developments of ACSSP approaches will have been realized. In the context of our descriptions in this paper a definition of ACSSP is now given by this specification of statistical procedure in combination with the definition of ACSP which was already given in Section 2. In order to make clear the characteristic aspects of an ACSSP, we have to include the following paragraph concerning the integration of statistical recognitions obtained from our ACSSP.

(6.4) *Integrated recognition.* Statistical recognitions belong to the realm of inductive recognitions, but the logic of induction could have its sound foundation, at least in our statistics, only after statisticians established mathematical formulations of statistical problems beginning with the epoch making work of R. A. FISHER. The achievements of modern statistics in this direction and particularly the Neyman-Pearson formulation of statistical problems including estimation and testing hypothesis could hardly be overestimated with respect to their significant contribution to the logic of induction.

However it should be noted that all learning processes of human beings in the light of data cannot be studied in the framework of estimation and testing hypotheses about a set of unknown parameter(s) in a prescribed mathematical model, because 1. our mathematical model may be tentatively and incompletely specified and 2. our uses of information obtained from data includes storage of information, and pattern recognition, as well as operational uses. These two reasons 1 and 2 are crucially important at the present stage of development of statistics in which an automatic learning process for handling mass data is urgently needed, as we have explained in Section 4. Data in themselves are indeed a bulk accumulation of information, and statistical recognition obtained from data should have a certain integration pattern.

Generally speaking, such an *integration pattern* of statistical recognitions can be defined with reference to two schemes which we call information summary scheme and information evaluation scheme respectively. An *information summary scheme* is a recognition pattern by which information obtained from data is stored and arranged. *An information evaluation scheme* is a set of utility functions associated with an information summary scheme, by which the utility of information stored and arranged in the information summary scheme is measured.

An integration pattern of statistical recognitions in this sense may be said to have been proposed by WALD [36] when he aimed to establish a unified theory of statistics by introducing a notion of risk function and developing the statistical decision function approach. His approach was quite successful in realizing this aim so far as the classical Neyman-Pearson theory and his sequential analysis were concerned. On the other hand a serious objection to his approach was raised by the Fisherian school pointing out the non-universal validity of the risk function approach to general scientific research work. The author of the present paper also referred to some limitations inherent in the Wald approach, both from the standpoint of the successive process of statistical inference as in [22] and from that of the relativistic logic of mutual specification as in [25].

In the Wald formulation a utility function of information may be said to be introduced in the form of a risk function which is the sum of a loss at the terminal decision and a cost of experiments performed before the experimenter reaches the terminal stage. By making use of the terminology introduced just now we may say that an information summary scheme in the Wald formulation is reduced to a terminal decision, while an information evaluation scheme is concerned with the evaluation of the terminal decision taking into account the cost of experiments performed before reaching the terminal decision. Our critique of the Wald formulation amounts therefore to the following objections:

1. The information summary pattern so far developed in the formulations of WALD and his successors is not broad enough to cover all three aspects of information, namely a) storage, b) pattern recognition and c) operational use, simply because of their limitation of a summary pattern to terminal decisions without considering the intermediate stages of their decision process.

2. In combination with the situation enunciated in paragraph 1 just now, their information evaluation schemes limit the statistician's concern to the operational uses of information c), with little attention paid to a) and b).

3. The tentativeness and incompleteness of mathematical models to be used as tools for statistical analysis are not duly taken into consideration in the Wald formulation, because that approach concentrates on the utility of terminal decisions based on a mathematical model whose validity has already been assumed.

It would therefore be an important responsibility for those of us who advocate an ACSSP approach to introduce an integrated pattern of statistical recognitions in terms of information summary and evaluation schemes, which can be incorporated with a broader range of information use a), b), and c), and which is free from the three critiques given in 1, 2 and 3. In what follows we shall outline our plan for realizing this aim, leaving until another occasion a discussion of the technical details of our formulation.

In analyzing the three objections 1, 2 and 3 just presented, it seems to us that the basic limitation of the statistical decision function approach can be handled by a decomposition of the whole process of statistical decision into a set of subprocesses of statistical decisions for each of which an individual risk function will be assumed. Under this situation our integration pattern may not necessarily be wholly integrated into one system, but it has a set of subsystems each of which is integrated with reference to its respective summary and evaluation schemes. Then there may arise naturally another problem for an integration pattern of statistical recognitions, namely how to coordinate these subsystems into a whole. A coordination principle may not necessarily belong to the same category as the utilities which define a set of risk functions for each subsystem, and it would in general be safe to assume that quite different considerations are important in evaluating coordinating subsystems. From the standpoint of statistical decision functions such an ACSSP must be condemned as disintegrated in itself and hence to be excluded by their approach. However we believe that such a disintegration provides a better model of general statistical recognitions and that it has a flexibility of introducing an integration pattern as a whole as our information from data accumulates. One of the main points we wish to emphasize is that such an integration pattern should emerge as a result of our learning process and hence it should admit of further elaboration and reconstruction in the course of our studies. On the other hand we shall make use of the risk function formulation in each subsystem so far as it is useful, in view of the mutual dependence of such subsystems.

For example, let us assume that any information to be expected from a subsystem S at any possible stage of our statistical recognitions is entirely limited to the mean value of data associated with the subsystem. In such a case, data in the subsystem S can be reduced to one quantity, the mean value of the subsystem S, and no further description of the original accumulation of data will be required at any later stage.

In defining our ACSSP, we appeal to the notion of tree and paths within a tree. These notions will lead us to an information set defined for our tree, and hence our information summary scheme can be defined as a certain subset of the set of couples consisting of each branch point and variables associated with that branch point in the tree.

An information set inherited in a tree of an ACSSP is a notion which has some resemblance to that of a game tree, with the difference that it has a set of variables associated with each individual branch point (vertex point). From our standpoint the critiques given by the Fisherian school about the non-universal validity of risk functions can be met abandoning the a prior setup of a risk function associated with one terminal decision in place of a decomposition into a disintegrated set of subsystems. However we believe it is realistic and useful to assume a risk function approach to each of these subsystems under a suitable decomposition of our whole system. A coordination principle can be developed in view of data accumulated, and the tentativeness and the incompleteness of our mathematical models can be discussed through the process of establishing a coordination principle and of reusing risk functions of each individual subsystems if it be required in the face of data.

These remarks mean to be entirely illustrative to explain our broad idea of how to make use of our ACSSP approach.

Further technical formulation is now in preparation to be published in the future.

References

[1] Anscombe, F. J., and J. W. Tukey: The Criterium of Transformation, 1954, Unpub. M.S.

[2] Bowman, R. T., and M. E. Martin: Changing Tasks in Official Statistics, The 34th Session of the International Statistical Institute, at Ottawa, 1963, August.

[3] Box, G. E. P., and J. S. Hunter: Multifactor experimental designs for exploring response surfaces, Ann. Math. Statist. **27**, 1017 (1956).

[4] — — Condensed Calculations for evolutionary operation programs. Technometrics **I**, 77 (1959).

[5] Bozivich, B., T. A. Bancroft, and H. O. Hartley: Power of analysis of variance test procedures for certain incompletely specified models, I. Ann. Math. Statist. **27**, 1017 (1956).

[6] Cooper, B. E.: Designing the Data Presentation of Statistical Program for the Experimentalist, The 34th Session of the International Statistical Institute at Ottawa, 1963, August.

[7] —, and Mrs. C. M. Whiteside: The Presentation of Experimental Data to Computer. Report United Kingdom Atomic Energy Authority Research Group Report AERE-R 4250, 1963.

[8] Deming, W. E. (1944): Some Theory of Sampling. New York: John Wiley and Sons 1950.

[9] Dodge, H. F.: I. A sampling plans for continuous production. Ann. Math. Statist. **14**, 264 (1943).

[10] —, and H. G. Romig: I. A method of sampling inspection. Bell System techn. J. **8**, 613 (1927).

[11] Douglas, A. S., and A. J. Mitchel: AUTO STAT: A Language for Statistical Data Processing. The Computer Journal **3**, 61 (1960).

[12] Fürst, G.: Changing Tasks in Official Statistics. The 34th Session of the International Statistical Institute at Ottawa, 1963, August.

[13] Goto, M.: Changing Tasks in Official Statistics in Japan. The 34th Session of the International Statistical at Ottawa, 1963, August.

[14] Hansen, H. M., W. N. Hurwitz, E. S. Marks, and W. P. Mauldin (1951): Respons errors in surveys. J. Amer. Stat. Ass. **46**, 147 (1951).

[15] Hayward, Lynn C.: BIMD Computer Programs Manual, UCLA Student Store, Los Angeles.

[16] Kendall, M. G., and P. Wegner: An Introduction to Statistical Programming. The 34th Session of the International Statistical Institute at Ottawa, 1963, August.

[17] Kitagawa, T.: Successive process of statistical inferences, Mem. Fac. Sci. Kyushu University, A. **15**, 139 (1950).

[18] — Some contributions to the design of sample surveys. Sankhya **17**, Part 4 to 6, 1 (1956).

[19] — Some aspects of stochastically approximative analysis. Bull. Math. Statist. **6**, 109 (1956).

[20] — Successive process of statistical inferences applied to linear regression analysis and its specifications to response surface analysis. Bull. Math. Statist. **8**, 80 (1959).

[21] — A mathematical formulation of the evolutionary operation program. Mem. Fac. Sci. Kyushu University, A. **15**, 21 (1961).

[22] — The logical aspects of successive processes of statistical inferences and controls, Bulletin de L'Institut International De Statistique, Actes de la 32e Session, Tokyo 1960, Tome 38, 4e Livraison, p. 152, Tokyo 1961.

[23] — The present problems of statistical inferences (Memorial Lecture on the Celebration of Three Hundred Years of Statistics delivered at the Annual Meeting of Japanese Society of Statistics) (in Japanese) Tokei, Vol. **13**, 11, 9 (1962).

[24] — Estimation after preliminary test of significance, University of California Publications in Statistics. **3**, No. 4, 147 (1963).

[25] — The relativistic logic of mutual specification in statistics. Mem. Fac. Sci. Kyushu University, A. **17**, 76 (1963).

[26] — Automatically controlled sequence of statistical procedures in data analysis. Mem. Fac. Sci. Kyushu University, A. **17**, 106 (1963).

[27] Leone, F. C.: Abstracts of Statistical Computer Routines. Statistical Laboratory, Case Institute of Technology, Ohio.

[28] Mahalanobis, P. C.: On large-scale sample surveys. Phil. Trans. Royal Soc., B. **231**, 329 (1944).

[29] — Recent experiments in statistical sampling in the Indian Statistical Institute. J. Roy. Statist. Soc. **109**, 325 (1946).

[30] MIL-STD-105 B: I. Sampling procedures and tables for inspection by attributes, 1958.

[31] Neyman, J., and E. S. Pearson: I. The testing of statistical hypotheses in relation to probabilities a priori. Proc. Cambridge Phil. Soc. **29**, 492 (1948).

[32] Pearson, K.: The grammar of Science, 1896.

[33] Miss Reed, S. J.: Screening Rules. M. S. Thesis. Rutgers University 1959.

[34] Terry, M. E.: The principles of statistical analysis using large electronic computers. The 34th Session of the International Statistical Institute at Ottawa, 1963, August.

[35] Tukey, J. W.: The Future of Data Analysis. Ann. Math. Statist. **33**, 1 (1962).

[36] Wald, A.: I. Statistical decision functions. New York: John Wiley and Sons 1950.

[37] Wiener, N.: Cybernetics, 2nd Edition. New York: John Wiley and Sons 1961.

[38] Yates, F., and H. R. Simpson: A General Program for the Analysis of Surveys. The Computer Journal. **3**, 136 (1960).

[39] — Computers in research-promise and performance. The Presidential Address. The Computer Journal **4**, 273 (1961).

[40] —, and H. R. Simpson: The Analysis of Surveys: Processing and Printing the Basic Tables. The Computer Journal **4**, 20 (1961).

On the Distribution of Sums of Independent Random Variables*

By **Lucien LeCam**

University of California, Berkeley

1. Introduction

Let $\{X_j; j = 1, 2, \ldots\}$ be a finite sequence of independent random variables. Let $S = \Sigma X_j$ be their sum, and let P_j be the distribution of X_j. Let M be the measure defined on the line deprived of its origin by $M(A) = \Sigma_j P_j\{A \cap \{0\}^c\}$. The purpose of the present paper is to develop certain results on the approximation of the distribution $\mathscr{L}(S)$ of S by the accompanying infinitely divisible distribution which has for Paul Lévy measure the measure M itself. If $\lambda = \|M\|$ is the total mass of M then $V = M/\lambda$ is a probability measure. Let $\{Z_k; k = 1, 2, \ldots\}$ be an independent sequence of random variables having common distribution V. Let N be a Poisson variable independent of the Z_k and such that $EN = \lambda$. A "natural" infinitely divisible approximation to the distribution of S is the distribution of $T = \sum_{k=0}^{N} Z_k$ with $Z_0 = 0$. If μ is a signed measure, let $\|\mu\|$ be its norm, equal to the total mass $\|\mu\| = \|\mu^+\| + \|\mu^-\|$. It can be shown that in some cases the approximation of $\mathscr{L}(S)$ by $\mathscr{L}(T)$ is good in the sense that $\|\mathscr{L}(S) - \mathscr{L}(T)\|$ is small. More generally it will be shown that the Kolmogorov-Smirnov distance $\varrho[\mathscr{L}(S), \mathscr{L}(T)]$ is small. This distance is defined by

$$\varrho(\mu, \nu) = \sup_x |\mu\{(-\infty, x]\} - \nu\{(-\infty, x]\}|$$

for any two signed measures μ and ν. One could also use Paul Lévy's diagonal distance $\varLambda[\mu, \nu]$ defined as the infimum of numbers α such that

$$\nu\{(-\infty, x - \alpha]\} - \alpha \leq \mu\{(-\infty, x]\} \leq \nu\{[-\infty, x + \alpha]\} + \alpha$$

for every value of x. However, since $\varLambda$ is not invariant under scale changes, approximations in this sense are not always entirely satisfactory.

One possible description of the theorems stated below is the following. Finite signed measures on the real line form a commutative Banach algebra for the convolution operation. In this algebra the distribution $\mathscr{L}(S)$ is simply the product $\prod_{j=1}^{n} P_j$. The distribution of T is the ex-

---

* This paper was prepared with the partial support of the United States Army Research Office (Durham), grant DA-ARO(D)-31-124-G 83.

ponential $\mathscr{L}(T) = \exp\left\{\sum_{j=1}^{n}(P_j - I)\right\}$ where I is the identity of the algebra, that is, the probability measure assigning mass unity to the origin. Letting $\Delta_j = P_j - I$, the theorems of the present paper are expressions of the fact that when the Δ_j are "small" the product $\Pi\,(I + \Delta_j)$ differs little from the exponential $\exp[\Sigma\,\Delta_j]$. It is easy to construct examples where each one of the Δ_j has small norm but where $\|\Pi\,(I + \Delta_j) - \exp\{\Sigma\,\Delta_j\}\|$ is large. However, it will be shown that when all the $\|\Delta_j\|$ are small the Kolmogorov distance $\varrho\{\Pi\,(I + \Delta_j) - \exp[\Sigma\,\Delta_j]\}$ is also small. Furthermore, it will be shown that when the variables X_j are suitably centered and small compared to their sum, a similar result is again available. In this case the description of "small" involves the use of Paul Lévy's concentration function.

The concentration function of a random variable X is defined at τ by $C_X(\tau) = \sup\{Pr\,[X \in J]\}$ where the supremum is taken over all intervals J of length at most equal to τ. We shall interpret the statement that X_j is small compared to S as meaning that there is a $\tau \geq 0$ such that $C_{X_j}(\tau)$ is close to unity and $C_S(\tau)$ is close to zero. The statement of the theorem is then that for suitably centered variables $\varrho\,[\mathscr{L}(S), \mathscr{L}(T)]$ is smaller than a certain increasing function of $C_S(\tau)$ and $\sup_j[1 - C_{X_j}(\tau)]$. By comparison, Paul Lévy's form of the usual Normal approximation theorem is that $\mathscr{L}(S)$ is close to a normal distribution whenever $C_S(\tau)$ and $\Sigma_j[1 - C_{X_j}(\tau)]$ are small. As is well known, this Normal approximation theorem possesses a converse.

We have been unable so far to prove or formulate an adequate converse for the approximation of $\mathscr{L}(S)$ by $\mathscr{L}(T)$.

The proofs given below are directly inspired from the work of Kolmogorov. In fact, they were obtained in an attempt to reproduce the results of [1]. It happened that in the process, we followed a slightly different path with a different end product.

In the meantime, Kolmogorov had also obtained the refinement leading to the replacement of the exponent 1/5 of [1] by the exponent (1/3) of [2]. It is a pleasure to acknowledge the fact that we received advanced notification of this at the time when we still had doubts about the correctness of a preliminary version of the present paper.

The most important differences between the results of [2] and the present ones seem to be the following. Theorem 3 relative to the case where $\sup\|\Delta_j\|$ is small does not appear to be a consequence of [2]. Furthermore, we have insisted here on the approximation of $\Pi\,(1 + \Delta_j)$ by $\exp\Sigma\,\Delta_j$ itself. Theorem 3 does not even involve any recentering possibility of the X_j. As for Theorem 4 appropriate centering seems to be a necessity, but this is the only modification to be made. Even there some flexibility remains, as shown by Theorem 5.

The approximation $\exp\{\Sigma_j \Delta_j\}$ which never possesses any normal component avoids the computation of a truncated variance and the introduction of the corresponding Gaussian component.

Further, one may conjecture that the approximability of $\Pi (1 + \Delta_j)$ by $\exp \Sigma \Delta_j$ is not often a consequence of the approximability of $\Pi (1 + \Delta_j)$ by an infinitely divisible distribution. This conjecture is supported by the observation that if the X_j are independent identically distributed, taking values -1 and $+1$ with probability one-half, and $S_n = \sum_{j=1}^{n} X_j$, there are infinitely divisible distributions F_n such that $\| \mathscr{L} (S_n) - F_n \| \to 0$ as $n \to \infty$ (see [3]); however, $\mathscr{L} (S_n) - \exp \left[\sum_{j=1}^{n} \Delta_j \right]$ does not even tend to zero for the weak convergence induced by the bounded continuous functions.

The theorems given below include mention of certain constants which can hardly be close to the best possible ones. Also, they involve an exponent $(1/3)$ which is not necessarily the best possible as indicated in [4].

2. Modulus of continuity and concentration functions

Let μ be a finite measure on the line and let τ be a nonnegative number. Let ϱ be the Kolmogorov-Smirnov seminorm, defined by $\varrho (\mu) = \sup\{\mu (-\infty, x]; -\infty < x < +\infty\}$. Let $S^{\alpha}\mu$ be the measure μ shifted by an amount α. If μ is the distribution of a variable X then $S^{\alpha}\mu$ is the distribution of $X + \alpha$. We shall call modulus of continuity under shift the function defined for all nonnegative values of τ by

$$\Gamma (\mu, \tau) = \sup\{\varrho [\mu - S^{\alpha}\mu]; \, | \alpha | \leq \tau\}.$$

For a probability measure P the modulus $\Gamma (P, \tau)$ is simply $\Gamma [P, \tau] = \sup\{P\{(x, x + \tau]\}; x \in (-\infty, +\infty)$. The concentration function C_μ of the measure μ is defined similarly by

$$C_\mu (\tau) = \sup\{P\{[x, x + \tau]\}; x \in (-\infty, +\infty)\}.$$

It follows that for positive measures $\Gamma (\mu, \tau) \leq C_\mu (\tau)$ and $C_\mu (\tau) \leq \leq \Gamma (\mu, \tau')$ for $\tau < \tau'$. Although similar moduli of continuity under shift may be defined for seminorms other than the Komogorov-Smirnov seminorm, the properties given below appear to be special properties of the Kolmogorov-Smirnov distance.

When X is a random variable having distribution P it will be convenient to use symbols such as $\Gamma (X, \tau)$ or $C_X (\tau)$ instead of $\Gamma (P, \tau)$ and $C_P (\tau)$. The following propositions summarize some of the important known properties of concentration functions which will be used in subsequent sections.

Proposition 1 (P. Lévy). *If X and Y are independent then $C_{X+Y} \leq$
$\leq \min [C_X, C_Y]$ and $\Gamma (X + Y) \leq \min [\Gamma (X), \Gamma (Y)]$.*

Proof. If J is an interval let $J - a = \{x: x + a \in J\}$. With this
notation

$$Pr [X + Y \in J] = E \{ Pr [X \in J - Y \mid Y] \} .$$

Proposition 2 (Kolmogorov-Lévy). *Let X and Y be two random variables having distributions $P = \mathscr{L} (X)$ and $Q = \mathscr{L} (Y)$. Then for every $\tau \geq 0$ the following inequality holds:*

$$\varrho (P, Q) \leq Pr \{ \mid X - Y \mid > \tau \} + \min \{ \Gamma (X, \tau), \Gamma (Y, \tau) \}.$$

Proof. This follows from a combination of four inequalities of the type

$$Pr [Y \leq x] \leq Pr [X \leq x + \tau] + Pr [\mid X - Y \mid > \tau]$$
$$\leq Pr [X \leq x] + \Gamma (X, \tau) + Pr [\mid X - Y \mid > \tau].$$

Proposition 3. *Let P, Q and W be three probability measures. Let PW and QW be the convolution products of P and Q by W. Let γ be the minimum of the moduli continuity of $[P - Q]^+$ and $[P - Q]^-$. Then for every $\tau \geq 0$*

$$[2 C_W (\tau) - 1] \varrho [P, Q] \leq \varrho [PW, QW] + C_W (\tau) \gamma (\tau) .$$

Proof. Let $F (x) = \mu_1 \{ (-\infty, x] \}$ and let $G (x) = \mu_2 \{ (-\infty, x] \}$ where μ_1 and μ_2 are the positive and the negative parts $\mu_1 = [P - Q]^+$ and $\mu_2 = [P - Q]^-$ of the measure $P - Q$. Let $H = F - G$. If H is identically zero then $P = Q$ and the inequality is satisfied. Otherwise, suppose that for some particular $x \in (-\infty, +\infty)$ one has $H (x) > \delta > 0$. Then, for $u \geq 0$ one can write

$$F (x + u) - G (x + u) = [F (x + u) - F (x)] + [F (x) - G (x)] -$$
$$- [G (x + u) - G (x)] > \delta - [G (x + u) - G (x)] \geq \delta - \Gamma (\mu_2, u).$$

Also

$$F (x - u) - G (x - u) = F (x) - G (x) +$$
$$+ [G (x) - G (x - u)] - [F (x) - F (x - u)] > \delta - \Gamma (\mu_1, u) .$$

In both cases there is an interval of length at least equal to τ in which H is larger than $\delta - \gamma (\tau)$. This implies the existence of a number y such that

$$\int H (y + u) W (du) \geq [\delta - \gamma (\tau)] C_W (\tau) - \{ \sup_x \mid H (x) \mid \} [1 - C_W (\tau)] .$$

Hence

$$\varrho [PW, QW] \geq [\delta - \gamma (\tau)] C_W (\tau) - \varrho (P, Q) [1 - C_W (\tau)] .$$

The desired result follows by letting δ tend to $\varrho (P, Q)$ at least whenever $\varrho (P, Q) = \sup_x H (x)$. If, on the contrary, $\varrho (P, Q) = \sup_x [- H (x)]$ the result is obtainable by interchanging the roles of P and Q in the above argument.

Note that $\gamma (\tau) \leq \min \{ \Gamma (P, \tau), \Gamma (Q, \tau) \}$. This will often be a usable upper bound for $\gamma (\tau)$.

One of the most important and most remarkable results on concentration functions is the following inequality of KOLMOGOROV [1], [5].

Proposition 4. *Let* $\{X_k\}$ *be a finite sequence of independent random variables. Let* $\gamma > 0$ *and* $\lambda \geq 0$ *be two numbers. Assume that for each k there are numbers b_k and α_k such that*

$$Pr\left[X_k \leq b_k - \gamma\right] \geq \alpha_k$$

and

$$Pr\left[X_k \geq b_k + \gamma\right] \geq \alpha_k .$$

Let $S = \Sigma X_k$ *then*

$$C_S\left(2\,\lambda\right) \leq \frac{1.3}{\sqrt{s}}\,\mathrm{Int}\left[1 + \frac{\lambda}{\gamma}\right]$$

with $s = \Sigma\,\alpha_k$ *and with* $\mathrm{Int}\,[x]$ *equal to the largest integer which does not exceed x.*

Corollary. *Let* $\{X_k;\ k = 1, 2, \ldots\}$ *be independent random variables and let* $S = \Sigma_k\,X_k$. *If γ and λ are two positive numbers then*

$$\left\{\sum_k [1 - C_{X_k}\left(\gamma\right)]\right\}\,[C_S\left(\lambda\right)]^2 \leq 4\left\{\mathrm{Int}\left[1 + \frac{\lambda}{\gamma}\right]\right\}^2.$$

A proof can be constructed as follows [5], [6]. Each X_k can be represented as a nondecreasing function $X_k = f_k\left(\eta_k\right)$ of a random variable η_k which is uniformly distributed on $[0, 1]$. Letting $2\,a_k\left(\theta\right) = f_k\left(\frac{1}{2} + \theta\right) + f_k\left(\frac{1}{2} - \theta\right)$ and $2\,D_k\left(\theta\right) = f_k\left(\frac{1}{2} + \theta\right) - f\left(\frac{1}{2} - \theta\right)$ for $\theta \in [\frac{1}{2}, 1]$ one may also replace each X_k by a variable of the type $X_k' = a_k\left(\theta_k\right) + \xi_k D_k\left(\theta_k\right)$ where ξ_k takes values (-1) and $(+1)$ with probability one-half and where θ_k is uniformly distributed on $[\frac{1}{2}, 1]$. Considering the problem conditionally for fixed values of the θ_k one is reduced to the special case covered by the following lemma of ERDÖS [7].

Let $S = \sum_{k=1}^{m} \xi_k\,x_k$ with $x_k \geq \gamma$ and with $\{\xi_k\}$ a sequence of independent random variables taking values $(+1)$ and (-1) with probability one-half. Then

$$Pr\left[a < S \leq a + 2\,\gamma\right] \leq 2^{-m}\,\binom{m}{p}$$

where p is the integer part $p = \mathrm{Int}\,(m/2)$ of $(m/2)$.

ERDÖS' result is a consequence of the fact that if two sums $\Sigma\,\varepsilon_j\,x_j$ and $\Sigma\,\varepsilon_j'\,x_j$ with ε_j and ε_j' equal to $+1$ or -1 fall in the interval $(a, a + 2\gamma]$ then the sets of indices $A = \{j;\ \varepsilon_j = +1\}$ and $A' = \{j;\ \varepsilon_j' = +1\}$ cannot be comparable.

We shall also need a bound on the concentration of infinitely divisible distributions as follows.

Proposition 5. *Let P be a probability measure whose characteristic function has the form*

$$\log\left\{\int e^{itx}\,P\left(dx\right)\right\} = \alpha it - \frac{1}{2}\,\sigma^2 t^2 + \int\left[e^{itx} - 1 - \frac{itx}{1 + x^2}\right] M\left(dx\right)$$

where M is a positive measure on the line deprived of its origin.

Let γ and λ be positive numbers and let

$$s\,(\gamma) = M\left\{(-\infty, -\gamma]\right\} + M\left\{[\gamma, +\infty)\right\}$$

$$D^2\,(\gamma) = \frac{\sigma^2}{\gamma^2} + \int \min\left[1, \frac{x^2}{\gamma^2}\right] M\,(dx)\ .$$

Then

$$\sqrt{s\,(\gamma)}\ C_P\,(\lambda) \leqq (1.2)\ \mathrm{Int}\left[1 + \frac{\lambda}{\gamma}\right]$$

and

$$D\,(\gamma)\ C_P\,(\gamma) \leqq 6\ .$$

Proof. Let $s_1 = M\left\{(-\infty, -\gamma]\right\}$ and $s_2 = M\left\{[\gamma, +\infty)\right\}$. Further, let $m = \mathrm{Int}\,(s_1/\log 2)$ and let $n = \mathrm{Int}\,(s_2/\log 2)$. With this notation M can be written in the form $M = mF + nG + H$ where F, G and H are positive measures, F carried by $(-\infty, -\gamma]$ and G by $[\gamma, \infty)$ and where $\|F\| = \|G\| = \log 2$. It follows that the concentration of P is smaller than the concentration of a random variable $T = \sum_{j=1}^{m+n} T_j$ where the T_j are independent and

$$\mathscr{L}\,(T_j) = \exp\left\{F - \|F\|\,I\right\}, \quad \text{for } j = 1, 2, \ldots, m\ ,$$

$$\mathscr{L}\,(T_j) = \exp\left\{G - \|G\|\,I\right\}, \quad \text{for } j = m + 1, \ldots, m + n\ .$$

Also $Pr\,(T_j = 0) = \exp\left[-\log 2\right] = Pr\left[\,|\,T_j\,| \geqq \gamma\right]$.

Let $\left\{\xi_j; j = 1, 2, \ldots, m + n\right\}$ be independent random variables taking values 0 and 1 with probability $1/2$. Then T has the same distribution as $\Sigma\left[\xi_j + (1 - \xi_j)\,U_j\right]$ where the U_j are independent random variables such that $|\,U_j\,| \geqq \gamma$. Consider a particular set $\left\{u_j; j = 1, 2, \ldots, m + n\right\}$ of values of the U_j and two possible sets of values $\left\{\varepsilon_j\right\}$ and $\left\{\varepsilon_j'\right\}$ of the ξ_j. Let $A = \left\{j; \varepsilon_j = 0, j = 1, 2, \ldots, m\right\}\left\{j; \varepsilon_j = 1, j = m + 1, \ldots, m + n\right\}$ and let A' be the corresponding set for the values ε_j'. If

$$b < \Sigma\left[\varepsilon_j + (1 - \varepsilon_j)\,u_j\right] \leqq b + \gamma$$

and

$$b < \Sigma\left[\varepsilon_j' + (1 - \varepsilon_j')\,u_j\right] \leqq b + \gamma,$$

the two sets A and A' are not comparable. Therefore

$$Pr\left\{b < T \leqq b + \gamma\right\} \leqq 2^{-(m+n)}\,\tbinom{m+n}{p}$$

with $p = \mathrm{Int}\,(m + n/2)$. The first inequality follows.

For the second inequality let V be the measure defined by $V\,(B) = M\left\{B \cap [-\gamma, +\gamma]\right\}$. The measure P is less concentrated than the measure Q, having for characteristic function the expression

$$\exp\left\{-\tfrac{1}{2}\,\sigma^2\,t^2 + \int\left[e^{itx} - 1 - \frac{itx}{1 + x^2}\right] V\,(dx)\right\}\ .$$

Assume first that V is a finite measure and let
$$\tau^2 = \sigma^2 + \int x^2\, V\,(dx)\,.$$

The usual Berry-Esseen type computation on characteristic functions shows that the Kolmogorov-Smirnov distance $\varrho\,(Q,\,W)$ between Q and a suitably centered normal distribution $W = \mathcal{N}\,(\mu,\,\tau^2)$ satisfies the inequality
$$\varrho\,[Q,\,W] \leqq \frac{5}{2}\,\frac{\gamma}{\tau}\,.$$

It follows that
$$C_P\,(\gamma) \leqq C_Q\,(\gamma) \leqq 5\,\frac{\gamma}{\tau} + \frac{\gamma}{\tau\,\sqrt{2\pi}}\,.$$

Combining this with the first inequality gives
$$\left[s\,(\gamma) + \frac{\tau^2}{\gamma^2}\right] C_P^2\,(\gamma) \leqq (2.4)^2 + \left(5 + \frac{1}{\sqrt{2\pi}}\right)^2\,.$$

This gives the desired result provided V be finite. The general result is an immediate consequence obtainable by taking a sequence V_n of finite measures which increases to V.

Note 1. It has been assumed here that $\gamma > 0$. The result is still valid for $\gamma = 0$ since M is assumed to be a measure on the line deprived of its origin. In this case, if $\sigma^2 = 0$, then $D^2\,(0)$ is simply the total mass of M. If $\sigma^2 > 0$ the distribution has a normal component. This obviously implies $C_P\,(0) = 0$.

Note 2. It is easily verified that for $\gamma > 0$ the concentration $C_P\,(\gamma)$ is always larger than
$$\frac{1}{2}\,e^{-s\,(\gamma)}\,\frac{\gamma}{\tau\,\sqrt{2\pi}}$$
with
$$\tau^2 = \int\limits_{-\gamma}^{+\gamma} x^2\,M\,(dx)\,.$$

In particular $C_P\,(\gamma)$ cannot be close to zero unless $D\,(\gamma)$ is large.

3. Approximation of a measure by its exponential

Let X be a random variable having for distribution the measure P. If $\alpha \in [0,\,1]$ is the probability that X be different from zero, one may write
$$P = (1 - \alpha)\,I + \alpha M = I + \alpha\,(M - I)\,,$$

where M is also a probability measure and I is the probability measure giving mass unity to the origin. Let Q be the convolution exponential $Q = \exp\{\alpha\,(M - I)\} = \exp\{P - I\}$. Such an exponential can be expanded in the form
$$Q = e^{-\alpha} \sum_{k=0}^{\infty} \frac{\alpha^k}{k!}\,M^k,$$

which shows that Q is the distribution of a sum $\sum_{k=1}^{N} X_k$ of N independent random variables X_k having distribution M. The number of terms in the sum is a Poisson random variable N which is independent of the X_k and has expectation α. Because of this interpretation and to simplify further formulae, we shall use the following notational convention. If μ is a finite positive measure then the exponential $\exp\{\mu - \|\mu\| I\}$ will be denoted pois μ.

The following properties of P and Q are well known and easily checked.

a) The minimum $P \wedge Q = W$ of the measures P and Q is at least equal to $(1 - \alpha) I + e^{-\alpha} M$. Therefore, there exist positive measures W' and W'' such that

$$\| W' \| = \| W'' \| \leq \alpha (1 - e^{-\alpha}) \leq \alpha^2$$

and $P = W + W'$ and $Q = W + W''$. In other terms it is possible to find a joint distribution of a pair (X, Y) on the plane such that $\mathscr{L}(X) = P$ and $\mathscr{L}(Y) = Q$ and $Pr[X \neq Y] \leq \alpha (1 - e^{-\alpha})$.

b) The expectations $\int x\, P\,(dx)$ and $\int x\, Q\,(dx)$ are equal.

c) If $\int x\, P\,(dx) = 0$ then P and Q have the same second and third moments. Furthermore $\int x^4 Q\,(dx) = 3\sigma^4 + \int x^4 P\,(dx)$ with $\sigma^2 = \int x^2 P\,(dx)$.

From these considerations it appears that Q may be a usable approximation to P whenever α is small. When α is not small but when the maximum possible value of $|X|$ is small, moments may be employed to indicate the structure of P. In this case property c) shows that Q is again a reasonable approximation to P, provided that $\int x\, P\,(dx) = 0$.

Instead of a single random variable X, consider a sequence $\{X_j;\ j = 1, 2, \ldots\}$ of independent random variables. Assume that $P_j = \mathscr{L}(X_j) = (1 - \alpha_j) I + \alpha_j M_j$ and let $Q_j = \text{pois}\,(P_j - I)$. Finally let $P = \Pi_j P_j$ and let $Q = \Pi_j Q_j = \exp\{\Sigma_j (P_j - I)\} = \text{pois}\, M$, with $M = \Sigma \alpha_j M_j$. The simplest results relative to the approximation of P by Q are probably the following.

Theorem 1. *If $P_j = (1 - \alpha_j) I + \alpha_j M_j$ where $\alpha_j \in [0, 1]$ and where M_j is a probability measure then*

$$\left\| \prod_j P_j - \text{pois}\,\left(\sum_j P_j\right) \right\| \leq 2\left\{1 - \prod_j (1 - \beta_j)\right\} \leq 2 \Sigma \alpha_j^2$$

with $\beta_j = \alpha_j (1 - e^{-\alpha_j})$.

Proof. This theorem is essentially due to Khintchin [8]. It can be proved as follows. Consider pairs (X_j, Y_j) where $\mathscr{L}(X_j) = P_j$ and $\mathscr{L} Y_j = \text{pois}\, P_j$. According to property a) above it is possible to select for (X_j, Y_j) a joint distribution such that $Pr[X_j \neq Y_j] \leq \alpha_j (1 - e^{-\alpha_j}) = \beta_j$. If $S = \Sigma X_j$ and $T = \Sigma Y_j$, then $Pr[S \neq T] \leq 1 - \Pi_j (1 - \beta_j)$. This implies the desired result.

Theorem 2 (PROHOROV). *If $P_j = (1 - \alpha) I + \alpha M$ where $\alpha \in [0, 1]$ and where M is a probability measure, then*

$$\left\| \prod_{j=1}^{n} P_j - \text{pois} \, \Sigma \, P_j \right\| \leqq 3 \, \alpha \, .$$

Proof. According to Theorem 1 it is sufficient to prove the result for $2 \, n\alpha > 3$. Furthermore, if the result is valid for the measure M which gives mass unity to the point $+ 1$, then it is valid for an arbitrary M. For the special choice of M, the result to be proved is that

$$\sum_{k=0}^{\infty} | b_k - p_k | \leqq 3 \, \alpha$$

with

$$b_k = \binom{n}{k} \alpha^k (1 - \alpha)^{n-k},$$

$$p_k = e^{-n\alpha} \frac{(n\alpha)^k}{k!} \, .$$

A simple proof, giving a coefficient 4 instead of 3, can be carried out by evaluating the maximum of the ratio b_k/p_k. Indeed, letting $q = 1 - \alpha$ and $\varphi (\nu) = (p_{n-\nu}/b_{n-\nu})$ or infinity according to whether $b_{n-\nu} > 0$ or $b_{n-\nu} = 0$, the logarithm of $\varphi (\nu)$ can be written

$$\log \varphi (\nu) = \frac{1}{2} \log \left(\frac{\nu}{n} \right) - \nu \left[\log \frac{nq}{\nu} - \frac{nq}{\nu} - 1 \right] + A (\nu, n)$$

where $A (\nu, n) \geq 0$. Further $\varphi (\nu)$ reaches a minimum for ν equal to the integer m which satisfies $nq - 1 < m \leqq nq$. It follows that $\varphi (\nu) \geq (m/n)^{1/2} = 1 - c$ (say). Let A be the set of integers k such that $b_k > p_k$. If P is the binomial distribution and Q is the Poisson distribution one can write

$$\| P - Q \| \leqq 2 \sum_{k \in A} (b_k - p_k) = 2 \sum_{k \in A} [(1 - c) \, b_k - p_k] + 2c \sum_{k \in A} b_k$$

$$\leqq 2 \, c \, P \, (A) \, .$$

Therefore $\| P - Q \| \leqq 2 (1 - \sqrt{m/n}) \, P \, (A)$. Evaluation of m gives $\| P - Q \| \leqq 4 \, \alpha$. To obtain a coefficient 3 instead of 4 it is sufficient to take into account not only the minimum $\varphi (m)$ but also the adjoining terms $\varphi (m - 1)$ and $\varphi (m + 1)$. The algebra becomes heavier and will be omitted.

The proof given by PROHOROV [9] is more elaborate and gives a much better evaluation of the difference $P - Q$ for n large.

Unfortunately the result indicated by Theorem 2 does not remain correct if $P_j = (1 - \alpha_j) I + \alpha_j M_j$ is allowed to vary with the index j. A similar inequality remains true with $\alpha = \sup \alpha_j$ if $P_j = (1 - \alpha_j) I + \alpha_j M$ with M fixed. The proof of this given in [10], is probably much too elaborate.

It should be possible to deduce this from PROHOROV's result by use of simple inequalities. If M_j concentrates all its mass at a single point $a_j \neq 0$ and if the a_j are rationally independent, then $\| P - Q \| \geq 1 - \Pi_j$ $[1 - \frac{1}{2} \alpha_j^2 (1 - \alpha_j)]$. For $\alpha = \sup \alpha_j$ small this can be made close to unity by taking $\Sigma \alpha_j^2$ large.

Such considerations show that approximation in the sense of the norm will be possible only in very special cases. The purpose of the next sections is precisely to show that on the contrary, approximations in the sense of the Kolmogorov-Smirnov distance are very often acceptable.

4. An approximation theorem for variables which are rarely different from zero

Let $\{P_j; j = 1, 2, \ldots\}$ be a finite sequence of probability measures on the line. Let $P = \Pi_j P_j$ be their convolution product and let $Q = \exp\{\Sigma_j (P_j - I)\}$ be the corresponding Poisson exponential. The main result of the present section is the following theorem.

Theorem 3. *For each j let α_j be the probability $\alpha_j = P_j\{\{0\}^c\}$ that a variable having distribution P_j be different from zero. Let $\alpha = \sup \alpha_j$. Let $\varrho(P, Q)$ be the Kolmogorov-Smirnov distance between P and Q. Then*

$$\varrho(P, Q) \leq K \alpha^{\frac{1}{3}}$$

with $K \leq 25$.

Since $\varrho(P, Q)$ is always less than or equal to unity one may assume $\alpha < K^{-3}$. Also, since Theorem 1 gives $\varrho(P, Q) \leq \Sigma \alpha_j^2 \leq \alpha (\Sigma_j \alpha_j)$ it is sufficient to prove the result for the case where $\Sigma_j \alpha_j$ is larger than $\alpha^{-(2/3)}$. In particular, letting $p_j = \alpha_j (1 - \alpha_j)^{-1}$ one may assume $p = \sup_j p_j < 1$.

The proof of the theorem will be divided in several lemmas outlined below. First let us note the following. The definition of α_j implies the existence of a probability measure M_j such that

$$P_j = (1 - \alpha_j) I + \alpha_j M_j = (1 - \alpha_j) [I + p_j M_j] .$$

Lemma 1. *Assume that $P_j = (1 - \alpha_j) [I + p_j M_j]$ and that $p_j < 1$. Let A_j and B_j be the positive measures defined by*

$$A_j = \sum_{k=0}^{\infty} \frac{p_j^{2k+1}}{2k+1} M_j^{2k+1} ,$$

$$B_j = \sum_{k=0}^{\infty} \frac{p_j^{2k+2}}{2k+2} M_j^{2k+2} .$$

Let $A = \Sigma_j A_j$ and $B = \Sigma B_j$. Then

$$P \text{ pois } B = \text{pois } A .$$

Proof. Since $p_j < 1$ the measure $I + p_j M_j$ possesses a logarithm obtainable by series expansions. This gives

$$
\begin{aligned}
\log P_j &= [\log (1 - \alpha_j)] \, I + \log [I + p_j M] \\
&= [\log (1 - \alpha_j)] \, I + A_j - B_j \\
&= \{ A_j - \| A_j \| \, I \} - \{ B_j - \| B_j \| \, I \},
\end{aligned}
$$

hence the result.

The proof of Theorem 3 proceeds by using repeatedly the fact that higher powers of p_j are small compared to p_j. The first step consists in replacing the measure B, which involves even powers of the M_j by a measure $H + K$ with

$$
H = \Sigma_j \, \beta_j \, M_j, \quad \beta_j = - \tfrac{1}{2} \log (1 - p_j^2) \, ,
$$

$$
K = \sum_j \sum_{k=0}^{\infty} \frac{p_j^{2k+2}}{2k+2} \, M_j^{2k+1} \, .
$$

The measure $H + K$ involves then only odd powers of the M_j. A second step consists in replacing M_j^{2k+1} by $(2k+1) \, M_j$ in the expansion of K. This replaces $H + K$ by a measure

$$
R = \sum_j \frac{p_j^2}{1 - p_j^2} \, M_j \, .
$$

The third step consists in replacing M_j^{2k+1} by $(2k+1) \, M_j$ in the expansion of A itself. This replaces A by $M + R$ with $M = \Sigma \, \alpha_j M_j$.

It follows then from the appropriate lemmas that the distance $\varrho \, [P \text{ pois } R, Q \text{ pois } R]$ is small, and since R is small compared to M one concludes that $\varrho \, (P, Q)$ is small.

The following lemma is intended to justify the substitution of $(2k+1) \, M_j$ to M_j^{2k+1}.

Lemma 2. *Let m be the largest integer such that $mp \leqq 1$ and let q be defined by $mq = 1$. Let*

$$
\delta (q) = 10 \, q^{\frac{2}{3}} \, [1 - q^{\frac{2}{3}}]^{-1} \, [1 - q^2]^{-\frac{1}{3}} \, .
$$

For each j let $\{ a_{j, k}; \, k = 0, 1, 2, \ldots \}$ be a sequence of nonnegative numbers satisfying for $k \geq 1$ the inequalities

$$
a_{j, 0} \, p^{2k} \geqq a_{j, k} \, .
$$

Let F and G be the measures

$$
F = \sum_j \sum_{k=0}^{\infty} (2k+1) \, a_{j, k} \, M_j \, ,
$$

$$
G = \sum_j \sum_{k=0}^{\infty} a_{j, k} \, M_j^{2k+1} \, .
$$

Then

$$
\varrho \, [\text{pois } F, \text{ pois } G] \leqq \delta (q) \, .
$$

Proof. Let $v_{j, k}; \, j = 1, 2, \ldots, k = 0, 1, 2, \ldots$ be independent random variables having Poisson distributions with

$$
E \, v_{j, k} = a_{j, k} \, .
$$

For $i = 1, 2, \ldots;\ j = 1, 2, \ldots;\ r = 0, 1, 2, \ldots$ let $\{Z_{i,j,r}\}$ be independent random variables independent of the $\{v_{j,k}\}$ and such that the distribution of $Z_{i,j,r}$ is M_j. Let $v_j = \sum_{k=0}^{\infty} (2k + 1)\, v_{j\,k}$. Then pois G is simply the distribution of

$$\sum_{j} \sum_{r=1}^{v_j} Z_{i,j,r}.$$

Let

$$n_{j,r} = \sum_{k=r}^{\infty} v_{j,k}.$$

The variable v_j may also be written

$$v_j = n_{j,0} + 2\, n_{j,1} + \ldots + 2\, n_{j,r} + \ldots.$$

Further, $n_{j,r}$ is a Poisson variable whose expectation is

$$E\, n_{j,r} = \sum_{k=r}^{\infty} a_{j,k},$$

and pois G is the distribution of

$$X_0 + (X_1 + Y_1) + (X_2 + Y_2) + \ldots + (X_r + Y_r) + \ldots$$

with

$$X_r = \sum_{j} \sum_{\xi=1}^{n_{j,r}} Z_{1,r,\xi},$$

$$Y_r = \sum_{j} \sum_{\xi=1}^{n_{j,r}} Z_{2,j,\xi}.$$

Let $\{n'_{j,r}\}$ and $\{n''_{j,r}\}$ be independent sequences of variables, independent of all the preceding ones, which have Poisson distributions with $E\, n'_{j,r} = E\, n''_{j,r} = E\, n_{j,r}$. Let X'_r and Y'_r be defined by

$$X'_r = \sum_{j} \sum_{\xi=1}^{n'_{j,r}} Z_{3,j,\xi}$$

$$Y'_r = \sum_{j} \sum_{\xi=1}^{n''_{j,r}} Z_{4,j,\xi}.$$

The set of $\{X'_r\}$ and $\{Y'_r\}$ is a set of independent variables. Further, $\mathscr{L}(X_r) = \mathscr{L}(Y_r) = \mathscr{L}(X'_r) = \mathscr{L}(Y'_r)$. Note that the X_r and Y_r are not independent and that pois F is the distribution of

$$X_0 + (X'_1 + Y'_1) + \ldots + (X'_r + Y'_r) + \ldots.$$

For $r > 0$ let S_r and T_r be respectively the sums

$$S_r = X_0 + (X'_1 + Y'_1) + \ldots + (X'_{r-1} + Y'_{r-1}) +$$
$$+ (X_r + Y_r) + (X_{r+1} + Y_{r+1}) + \ldots$$

$$T_r = X_0 + (X'_1 + Y'_1) + \ldots + (X'_{r-1} + Y'_{r-1}) +$$
$$+ (X'_r + Y'_r) + (X_{r+1} + Y_{r+1}) + \ldots.$$

The two sums S_r and T_r differ only at the rth place so that

$$S_r - T_r = \{(X_r + Y_r) - (X'_r + Y'_r)\} = (X_r - X'_r) + (Y_r - Y'_r).$$

Referring back to the definition of $n_{j,r}$ note that the part of the sum S_r which does not involve the primed variables has for distribution an exponential pois L where L is the measure

$$L = \sum_j a_{j,0} M_j + \sum_j \sum_{k=1}^{\infty} a_{j,k} M_j^{\sigma_k},$$

and where the exponents σ_k are positive. In particular

$$L \geq \sum_j a_{j,0} M_j.$$

For $r \geq 1$ the variable X_r has an exponential distribution pois L_r with

$$L_r = \sum_j b_{j,r} M_j$$

$$b_{j,r} = \sum_{k=r}^{\infty} a_{j,k} \leq \sum_{k=r}^{\infty} (a_{j,0}) \, p^{2k} = a_{j,0} \frac{p^{2r}}{1-p^2}.$$

Let m_r be the largest integer such that

$$m_r \frac{p^{2r}}{1-p^2} \leq 1.$$

It follows from these inequalities that the concentration function Γ_r of S_r is smaller than the concentration function of a sum of m_r-independent variables having the same distribution as X_r.

Let $\varepsilon > 0$ be a positive number such that $[1 - C_r(0)] > \varepsilon$. There is a number $\tau > 0$ such that

$$1 - C_r(\gamma) \geq \varepsilon \text{ for } \gamma < \tau$$

$$1 - C_r(\lambda) \leq \varepsilon \text{ for } \lambda > \tau.$$

Fixing γ and λ such that $0 < \gamma < \tau < \lambda < (3/2)\gamma$ Kolmogorov's inequality gives

$$\Gamma_r(2\lambda) \leq \frac{2}{\sqrt{m_r}} \operatorname{Int}\left[1 + \frac{\lambda}{\gamma}\right]\left[1 - C_r(\gamma)\right]^{-\frac{1}{2}} \leq \frac{6}{\sqrt{m_r \varepsilon}}.$$

Also, there is a number x such that

$$\operatorname{Prob}\left\{|X_r - x| > \frac{\lambda}{2}\right\} \leq \varepsilon.$$

This implies

$$\operatorname{Prob}\left\{|X_r - Y_r| > 2\lambda\right\} \leq 2\varepsilon.$$

Hence

$$\operatorname{Prob}\left\{|S_r - T_r| > 2\lambda\right\} \leq 4\varepsilon.$$

192 LUCIEN LE CAM

Thus

$$\varrho \left[\mathscr{L} (S_r), \mathscr{L} (T_r) \right] \leq 4\,\varepsilon + \Gamma_r (2\,\lambda)$$

$$\leq 4\,\varepsilon + \frac{6}{\sqrt{m_r \varepsilon}}\,.$$

If $m_r \left[1 - C_r (0) \right]^3 > 1$, take the value $\varepsilon = 1 - C_r (0)$. If on the contrary $m_r \left[1 - C_r (0) \right]^3 \leq 1$ there is a point x such that $m_r \{ Pr \{ \, | X_r - - x | > > \{ \, 0 \, \}^3. \, \leq 1$ This implies

$$\| \mathscr{L} (S_r) - \mathscr{L} (T_r) \| \leq \frac{8}{m_r^{\frac{1}{3}}}\,.$$

Therefore, in all cases

$$\varrho \left[\mathscr{L} (S_r);\, \mathscr{L} (T_r) \right] \leq 10\,\frac{1}{m_r^{\frac{1}{3}}}\,.$$

Applying this successively to S_1 and T_1, then $S_2 = T_1$ and T_2, and so forth, one obtains

$$\varrho \left[\mathscr{L} (S_1),\, \mathscr{L} (T_r) \right] \leq 10 \sum_{r=1}^{\infty} \left(\frac{1}{m_r} \right)^{\frac{1}{3}}\,.$$

Hence also

$$\varrho \, (\text{pois } G,\, \text{pois } F) \leq 10 \sum_{r=1}^{\infty} \left(\frac{1}{m_r} \right)^{\frac{1}{3}}\,.$$

The integer m_r is certainly as large as the largest integer m_r' satisfying $m_r'\, q^{2r} \leq 1 - q^2$. This inequality can also be written $m_r' \leq m^{2r} - m^{2r-2}$. Since this last expression is an integer it follows that $m_r \geq m^{2r} - m^{2r-2}$. Therefore,

$$\frac{1}{m_r} \leq \frac{1}{1 - q^2}\, q^{2r}$$

and finally

$$\sum_{r=1}^{\infty} \left(\frac{1}{m_r} \right)^{\frac{1}{3}} \leq \frac{1}{(1 - q^2)^{\frac{1}{3}}}\, \frac{q^{\frac{2}{3}}}{1 - q^{\frac{2}{3}}}\,.$$

This completes the proof of the lemma.

Lemma 3. *Let m be the largest integer such that $mp \leq 1$ and let $q = m^{-1}$. Let H and K be the measures*

$$K = \sum_{j} \sum_{k=0}^{\infty} \frac{p_j^{2k+2}}{2k+2}\, M_j^{2k+1}$$

$$H = \Sigma\, \beta_j\, M_j$$

with

$$\beta_j = -\tfrac{1}{2} \log (1 - p_j^2)\,.$$

Then

$$\varrho \left\{ P \text{ pois } B,\, P \text{ pois } (H + K) \right\} \leq 6\, q^{\frac{1}{3}}\,.$$

Proof. For $i = 1, 2, \ldots;\ j = 1, 2, \ldots;\ k = 1, 2, \ldots$, let $\{Z_{i,j,k}\}$ be independent variables such that $\mathscr{L}(Z_{i,j,k}) = M_j$. Let $\{v_{j,k}\}$ be independent Poisson variables independent of the $\{Z_{i,j,k}\}$ with expectations

$$E\, v_{j,k} = \frac{p_j^{2k+2}}{2k+2}.$$

Let

$$v_j' = \sum_{k=0}^{\infty} v_{j,k},$$

$$v_j'' = \sum_{k=0}^{\infty} (2k+1)\, v_{j,k},$$

$$X = \sum_{j} \sum_{r=1}^{v_j'} Z_{1,j,r},$$

$$Y = \sum_{j} \sum_{r=1}^{v_j''} Z_{2,j,r}.$$

The distribution of Y is $\mathscr{L}(Y) = \text{pois}\, K$ while the distribution of $X + Y$ is pois B.

Let S and V be two other independent variables independent of all the $v_{j,k}$ and $Z_{i,j,k}$ but such that

$$\mathscr{L}(S) - P$$

$$\mathscr{L}(V) = \mathscr{L}(Y) = \text{pois}\, K.$$

One can write

$$\text{pois}\, A = P\, \text{pois}\, B = \mathscr{L}(S + X + Y),$$

$$P\, \text{pois}\, (H + K) = \mathscr{L}(S + X + V).$$

In addition

$$K = \sum_{j} p_j \sum_{k=0}^{\infty} \frac{2k+1}{2k+2} \left(\frac{p_j^{2k+1}}{2k+1}\, M_j^{2k+1} \right) \leqq \sum_{j} p_j\, A_j \leqq pA.$$

Therefore pois K is more concentrated than pois pA. Let C be the concentration function of pois K and let Γ be the concentration function of pois A. Let $\varepsilon > 0$ be a positive number. If $C(0) \geqq 1 - \varepsilon$, there is a number a such that

$$Pr\,[Y \neq a] \leqq \varepsilon \ \text{and}\ Pr\,[V \neq a] \leqq \varepsilon.$$

Therefore

$$\| \mathscr{L}(S + X + V) - \mathscr{L}(S + X + Y) \| \leqq 4\,\varepsilon.$$

If on the contrary $C(0) < 1 - \varepsilon$ there is a number $\tau > 0$ such that

$$[1 - C(\gamma)] \geqq \varepsilon \ \text{for}\ \gamma < \tau,$$

$$1 - C(\lambda) \leqq \varepsilon \ \text{for}\ \lambda > \tau.$$

Fix γ and λ such that $\gamma < \tau < \lambda < 2\gamma$. There exists numbers b such that

$$Pr\left\{ |Y - b| > \frac{\lambda}{2} \right\} \leqq [1 - C(\lambda)] \leqq \varepsilon.$$

Hence

$$Pr\{\,|\,Y - V\,|\, > \lambda\} \leq 2\,\varepsilon\,.$$

By Kolmogorov's inequality

$$\Gamma(\lambda) \leq 2\,q^{\frac{1}{2}}\,\mathrm{Int}\left(1 + \frac{\lambda}{\gamma}\right)[1 - C\,(\gamma)]^{-\frac{1}{2}}$$

$$\leq 4\,q^{\frac{1}{2}}\,\varepsilon^{-\frac{1}{2}}\,.$$

An application of Proposition 2 gives

$$\varrho\left\{\mathscr{L}\,(S + X + V),\,\mathscr{L}\,(S + X + Y)\right\} \leq \varepsilon + 4\,q^{\frac{1}{2}}\,\varepsilon^{-\frac{1}{2}}\,.$$

Taking $\varepsilon = q^{1/3}$ gives

$$\varrho\left\{\mathscr{L}\,(S + X + V),\,\mathscr{L}\,(S + X + Y)\right\} \leq 6\,q^{\frac{1}{3}}\,,$$

and completes the proof of the lemma.

Finally, a combination of the preceding lemmas gives the following.

Lemma 4. *Let m be the largest integer such that $mp \leq 1$ and let q be defined by the equality $mq = 1$. Let $\delta\,(q)$ be the function defined in Lemma 2. Let R be the measure*

$$R = \sum_j \frac{p_j^2}{1 - p_j^2}\,M_j\,.$$

Then

$$\varrho\,[P \text{ pois } R,\, Q \text{ pois } R] \leq 2\,\delta\,(q) + 6\,q^{\frac{1}{3}}\,.$$

Proof. The measure $H + K$ of Lemma 3 involves only odd powers of the M_j. Furthermore, the coefficient of M_j is $\beta_j + (1/2)\,p_j^2$ and the coefficient of M_j^{2k+1} for $k \geq 1$ is only $[p_j^{2k+2}]\,[2\,k + 2]^{-1} \leq p^{2k}\,[\beta_j + (1/2)\,p_j^2$. It follows from Lemma 2 that

$$\varrho\,[\text{pois}\,(H + K),\,\text{pois } R] \leq \delta\,(q)\,.$$

Similarly the measure A involves only odd powers of the M_j and the coefficients satisfy the assumptions of Lemma 2. It follows that $\varrho\,[\text{pois } A,\,\text{pois } R_1] \leq \delta\,(q)$ for the measure R_1 defined by

$$R_1 = \sum_j \left(\sum_{k=0}^{\infty} p_j^{2k+1}\right) M_j = M + R\,.$$

Since $Q = \text{pois } M$, the result follows.

The proof of Theorem 3 can now be completed as follows.

Proof of Theorem 3. Let Γ be the concentration function of Q and let c be the concentration function of pois R. The measures M and R satisfy the inequality $nR \leq M$ for the integer $n = m - 1$, the largest such that $np \leq 1 - p$. Let $\varepsilon^3 = n^{-1}$. If $c\,(0) \geq 1 - \varepsilon$ then

$$\varrho\,[P,\,Q] \leq \varrho\,[P \text{ pois } R,\, Q \text{ pois } R] + 4\,\varepsilon\,.$$

If on the contrary $c\,(0) < 1 - \varepsilon$, let τ be the infimum of numbers x such that $c\,(x) \geq 1 - \varepsilon$. Let γ be a number such that $0 < \gamma < \tau < 2\,\gamma$. Kolmogorov's inequality implies $\Gamma(\tau) < 4\,\varepsilon$. Furthermore, according to Proposition 3 one can write

$$[2\,c\,(\tau) - 1]\,\varrho\,(P, Q) \leq \varrho\,[P\ \text{pois}\ R, Q\ \text{pois}\ R] + \Gamma(\tau)\,c\,(\tau)\,.$$

This inequality and Lemma 4 can be combined to obtain

$$[1 - 2\,\varepsilon]\,\varrho\,[P, Q] \leq 2\,\delta\,(q) + 6\,q^{\frac{1}{3}} + 4\,\varepsilon\,.$$

Since q is related to ε by the formula $(1 + \varepsilon^3)\,q = \varepsilon^3$ a simple computation yields the desired result.

In the following section we shall use a "splitting" technique to obtain results relative to the general case where the random variables may be often different from zero but "usually small". As part of the process it will be found convenient to separate the small values from the large ones and approximate a convolution product $\Pi\,[(1 - \alpha_k)\,M_k + \alpha_k\,N_k]$ by a product of the type $\Pi\,[(1 - \alpha_k)\,M_k + \alpha_k I]\,[(1 - \alpha_k)\,I + \alpha_k N_k]$. The method of proof of Theorem 3 shows that such an approximation will often be reasonable. More precisely one can prove the following result.

Proposition 6. *Let* $\{\alpha_k; k = 1, 2, \ldots\}$ *be a finite sequence of numbers,* $0 \leq \alpha_k \leq \alpha < 1$. *For each k let M_k and N_k be two probability measures and let*

$$L_k = (1 - \alpha_k)\,M_k + \alpha_k N_k\,,$$
$$L_k' = [(1 - \alpha_k)\,M_k + \alpha_k I]\,[(1 - \alpha_k)\,I + \alpha_k N_k]\,.$$

Let $P = \Pi L_k$ *and* $P' = \Pi L_k'$ *and let* Q *be the Poisson exponential* $Q = \text{pois}\ \Sigma\,L_k$. *Further, let* $\beta = \sup_\tau\,[\gamma\,(\tau) - C_Q\,(\tau)]^+$, *where* $\gamma\,(\tau)$ *is the minimum* $\gamma\,(\tau) = \min\,[\Gamma_P\,(\tau), \Gamma_P\,(\tau)]$ *of the moduli of continuity of P and P'. Then*

$$\varrho\,(P, P') \leq \beta + 12\,\alpha^{\frac{1}{3}}\,.$$

Proof. Let $\xi_k, \eta_k, U_k, V_k; k = 1, 2, \ldots$ be independent random variables such that $\mathscr{L}\,(U_k) = M_k$ and $\mathscr{L}\,(V_k) = N_k$. Assume also that $\mathscr{L}\,(\xi_k) = \mathscr{L}\,(\eta_k)$ and that $Pr\,[\xi_k = 1] = 1 - Pr\,[\xi_k = 0] = \alpha_k$. The measure P is the distribution of a sum $S = \Sigma\,[(1 - \xi_k)\,U_k + \xi_k V_k]$. Similarly the measure P' is the distribution of a sum $S' = \Sigma\,[(1 - \eta_k)\,U_k + \xi_k V_k]$. Let $p_j = \alpha_j\,(1 - \alpha_j)^{-1}$ and let A and B be the measures introduced in Lemma 1. If $T = \Sigma\,\xi_k U_k$, Lemma 1 gives the equality $\mathscr{L}\,(T)$ pois $B = $ pois A. Thus, the concentration functions C of pois A and $C_T)$ of $\mathscr{L}\,(T)$ satisfy the inequality $C_T \geq C$. In addition, with the notation of Lemma 2 and for $F = \Sigma\,\{p_j/(1 - p_j^2)\}\,M_j$ one can write

$$\varrho\,[\text{pois}\ F, \text{pois}\ A] \leq \delta\,(q)\,.$$

Letting C' be the concentration function of pois F, this gives

$$C_T \geq C' - 2\,\delta\,(q)\,.$$

Let $Y = \Sigma\, \xi_k U_k - \Sigma\, \eta_k U_k = \Sigma\, (\xi_k - \eta_k)\, U_k$. The foregoing inequalities imply

$$Pr\big\{\,|\,Y\,| > \lambda\big\} \leqq 2\,[1 - C_T\,(\lambda)] \leqq 2\,[1 - C'\,(\lambda)] + 4\,\delta\,(q)\ .$$

Let m_1 be the largest integer such that $m_1\,\alpha \leqq (1 - \alpha)\,(1 - 2\,\alpha)$ and let $\beta = \sup_\tau\,[\gamma\,(\tau) - C_Q\,(\tau)]^+$. Kolmogorov's inequality and Proposition 3 imply

$$\varrho\,(P, P') \leqq \beta + Pr\,[\,|\,Y\,| > \lambda] + \min\,[\Gamma_P\,(\lambda),\, \Gamma_{P'}\,(\lambda)]$$

$$\leqq \beta + 2\,[1 - C'\,(\lambda)] + 4\,\delta\,(q)$$

$$+ \frac{2}{\sqrt{m_1}}\ \mathrm{Int}\,[1 + \lambda/\gamma]\,[1 - C'\,(\gamma)]^{-\frac{1}{2}}\ .$$

Let $\varepsilon = m_1^{-1/3}$. If $C'\,(0) > 1 - \varepsilon$ take $\lambda = 0$. Otherwise select γ and λ in such a way that $\gamma < \lambda < 2\,\gamma$ and $1 - C'\,(\lambda) \leqq \varepsilon$ but $1 - C'\,(\gamma) \geqq \varepsilon$. Then

$$\varrho\,(P, P') \leqq \beta + 6\,\varepsilon + 4\,\delta\,(q)\ .$$

For the values of α for which the inequality is relevant this gives

$$\varrho\,(P, P') \leqq \beta + 12\,\alpha^{\frac{1}{3}}$$

and concludes the proof of the proposition.

5. Sums of independent variables which are usually small

In the preceding section we considered sequences $\big\{X_j;\, j = 1, 2, \ldots\big\}$ of independent random variables which differ from zero only very rarely. The purpose of the present section is to show that similar results can be obtained whenever the X_j are suitably centered and "usually small" in the sense that there is some number $\theta \geqq 0$ such that both $\sup\big\{Pr\,[|\,X_j\,| > \\ > \theta]\big\}$ and the concentration of $\Sigma\,X_j$ at θ are small.

Theorem 3 did not involve any centering of the individual variables or of their sum. In the present case, centering seems to be a necessity. We do not know which centering process gives the best results. However, the following "equal tails" centering can always be used. Let Y be an arbitrary random variable. Then Y has the same distribution as $f\,(\eta)$ where η is uniformly distributed on $[0, 1]$ and where f is a nondecreasing function continuous from the left for $\eta > 1/2$ and from the right for $\eta < 1/2$. Let α be a given number $\alpha\,\varepsilon\,(0, 1/2)$ and let a be the conditional expectation defined by

$$(1 - \alpha)\,a = \int_{\alpha/2}^{1-(\alpha/2)} f\,(\xi)\,d\xi\ .$$

Further, let $\tau\,(\alpha) = \max\big\{f\,(1 - \alpha/2) - a,\, a - f\,(\alpha/2)\big\}$ and let X be the random variable $X = Y - a$. Then the distribution of X can be written in the form

$$\mathscr{L}\,(X) = (1 - \alpha)\,M + \alpha N$$

where M is the conditional distribution of $f(\eta) - a$ given that $\alpha < 2\eta <$ $< 2 - \alpha$ and where N is the conditional distribution of $f(\eta) - a$ given that $2 \min(\eta, 1 - \eta) < \alpha$. It follows from the construction that

$$M\left[-\tau(\alpha), +\tau(\alpha)\right] = 1$$

and

$$\int x\, M\,(dx) = 0 .$$

The main theorem of the present section does not make any particular reference to the centering system used so that the foregoing equal tails procedure is only one of the available possibilities.

Consider now a finite sequence $\{X_j; j = 1, 2, \ldots\}$ of independent random variables subject to the following restrictions

a) The distribution L_j of X_j has the form

$$L_j = (1 - \alpha_j)\, M_j + \alpha_j N_j$$

where $\alpha_j \in [0, 1]$ and where M_j and N_j are probability measures.

b) $M_j\left[-\theta, +\theta\right] = 1$ and $\int x\, M_j\,(dx) = 0$.

It will be convenient to use the following notation.

1. $M'_j = \alpha_j I + (1 - \alpha_j)\, M_j$; $N'_j = (1 - \alpha_j)\, I + \alpha_j N_j$.
2. $\sigma_j^2 = \int x^2\, M'_j\,(dx) = (1 - \alpha_j) \int x^2\, M_j\,(dx)$.
3. $\sigma^2 = \Sigma_j\, \sigma_j^2$.
4. $\sigma^4 \delta = \Sigma\, \sigma_j^4$.

5. τ is a nonnegative number and G is a probability measure whose characteristic function $\widetilde{G}$ satisfies the inequality

$$\left|\,\widetilde{G}(t)\,\right| \leq \exp\left[-\frac{\tau^2 t^2}{3}\right]$$

and vanishes for $|\,\theta t\,| > 1$. For instance, one may take

$$\widetilde{G}(t) = [1 - \theta\,|\,t\,|\,]^+ \exp\left[-\frac{\tau^2 t^2}{3}\right] .$$

Proposition 7. *Assume that the variables* $\{X_j; j = 1, 2, \ldots\}$ *satisfy conditions* a) *and* b). *Let* $P = \Pi_j L_j$ *and* $P' = \Pi_j M'_j N'_j$. *Further, let* $H = (\Pi_j N'_j)$ pois $[\Sigma_j M'_j]$ *and let* G *be a probability measure satisfying condition* 5. *Then*

$$\varrho\,[PG, P'\,G] \leq \frac{10}{3\pi}\, K\,\alpha$$

and

$$\varrho\,[P'\,G, HG] \leq \frac{5}{3\pi}\, K^2\,\delta ,$$

with $\sigma^2 = K\,[\sigma^2 + \tau^2]$ *and* $\alpha = \sup \alpha_j$.

Proof. Let $F_j(\lambda)$ be the measure

$$F_j(\lambda) = [(1 - \lambda)\, I + \lambda\, M'_j]\exp\{(1 - \lambda)\,(M'_j - I)\},$$

for $0 \leq \lambda \leq 1$. Then

$$H = (\prod_j N'_j) \, [\prod_j F_j (0)] \,,$$

$$P' = (\prod_j N'_j) \, [\prod_j F_j (1)] \,.$$

An application of Taylor's formula gives

$$H - P' = (\prod_j N'_j) \int_0^1 \lambda \{\sum_k S_k (\lambda) \, (M'_k - I)^2\} \, d\lambda$$

with

$$S_k (\lambda) = \{\prod_{j \neq k} F_j (\lambda)\} \exp \{(1 - \lambda) \, (M'_k - I\}.$$

Similarly,

$$P' - P = \Sigma \, \alpha_k \, (1 - \alpha_k) \, R_k \, (M_k - I) \, (N_k - I)$$

for measures R_k defined by

$$R_k = (\prod_{j \leq k-1} L_j) \, (\prod_{j \geq k+1} M'_j \, N'_j).$$

In this expression products with empty sets of indices are, as usual, taken equal to the identity.

Let $\widetilde{M}_j$ be the Fourier transform of the measure M_j. The usual Taylor expansion gives

$$\widetilde{M}_j (t) = 1 - \frac{1}{2} \, \sigma_j^2 \, t^2 \left[1 + \frac{1}{3} \, | \, \theta t \, | \, \varepsilon \, (t) \right]$$

where $| \, \varepsilon \, (t) \, | \leq 1$. Therefore, the range of values in which

$$| \widetilde{M}_j \, (t) \, | \leq 1 - \left(\frac{1}{3}\right) \sigma_j^2 \, t^2 \text{ and } | \, M_j \, (t) - 1 \, | \leq \left(\frac{2}{3}\right) \sigma_j^2 \, t^2$$

includes at least the interval $\{t : | \, \theta t \, | \leq 1\}$. In this same range of values both R_k and $S_k \, (\lambda)$ have Fourier transforms smaller in modulus than $\exp \, [- \, (1/3) \sum_{j \neq k} \sigma_j^2 t^2]$. Substitution in the expressions of $P' - P$ and $P' - H$ gives

$$[\widetilde{P} \, (t) - \widetilde{P}' \, (t) \, | \leq \frac{20}{9} \, \alpha \sigma^2 \, t^2 \exp \left[- \frac{1}{3} \, \sigma^2 \, t^2\right]$$

and

$$| \, \widetilde{P}' \, (t) - \widetilde{H} \, (t) \, | \leq \frac{10}{27} \, \sigma^4 \, t^4 \exp \left[- \frac{1}{3} \, \sigma^2 t^2\right]$$

for all values of t such that $| \, \theta t \, | \leq 1$.

Since the Fourier transform of PG and $P'G$ and HG are all integrable, direct application of the inversion formula gives

$$\varrho \, [PG, P'G] \leq \frac{20}{9 \, \pi} \, \alpha \int_0^\infty \exp \left[- \frac{1}{3} \, (\sigma^2 + \tau^2) \, t^2\right] \sigma^2 \, t \, dt$$

$$= \frac{10}{3 \, \pi} \, \alpha \, \frac{\sigma^2}{\sigma^2 + \tau^2}$$

Similarly

$$\varrho\,[P'G,\,HG] \leq \frac{10}{27\,\pi}\,\delta \int\limits_0^\infty \exp\left[-\frac{1}{3}\,(\sigma^2 + \tau^2)\,t^2\right]\sigma^4\,t^3\,dt$$

$$= \frac{5}{3\,\pi}\,\delta\left[\frac{\sigma^2}{\sigma^2 + \tau^2}\right].$$

This concludes the proof of Proposition 7.

Corollary. *Under the conditions of Proposition 7, let $Q = \mathrm{pois}\,(\Sigma\,L_j)$.*
Then

$$\varrho\,[P,\,Q] \leq 2\,\varrho\,[QG,\,HG] + \frac{10}{3\,\pi}\,[2\,\alpha K + \delta K^2] + \frac{3}{2}\,\Gamma\,[(12.8)\,\theta + (3.6)\,\tau]\,,$$

where Γ is the minimum of the moduli of continuity of P and Q.

Proof. Let G be the measure such that

$$\widetilde{G}\,(t) - [1 - |\,\theta t\,|\,]^+ \exp\left[-\frac{\tau^2\,t^2}{2}\right].$$

If λ is a number such that $C_G\,(\lambda) = 3/4$ then Propositions 3 and 7 give

$$\varrho\,(P,\,Q) \leq 2\,\varrho\,[QG,\,HG] + \frac{10}{3\,\pi}\,[2\,\alpha K + \delta K^2] + \frac{3}{2}\,\Gamma\,(\lambda)\,.$$

To obtain the result it is sufficient to check that the number λ defined
by $C_G\,(\lambda) = 3/4$ is smaller than $(12.8)\,\theta + (3.6)\,\tau$. This is easily verifiable.

Theorem 4. *Let $\{X_j;\,j = 1,\,2,\,\ldots\}$ be a finite sequence of independent
random variables. Let θ be a nonnegative number. Let L_j be the distribution
of X_j. Assume that there exist numbers $\alpha_j \in [0,\,1]$ and probability measures
M_j and N_j such that*

    1. *$L_j = (1 - \alpha_j)\,M_j + \alpha_j N_j$.*

    2. *$M_j\,[-\,\theta,\,+\,\theta] = 1$ and $\int x\,M_j\,(dx) = 0$.*

*Let $P = \Pi\,L_j$ be the distribution of the sum $\Sigma\,X_j$ and let Q be the accom-
panying exponential*

$$Q = \exp\{\Sigma_j\,(L_j - I)\}.$$

Then

$$\varrho\,[P,\,Q] \leq 51\,\alpha^{\frac{1}{3}} + 15\,\Gamma^{\frac{2}{3}}\,(\theta) \leq 51\,\{\alpha^{\frac{1}{3}} + [D\,(\theta)]^{-\frac{2}{3}}\}\,,$$

*where $\alpha = \sup \alpha_j$, where Γ is the minimum of the moduli of continuity of
P and Q and where*

$$D^2\,(\theta) = \sum_{j}\,\int\limits_{x \neq 0}\,\min\left[1,\,\frac{x^2}{\theta^2}\right]L_j\,(dx)\,.$$

Proof. In the inequality of the corollary to Proposition 7 substitute
for $\varrho\,[Q,\,H]$ the bound $25\,\alpha^{1/3}$ abtained in Theorem 3. The coefficient K
is smaller than unity and $\delta\,K^2$ is smaller than (θ^2/τ^2), since $\sigma_j^2 \leq \theta^2$.
To prove the first inequality it is sufficient to replace τ by a suitably
selected value, which renders (θ^2/τ^2) small without increasing $\Gamma\,[(12.8)\,\theta\,+$

$+ (3.6)\ \tau]$ excessively. The value defined by

$$\frac{\theta^3}{\tau^3} = (2.7)\ \Gamma(\theta)$$

gives the indicated result.

The second inequality follows from the first by direct application of Proposition 5. This concludes the proof of the theorem.

The inequality of the corollary to Proposition 7 can often be used directly to better advantage than final inequalities of Theorem 4. This will occur in particular if δ itself is already known to be small. In this case the introduction of the supplementary variable τ becomes unnecessary.

For example, the inequality implies the following result, originally due to PROHOROV [4].

Proposition 8. *Assume that the variables $\{Y_j;\ j = 1, 2, \ldots, n\}$ are independent and identically distributed. Let a be the centering constant obtained from the "equal tails" procedure corresponding to $\alpha = 2\ n^{-(1/3)}$. Let $X_j = Y_j - a$. Let $L_j = \mathscr{L}(X_j)$ and let $P = \Pi\ L_j$ and $Q = \mathrm{pois}\ [\Sigma\ L_j]$. Then*

$$\varrho\ [P, Q] \leq (132)\ n^{-\left(\frac{1}{3}\right)}.$$

Proof. In this case Theorem 2 implies that $\varrho\ [Q, H] \leq 3\ \alpha$. Also δ is an average of quantities (σ_j^2/σ^2) which are all equal to $(1/n)$. Thus the corollary of Proposition 7 implies

$$\varrho\ [P, Q] \leq 6\ \alpha + \frac{20}{3\pi}\ \alpha + \frac{10}{3\pi}\ \frac{1}{n} + \frac{3}{2}\ \Gamma\ [(12.8)\ \theta].$$

In addition $D^2(\theta)$ is at least equal to $\frac{1}{2}\ n\alpha = n^{2/3}$. Thus $\Gamma\ [(12.8)\ \theta] \leq$ $\leq 78\ n^{-|1|3|}$. The result follows by simple arithmetic.

The following theorem is intended to indicate that a slight change in the centering constants does not affect the possibility of exponential approximation to a very large extent. Here, "slight" will be interpreted to mean that the conditional truncated expectations a_k are such that the sum $\Sigma\ a_k^2$ is small compared to $\theta^2\ D^2(\theta)$.

Theorem 5. *Let $\{X_k;\ k = 1, 2, \ldots\}$ be a finite sequence of independent random variables having distributions $P_k = \mathscr{L}(X_k)$. Let θ be a nonnegative number. Assume that there exist numbers $\alpha_k \in [0, 1]$ and probability measures M_k and N_k such that $P_k = (1 - \alpha_k)\ M_k + \alpha_k N_k$ and such that M_k is entirely concentrated in an interval of length at most θ.*

Let $P = \Pi_k\ P_k = \mathscr{L}\ [\Sigma\ X_k]$ and let Q_0 be the exponential $Q_0 = \mathrm{pois}\ F$ with $F = \Sigma_k\ P_k$. Then

$$\varrho\ [P, Q_0] \leq (51)\ [\alpha^{\frac{1}{3}} + \beta^{\frac{1}{3}} + \varepsilon^{\frac{1}{3}}]$$

with $\alpha = \sup \alpha_k$ and with β and ε defined by $\beta D^2 = 1$ and $\varepsilon\theta^2 D^2 = \Sigma\ a_k^2$ for

$$D^2 = \sum_k \int_{x \neq 0} \min\left[1, \frac{x^2}{\theta^2}\right] P_k\ (dx)$$

and

$$a_k = \int x \, M_k \, (dx) \, .$$

Proof. Let $\{Y_{k,i}; k = 1, 2, \ldots; i = 1, 2, \ldots\}$ be independent random variables having distributions $\mathscr{L}(Y_{k,i}) = L_k = \mathscr{L}[X_k - a_k]$. Let $\{\nu_k; k = 1, 2, \ldots\}$ be ordinary Poisson random variables which are mutually independent and independent of the $Y_{k,i}$ and are such that $E \nu_k = 1$. Let $L = \Sigma_k L_k$. The Poisson exponential pois L is the distribution of

$$T = \sum_k \sum_{i=1}^{\lambda_k} Y_{k,i} \, .$$

The exponential Q_0 is the distribution of $T_0 = T + \varDelta$ for $\varDelta = \Sigma_k \nu_k a_k$. This variable $\varDelta$ itself has an exponential distribution such that

$$a = E \varDelta = \Sigma \, a_k \, ,$$

$$b^2 = E \, (\varDelta - a)^2 = \Sigma \, a_k^2 \, .$$

Let S be the shift of magnitude a and let G be the distribution used in the proof of Theorem 4. According to the inequalities of Proposition 7 and Theorem 3 one can write

$$\varrho \, [PG, SQG] \leqq 25 \, \alpha^{\frac{1}{3}} + \frac{5}{3\pi} \, \lceil 2 \, \alpha C + \delta C^2 \rceil \, .$$

According to Chebyshev's inequality $\lambda^2 \, Pr\{ \, |\varDelta - a \, | \geqq \lambda\} \leqq b^2$. Therefore

$$\varrho \, [SQG, Q_0 G] \leqq \frac{b^2}{\lambda^2} + C_{Q_0 G} \, (\lambda) \, .$$

An application of Proposition 3 yields

$$\varrho \, [P, Q_0] \leqq 50 \, \alpha^{\frac{1}{3}} + \frac{10}{3\pi} \, [2 \, \alpha C + \delta C^2] + \frac{3}{2} \, \varGamma_0 \, [(12.8) \, \theta + (3.6) \, \tau]$$

$$+ \, 2 \frac{b^2}{\lambda^2} + 2 \, C_{Q_0 G} \, (\lambda) \, ,$$

with $\varGamma_0$ equal to the minimum $\varGamma_0 = \min [\varGamma_P, \varGamma_{Q_0}]$ of the moduli of continuity of P and Q_0. The modulus $\varGamma_0 \leqq C_{Q_0}$ may be bounded through the use of Proposition 5. The result follows then by appropriate choice of τ and λ.

In the statement of Theorem 4, we have used the "modulus of continuity" instead of the more usual concentration function for the simple reason that, with this small precaution, Theorem 4 essentially reduces to Theorem 3 when θ is taken equal to zero.

To give an intuitive content to Theorem 4, let us assume for simplicity that the variables X_j are symmetrically distributed. In this case, recentering is unnecessary, so that for each $\theta \geqq 0$ the number $\alpha (\theta) = \sup Pr\{ \, |X_j| > \theta\}$ gives an indication of the spread of the variable X_j.

Thus, letting $\varepsilon\left[\{L_j\}\right]$ be the number

$$\varepsilon\left[\{L_j\}\right] = \inf_{\theta}\left\{51\left[\alpha\left(\theta\right)\right]^{\frac{1}{3}} + 15\,\Gamma^{\frac{2}{3}}\left(\theta\right)\right\},$$

Theorem 4, which states that

$$\varrho\left\{\prod_j L_j,\ \exp\left[\sum_j \left(L_j - I\right)\right]\right\} \leqq \varepsilon\left[\{L_j\}\right]$$

is an explicit expression of the fact that the convolution product $\Pi\, L_j$ is close to the corresponding exponential whenever each of the terms L_j contributes little to the product as a whole.

Although there is no hope that Theorem 4 be in any way the best of its kind one may inquire about the possibility that such a neglibility of each L_j compared to $\Pi\, L_j$ be necessary for the validity of the approximation. For instance, is it true that if the sequence $\left\{L_j; j = 1, 2, \ldots\right\}$ varies in such a way that $\varrho = \varrho\left\{\Pi\, L_j,\ \exp\left[\Sigma\left(L_j - I\right)\right]\right\}$ tends to zero then $\varepsilon\left[\{L_j\}\right]$ also converges to zero?

It will be shown, in a separate publication, that, for symmetric measures, the conditions $\varepsilon\left[\{L_j\}\right] \to 0$, $P\left\{\Pi L_j,\ \exp \Sigma\left(L_j - I\right)\right\} \to 0$ and $\sup_j P\left\{\prod_{k \neq j} L_k, \Pi\, L_k\right\}$ are all equivalent.

References

[1] Kolmogorov, A. N.: Deux théorèmes asymptotiques sur les sommes de variables indépendentes. Teor. Veroyatnost. i Primenen. 1, 427 (1956).

[2] — On the approximation of distributions of sums of independent summands by infinitely divisible distributions. Sankhya. 2, 159 (1963).

[3] Meshalkin, L. D.: On the approximation of distributions of sums by infinitely divisible laws. Teor. Veroyatnost. i Primenen. 6, 257 (1961).

[4] Prohorov, Yu. V.: On the uniform limit theorems of A. N. Kolmogorov. Teor. Veroyatnost. i Primenen. 5, 103 (1960).

[5] Kolmogorov, A. N.: Sur les propriétés des fonctions de concentration de M. P. Lévy. Ann. Inst. H. Poincaré. 16, 27 (1958).

[6] Rogozin, B. A.: On an estimate of the concentration function. Teor. Veroyatnost. i Primenen. 6, 103 (1961).

[7] Erdös, P.: On a lemma of Littlewood and Offord. Bull. Amer. Math. Soc. 51, 898 (1945).

[8] Khintchine, A. Ya: Asymptotische Gesetze der Wahrscheinlichkeitsrechnung. Ergebnisse der Mathematik und ihrer Grenzgebiete. Berlin: Springer 1933.

[9] Prohorov, Yu V.: Asymptotic behavior of the binomial distribution. Uspehi Mat. Nauk. 8, 135 (1953).

[10] Le Cam, L.: An approximation theorem for the Poisson binomial distribution. Pacific J. Math. 10, 1181 (1960).

Limit Solutions of Sequences of Statistical Games

By **Jerzy Łoś**

Mathematical Institute, Polish Academy of Sciences, Warszawa
and
University of California, Berkeley

0. The present paper deals with the so-called "empirical Bayes approach" proposed and studied by Herbert Robbins. His papers, especially paper [3], have inspired us to show the notion of "empirical Bayes approach" fits into the framework of game theory by means of simple operations on games. We show it by examples of testing two simple statistical hypotheses (Example 5; the original version of this example was presented by H. Robbins in [3]). At the end we prove some easy lemmas concerning the relations between Bayes, minimax and limit solutions.

1. Let $\Gamma = \langle S, T, a \rangle$ be a two-person game in the normal form.

We shall assume that the payoff function $a\,(s, t)$ is bounded on $S \times T$ and that the whole game Γ is positive for the first player who controls the strategies in S (i.e., the first player intends to maximize the payoff). In what follows we shall always discuss the game Γ from the point of view of the first player.

In order to avoid discussion of ε-approximation of optimalities, we shall always assume that on considered sets the inf and sup are attained at some elements. This allows us to write simply min and max. Also for reasons of simplicity, we shall not discuss the question of measurability of functions under consideration. We shall assume that in all sets under discussion (e.g., S and T), we have selected some fields of subsets, and all considered probability measures are defined on the corresponding fields. Also all considered functions will be assumed measurable with respect to the appropriate fields.

By a "valuation" we understand every real-valued, bounded function φ defined on S.

Let us mention here two important examples of valuations:

(i) the min valuation m, defined by $m\,(s) = \min_{t \in T} a\,(s, t)$; and

(ii) the ν-Bayes valuation β_ν, defined by $\beta_\nu\,(s) = \int_T a\,(s, t)\,d\nu\,(t)$, where ν is a probability measure in T.

For any given valuation φ, the φ-solution is defined as the element s_0 of S for which $\varphi\,(s_0) = \max_{s \in S} \varphi\,(s)$.

If the notion of a limit is introduced in S, so that S becomes at least an L-space, then the continuity of a valuation φ means simply the ordinary continuity in the sense of HEINE (for the definition of L-spaces see [7], p. 84).

Since to every s in S there corresponds a function $s(t) = a(s, t)$ on T, there are several possibilities of forming an L-space from S. Thus, for example, we may define $s_n \to s$ if $s_n(t) \to s(t)$ asymptotically, or almost everywhere with respect to a given measure v on T, or simply everywhere, or even uniformly in T. In each of these cases we obtain from S a different L-space.

By an S-sub-game of Γ (or simply sub-game) we understand a game $\Gamma_0 = \langle S_0, T, a \rangle$, where S_0 is a nonempty subset of S. Here T is the same set as in Γ and a is the same payoff function restricted to $S_0 \times T$.

If S is an L-space and $\Gamma_n = \langle S_n, T, a \rangle$ is a sequence of sub-games of Γ, then by the derived game we understand the game $\Gamma'' = \langle \{S_n\}', T, a \rangle$, where $\{S_n\}' = S' = \{s;$ there exists s_n in S_n with $s_n \to s\}$. The only condition for the existence of derived game Γ'' is that the set S' be not empty.

Lemma 1. *If S is an L-space and φ is a continuous valuation, then for every sequence of sub-games Γ_n for which the derived sub-game exists, we have*

$$\lim_{n \to \infty} \max_{s \in S_n} \varphi(s) \geq \max_{s \in S'} \varphi(s).$$

For the proof let us assume that $\varphi(s') = \max_{s \in S'} \varphi(s)$ and $s'_n \to s'$ with s'_n in S_n. Then $\max_{s \in S'} \varphi(s) \geq \varphi(s'_n) \to \varphi(s')$, which shows that the asserted relation holds.

Lemma 2. *Let S be an L-space and let φ be a continuous valuation. If the derived game of the sequence Γ_n exists, then for every sequence s_n of φ-solutions of the games Γ_n, the condition $s_n \to s$ implies that s is a φ-solution of Γ''.*

Lemma 2 is an easy corollary of Lemma 1.

The following lemmas refer to special valuations and to the concept of a dominant strategy. By the latter we mean a strategy d in S with $a(d, t) \geq a(s, t)$, for all s in S and all t in T.

Lemma 3. *If an L-space in S is introduced by means of the asymptotic convergence with respect to a given measure v in T, and the dominant strategy d of Γ belongs to the derived game Γ'' (i.e., to the set S'), then for the β_v-solutions s_n of the games Γ_n we have $s_n \to d$.*

Let us notice first that in this case β_v is a continuous valuation. From the assumption we have $\max_{s \in S_n} \beta_v(s) \leq \beta_v(d) = \max_{s \in S} \beta_v(s)$. Applying Lemma 1 we obtain

$$\lim_{n \to \infty} \max_{s \in S_n} \beta_v(s) = \lim_{n \to \infty} \beta_v(s_n) = \beta_v(d),$$

where the s_n are β_ν-solutions of Γ_n. This means that

$$\int_T a\,(s_n,\,t)\;d\nu\,(t) \to \int_T a\,(d,\,t)\;d\nu\,(t).$$

Since $a\,(s_n,\,t) \leqq a\,(d,\,t)$, for all t in T, we conclude that $a\,(s_n,\,t) \to a\,(d,\,t)$ asymptotically. This shows that $s_n \to d$, as required.

Now let us assume that there exists a dominant strategy d in the game Γ and let us denote by Δ the game $\langle S,\,T,\,a\,(s,\,t) - a\,(d,\,t)\rangle$. Following this notation, we denote by Δ_n the sub-game of Δ based on the same set of strategis as the sub-game Γ_n of Γ.

Lemma 4. *If an L-space is introduced in S by means of the uniform convergence and the dominant strategy d of Γ belongs to the derived game Γ'', then for m-solutions s_n of the games Δ_n, we have $s_n \to d$.*

For the proof let $s'_n \to d$, where s'_n are in S_n. Then

$$\max_{t \in T}\,[a\,(d,\,t) - a\,(s'_n,\,t)] \to 0\,.$$

Since

$$\max_{s \in S_n}\,\min_{t \in T}\,[a\,(s,\,t) - a\,(d,\,t)] = \min_{t \in T}\,[a\,(s_n,\,t) - a\,(d,\,t)] \geqq$$

$$\geqq \min_{t \in T}\,[a\,(s'_n,\,t) - a\,(d,\,t)]\,,$$

hence

$$\max_{t \in T}\,[a\,(d,\,t) - a\,(s_n,\,t)] \leq \max_{t \in T}\,[a\,(d,\,t) - a\,(s'_n,\,t)] \to 0\,.$$

This shows that $s_n \to d$, as asserted.

2. If $\Gamma = \langle S,\,T,\,a\rangle$ is a game, then by a statistical information for Γ we understand a function λ which assigns to every t in T a probability measure λ_t in a given set X. We assume that all λ_t are defined on the same field of subsets of X.

By the statistical game over Γ with information λ, we understand the game $\Gamma\,(\lambda) = \langle F,\,T,\,A\rangle$, where F is the set of all measurable functions $f\colon X \to S$ and the payoff function A is defined by

$$A\,(f,\,t) = \int_X a\,[f\,(x),\,t]\;d\lambda_t\,(x)\,.$$

Let us consider some examples of statistical games, which will be used in the sequel.

Example 1. If $X = T$ and $\lambda_t\,(X_0) = 1$ iff t is in X_0, then we say that $\Gamma\,(\lambda)$ is a game with perfect information. Every function $f\colon T \to S$ may be used as a strategy and $A\,(f,\,t) = a\,[f\,(t),\,t]$.

For every game with perfect information there exists a dominant strategy, namely a function $d\colon T \to S$ such that $a\,[d\,(t),\,t] = \max_{s \in S} a\,(s,\,t)$.

Example 2. Suppose that the first player, instead of playing pure strategies from S, uses mixed strategies. This means that he selects first a probability measure μ in S and then selects an element s in S at random according to μ. Now the original game changes into a randomized game $\langle M, T, b \rangle$ where M is the set of all probability measures μ in S and

$$b\,(\mu,\,t) = \int_S a\,(s,\,t)\,d\mu\,(s)\,.$$

When applying perfect information to this game we obtain the game $\langle \Psi, T, B \rangle$, where Ψ is the set of all mappings $\Psi\colon T \to M$ and

$$B\,(\Psi,\,t) = b\,(\Psi_t,\,t) = \int_S a\,(s,\,t)\,d\Psi_t\,(s)\,.$$

The game so defined will be called the universal statistical game over Γ.

Example 3. Let us assume now that $\Gamma = \langle S, T, a \rangle$ is a game with finite T and let us denote by N the set of all distributions $\nu = \langle \ldots, \nu_t, \ldots \rangle$ over T. If the second player uses mixed strategies, we have the game $\langle S, N, A \rangle$, where

$$A\,(s,\,\nu) = \sum_{t\in T} a\,(s,\,t)\,\nu_t\,.$$

Now let λ be a statistical information for this game, such that $X = T^k$ and $\lambda_\nu = \nu^k$ where T^k denotes the cartesian kth power of T and ν^k the independent kth power of ν in T^k. The statistical information λ transforms $\langle S, N, A \rangle$ into $\langle F, N, b \rangle$, where F is the set of all functions $f\colon T^k \to S$ and

$$\sum_{t_1,\,\ldots,\,t_k \in T^{k+1}} a\,[f\,(t_1,\,\ldots,\,t_k),\,t]\,\nu_{t_1},\,\ldots\,\nu_{t_k}\,\nu_t\,.$$

Example 4. If a simple hypothesis H_0 is to be tested against a simple alternative H_1, and s_0, s_1 are actions to be taken according to the outcome of the test, then we are concerned with a simple 2×2 game, provided that the losses due to errors $a\,(s_1,\,H_0) = -l_0$ and $a\,(s_0,\,H_1) = -l_1$ are given. The strategies are $S = (s_0,\,s_1)$, $T = (H_0,\,H_1)$ and the payoff function is defined by $l_0,\,l_1$ and $a\,(s_0,\,H_0) = a\,(s_1,\,H_1) = 0$. The application of statistical information $\lambda = (\lambda_0,\,\lambda_1)$ in a set X, to this game, transforms it into a statistical game with functions $f\colon X \to (s_0,\,s_1)$ as strategies of the first player ("Statistician") and the payoff function defined as $A\,(f,\,H_0) = -l_0\,\lambda_0\,[f^{-1}\,(s_1)]$, $A\,(f,\,H_1) = -l_1\,\lambda_1\,[f^{-1}\,(s_0)]$.

Example 5. Let us suppose that, as in Example 4, two hypotheses H_0 and H_1 are to be tested with statistical information $\lambda = (\lambda_0,\,\lambda_1)$ and the previously defined loss function a.

If we assume that the second player ("Nature") selects states H_0, H_1 at random with probabilities p for H_0 and $1-p$ for H_1, then the former game $\langle F, (H_0,\,H_1), A \rangle$ changes into another one which has mixed strategies for the second player, namely $\langle F, [0, 1], A] \rangle$, where

$$A\,(f,\,p) = -\left\{l_0\,p\lambda_0\,[f^{-1}\,(s_1)] + l_1\,(1-p)\,\lambda_1\,[f^{-1}\,(s_0)]\right\}\,.$$

Now let us suppose that the hypotheses H_0, H_1 are to be tested several times in a row, with the same mixed strategy p of Nature. Then, when testing for the $(n + 1)$st time, n independent outcomes of the experiment on X are known, each governed by the same probability measure $\Lambda_p = p\lambda_0 + (1 - p)\,\lambda_1$. In this case we have new information, which will be denoted by Λ_p^n; it has the form of nth independent power of Λ_p defined on X^n. The corresponding statistical game is $\langle F^{(n)}, [0, 1], B^{(n)}\rangle$ where F^n consists of the functions $f^{(n)}: X^n \to F$. Denoting $\langle x_1, \ldots, x_n\rangle$ by x and $f^n\,(x_1, \ldots, x_n)$ by $f_x^{(n)}$ we have

$$B^{(n)}\,(f^{(n)}, p) = \int_{X^n} (f_x^n, p)\, d\Lambda_p^n\,(x) = \int_{X^{(n)}} -\left\{l_0\,p\lambda_0\,[(f_x^{(n)})]^{-1}\,(s_1) + \right.$$
$$\left. + l_1\,(1 - p)\,\lambda_1\,[(f_x^{(n)})]^{-1}\,(s_0)\right\}\,d\Lambda_p^n\,(x)\ .$$

Let us remark once more that $\langle F^{(n)}, [0, 1], B^{(n)}\rangle$ is a statistical game over $\langle F, [0, 1], A\rangle$.

3. Let $\lambda^{(n)}$ be a sequence of information for a game $\Gamma = \langle S, T, a\rangle$, $\lambda^{(n)}$ being defined in $X^{(n)}$. Using $\lambda^{(n)}$ we may construct a sequence of statistical games over $\Gamma: \Gamma_n = \Gamma\,(\lambda^{(n)}) = \langle F^{(n)}, T, A^{(n)}\rangle$. By definition $F^{(n)}$ consists of the functions $f^{(n)}: X^{(n)} \to S$, and

$$A^{(n)}\,(f^{(n)}, t) = \int_{X^{(n)}} a\,[f^{(n)}\,(x, t)]\, d\lambda_t^{(n)}\,(x)\ .$$

We shall say that the sequence Γ_n has a limit solution, if there exists a sequence of functions $l^{(n)}$, each in the corresponding $F^{(n)}$, such that $A^{(n)}\,(l^{(n)}, t) \to \max_{s\,\in\,S} a\,(s, t)$. The last limit may be understood as an asymptotic limit with respect to a given measure ν in T, or a limit everywhere, or even a uniform limit. In each of these cases the sequence $l^{(n)}$ will be called respectively an asymptotic, everywhere or uniform limit solution.

Example 6. If λ is a statistical information for Γ in the set X, then taking for every natural k and t in T, the kth independent product of λ_t, we get a sequence of information λ^k in X^k, each of them for the same game Γ. This sequence yields a sequence of statistical games $\Gamma\,(\lambda^k)$ called the power sequence of λ over Γ.

Example 7. In Example 3 a power sequence over the game $\langle S, N, A\rangle$ was constructed. The initial information was $\lambda^{(1)} = \nu$ for every ν in N. This sequence has everywhere the limit solution $l^{(n)}\,(t_1, \ldots, t_n) = s_0$, where s_0 is defined as an element in S satisfying the condition $A\,(s_0, \eta) = \max_{s\,\in\,S} A\,(s, \eta)$ for the empirical distribution of $t_1, \ldots, t_n$

$$\eta_t = \frac{\text{number of } i\text{'s with } t_i = t,\ i = 1, \ldots, n}{n}\ .$$

Example 8. In Example 5 we also have a power sequence over the game $\langle F, [0, 1], A\rangle$. We start with information $\Lambda_p = p\lambda_0 + (1 - p)\,\lambda_1$. It has been shown by H. ROBBINS [3] that this sequence has an everywhere limit solution.

4. Let $\Gamma = \langle S, T, a \rangle$ be a game and let $\Gamma(\lambda) = \langle F, T, A \rangle$ be the statistical game over Γ with information λ in X.

By definition, elements of F are measurable functions $f: X \to S$ and $A(f, t) = \int_X a[f(x), t] \, d\lambda_t(x)$.

Every function f in F transforms every measure λ_t defined in X into a measure $\chi_t(f)$ defined in S, in such a way that for measurable subsets Y of S: $[\chi_t(f)](Y) = \lambda_t[f^{-1}(Y)]$. Thus $\chi(f)$ is a function which maps T into M, where M is the set of measure in S. It follows that $\chi(f)$ is a strategy of the universal statistical game $\Gamma: \langle \Psi, T, B \rangle$, defined in Example 2. Moreover, we have

$$A(f, t) = \int_X a[f(x), t] \, d\lambda_t(x) = \int_S a(s, t) \, d[\chi_t(f)](s) = B[\chi(f), t] \ .$$

In this way we embedded $\Gamma(\lambda)$ into the universal statistical game over Γ. Let us denote by $\Gamma(\lambda) = \langle \chi F, T, B \rangle$, the sub-game of the universal statistical game over Γ, which is the χ-image of $\Gamma(\lambda)$.

If we are concerned with a sequence of statistical games $\Gamma''(\lambda^{(n)})$ over Γ, then by embedding each game of this sequence into the universal statistical game, we obtain a sequence of sub-games $\chi\Gamma(\lambda^{(n)})$. We may apply all concepts and lemmas of Section 1 to this sequence. Thus the existence of a limit solution means that the dominant strategy belongs to the derived game of the sequence $\chi\Gamma(\lambda^{(n)})$. Here the limit solution is to be understood in the same sense as for the derived game.

Theorem 1. *If there exists a v-asymptotic limit solution of the sequence $\Gamma(\lambda^{(n)})$, then the sequence of β_v-solution of $\Gamma(\lambda^{(n)})$ has the same property.*

Applying Lemma 4 we obtain

Theorem 2. *If there exists a uniform limit solution of the sequence $\Gamma(\lambda^{(n)})$, then the sequence of m-solutions of the corresponding Δ_n is also a uniform limit solution for $\Gamma(\lambda^{(n)})$.*

Neither of these theorems inform us about the existence of limit solutions, but give us a method of finding them, provided they do exist. No analogous theorem is known for the everywhere limit solutions.

References

[1] KURATOWSKI, K.: Topologie 1, Monografie Matematyczne 1958.

[2] ROBBINS, H.: An empirical Bayes approach to statistics. Proc. Third Berkeley Symp. Math. Stat. Probab. **I**, 157 (1956).

[3] — The empirical Bayes approach to testing statistical hypotheses. Rev. Int. Stat. Inst. **31**, 195 (1963).

Some Remarks on Statistical Inference

By E. J. G. PITMAN

The Johns Hopkins University

Probability and Statistics have always been highly contentious parts of science. They have been the occasion of many disputes about principles, conducted not always with cool scientific detachment and decorum. What has been disputed about is not mathematics, not pure mathematics. It is true that, from time to time, mathematicians, even eminent mathematicians, have resisted or scoffed at the introduction of new ideas or new areas of investigation in pure mathematics, for example, continuous functions without derivatives, the Lebesgue integral, and a proper treatment of the real number system, and that even today some older applied mathematicians speak of such things as abstract algebra and topology as frills. But these are merely the grumbles of old age and laziness, and they never last long. It is in the applications of mathematics, in applied mathematics, especially where experimental testing or verification is difficult, or for the time being impossible, that the violent arguments take place. Relativity in its early days is an example. In probability it is the principles governing the applications, especially in statistics, about which there are great differences of opinion.

Since the publication in 1933 of KOLMOGOROFF's book on the fundamentals of probability, "Grundbegriffe der Wahrscheinlichkeitsrechnung," we have been able to separate off the pure mathematics of probability from the applied mathematics. We are agreed that as a branch of pure mathematics probability is concerned with a set of elements, a space Ω, with a measure P such that $P(\Omega) = 1$. A real random variable X is a real, measurable function defined on Ω, and the mathematical expectation or mean value of X is simply the integral of X over Ω with respect to the measure P. $E(X) = \int X \, dP$. We are by no means agreed as to the applications of this branch of pure mathematics.

The oldest and perhaps still the most important, application of probability theory is to repeated trials. It is this application that has been our main guide in the choice of the fundamental principles of probability, and it is the one we come back to when in doubt or difficulty. Here the practical or engineering definition of the probability of an event A is the expected or estimated relative frequency of A in a large number of trials. Sometimes we think of probability as something like strength of opinion about the truth of a proposition or an hypothesis. Here the repeated trials model seems not immediately appropriate. Can we combine the two uses?

There are great differences of opinion about the theory of estimation and the theory of statistical tests, which we now group together as the theory of statistical inference, the art or science of drawing valid conclusions from the observed values of random variables. In both the closely related operations of statistical estimation and statistical testing we pass from observed values of random variables to the distributions of those random variables. We attempt to make statements, or to test statements, about more or less unknown distributions. That is the abstract mathematical formulation of the problem. From the practical point of view, we may think of arguing from effects to their causes. Until this century the main method of dealing with such a problem was to calculate a conditional *a posteriori* distribution of causes (distributions) from a given or assumed *a priori* distribution. Now we have significance tests, confidence intervals, and fiducial distributions; but we have become dissatisfied with these more modern tools and we are giving serious consideration to the use of *a priori* distributions.

Of the three mathematicians whose achievements in probability we are celebrating in this seminar, it is BAYES who would best have appreciated our present-day perplexities. Let me remind you of what he did. His famous essay, published in the *Philosophical Transactions of the Royal Society*, is entitled:

> *An Essay towards solving a Problem in the Doctrine of Chances. By the late Rev. Mr. Bayes, F.R.S. communicated by Mr. Price, in a letter to John Canton, A.M. F.R.S.*

It was read on the 23rd of December, 1763, nearly two hundred years ago.

There is an introductory letter of six pages by RICHARD PRICE, in which he stresses the importance of the problem discussed by BAYES, and comments on his solution. Toward the end of the letter he writes "Mr. BAYES has thought fit to begin his work with a brief demonstration of the general laws of chance. His reason for doing this, as he says in his introduction, was not merely that his reader might not have the trouble of searching elsewhere for the principles on which he has argued, but because he did not know whither to refer him for a clear demonstration of them. He has also made an apology for the peculiar definition he has given of the word *chance* or *probability*. His design herein was to cut off all dispute about the meaning of the word, which in common language is used in different senses by persons of different opinions, and according as it is applied to *past* or *future* facts."

The essay itself starts with the heading

Problem

Given the number of times in which an unknown event has happened and failed: *Required* the chance that the probability of its happening in a single trial lies somewhere between any two degrees of probability that can be named.

It is divided into two sections. Section I is devoted to the general laws of chance. It consists of seven numbered definitions, and seven numbered propositions. The definitions are definitions of inconsistent, contrary, fail, determined, and independent as applied to events, and definitions of probability and chance. The important one is what PRICE refers to as "the peculiar definition he has given of the word *chance* or *probability*". This is Definition 5.

The *probability of any event* is the ratio between the value at which an expectation depending on the happening of the event ought to be computed, and the value of the thing expected upon its happening.

About BAYES' definition of probability PRICE writes as follows.

"But whatever different senses it may have, all (he observes) will allow that an expectation depending on the truth of any *past* fact, or the happening of any *future* event, ought to be estimated so much the more valuable as the fact is more likely to be true, or the event more likely to happen. Instead therefore, of the proper sense of the word *probability*, he has given that which all will allow to be its proper measure in every case where the word is used."

Thus if A is an event whose happening brings a prize a, and if the estimated value of this uncertain prize be b, then $P(A) = b/a$. If $a = 1$, $P(A) = b$, the "expectation" or estimated value of a unit prize. In other words $P(A)$ is the stake for a bet on A which brings a total return of unity if A happens. This definition has the great advantage that we can apply it to hypotheses as well as to events in repeated trials. Many statisticians are now adopting it, or something like it.

Of the seven propositions in Section I, the two most important are the third and the fifth.

Proposition 3. The probability that two subsequent events will both happen is a ratio compounded of the probability of the 1st, and the probability of the 2d on supposition the 1st happens.

As we write it

$$P(AB) = P(A) P(B \mid A).$$

Proposition 5. If there be two subsequent events, the probability of the 2d b/N and the probability of both together P/N, and it being 1st discovered that the 2d event has happened, from hence I guess that the 1st event has also happened, the probability that I am in the right is P/b.

As we would express it, this is

$$P\,(A \mid B) = P\,(AB)/P\,(B)\,.$$

This is Bayes' theorem. It is essentially the same as Proposition 3. Bayes seems to think that they require different proofs because of the order of events, A first and B second. Note that Price speaks of a past *fact* and a future *event*. Bayes seems reluctant to speak of the probability of a past event, and speaks of "the probability that I am in the right" if I guess that the first event has happened.

In Section II Bayes gives the solution of the proposed problem. He first describes a physical setup which will give rise to a problem of this kind.

Postulate I. I suppose the square table or plane $ABCD$ to be so made and leveled, that if either of the balls O or W be thrown upon it, there shall be the same probability that it rests upon any one equal part of the plane as another, and that it must necessarily rest somewhere upon it.

The ball W is thrown first, and a line QS drawn through the point where it rests, parallel to the edge AD. The ball O is then thrown n times. In any throw the event M happens if O rests between QS and AD. Thus we have an event M whose probability λ is truly a random variable with a uniform distribution from 0 to 1. Bayes obtains the *a posteriori* distribution of λ, that is, the conditional distribution of λ given that in n trials M has happened p times and failed q times.

He then goes on to argue in favor of using the same *a posteriori* distribution "in relation to any event concerning the probability of which nothing at all is known antecedently to any trials made or observed concerning it". This he does in a scholium, or explanatory remark.

Scholium. From the preceding proposition it is plain, that in the case of such an event as I there call M, from the number of times it happens and fails in a certain number of trials, without knowing anything more concerning it, one may give a guess whereabouts it's probability is, and, by the usual methods computing the magnitudes of the areas there mentioned, see the chance that the guess is right. And that the same rule is the proper one to be used in the case of an event concerning the probability of which we absolutely know nothing antecedently to any trials made concerning it, seems to appear from the following consideration; viz. that concerning such an event I have no reason to think that, in a certain number of trials, it should rather happen any one possible number of times than another. For, on this account, I may justly reason concerning it as if its probability had been at first unfixed, and then determined in such a manner as to give me no reason to think that, in a certain number of trials, it should rather happen any one possible number of times than another. But this is exactly the case of the event M. ...

This essay was a wonderful achievement. It contained a new and exciting definition of probability, BAYES' Theorem in Proposition 5, the solution of an interesting and carefully chosen problem on inverse or conditional probability, and the bold and frankly stated proposal to apply the same calculations to any event concerning the probability of which nothing is known. As to the last, we may not always agree to do as BAYES proposes; but we must admire his boldness and his scrupulous care in stating his assumptions.

In the last forty years *a priori* distributions of unknown parameters have been largely neglected because of the difficulty of justifying a particular choice. Sometimes, however, it is possible to estimate an *a priori* distribution. Let us consider a simple example of the sort of problem discussed by ROBBINS.

Suppose that

$$x_1, x_2, \ldots, x_n$$

are observed values of n independent random variables, each of which has a probability density function that is either f or g. We wish to decide for each x_r whether its density function is f or g, and in such a way that the expected number of correct decisions will be a maximum. We shall suppose that in each case the probabilities of the density being f or g are p, $1-p$ respectively. Let H_1 denote the statement that the density is f, and H_2 the statement that the density is g. Denote by

$$J_r, \qquad (r = 1, 2, 3, \ldots, 2^n)$$

the 2^n possible sequences of n symbols, each H_1 or H_2. The true state of affairs is specified by some J_r, and so is the compound decision. Let $N(J_r, J_s)$ denote the number of agreements between J_r and J_s.

Write

$$x = (x_1, x_2, \ldots, x_n),$$

and denote the decision by J_D. We want to define D as a function of x which takes the values $1, 2, \ldots, 2^n$, in such a way that

$$E\{N(J_D, J_A)\}$$

is a maximum, where J_A denotes the correct decision.

$$E\{N(J_D, J_A)\} = \sum_{r=1}^{2^n} P(J_r) \int N(J_D, J_r) \, \varnothing_r \, dx,$$

where $\varnothing_r$ is the density function of x under J_r,

$$= \int \sum_{r=1}^{2^n} P(J_r) N(J_D, J_r) \, \varnothing_r \, dx.$$

For any given value of x we must choose D so that

$$\sum_{r=1}^{2^n} P(J_r)\, N(J_D, J_r)\, \emptyset_r(x)$$

is a maximum. If we know p, this is possible, although likely to be tedious.

When we do not know p, we can easily estimate it. The observed values

$$x_1,\ x_2,\ \ldots,\ x_n$$

may be regarded as n independently observed values of a random variable with density function

$$pf + (1 - p)\, g\,.$$

The logarithm of the likelihood function is

$$L(p) = \sum_r \log\left\{pf(x_r) + (1 - p)\, g(x_r)\right\}.$$

Then

$$\frac{\partial L}{\partial p} = \sum_r \frac{f(x_r) - g(x_r)}{pf(x_r) + (1 - p)\, g(x_r)}\,,$$

which always decreases as p increases, because $\partial^2 L/\partial p^2 < 0$. Thus the maximum likelihood estimate is uniquely determined, and we may use it in the above calculations.

Statistics being, as Fisher said, essentially a branch of Applied Mathematics, we should be guided in our choice of principles and methods by the practical applications. All actual sample spaces are discrete, and all observable random variables have discrete distributions. The continuous distribution is a mathematical construction, suitable for mathematical treatment, but only an approximation. We develop our fundamental concepts, principles, and methods in the study of discrete distributions. In the case of a discrete sample space, it is easy to understand and appreciate the practical or experimental significance and value of conditional distributions, the likelihood principle, the principles of sufficiency and conditionality, and the method of maximum likelihood. These are then extended to more general distributions, and the extension is supported by mathematical theorems.

Let us consider the likelihood principle. If the sample space is discrete, its points may be enumerated,

$$x_1,\ x_2,\ \ldots\,.$$

Suppose that the probability of observing the point x_r is $f(x_r, \theta)$, and that θ is unknown. An experiment is performed, and the outcome is the point x of the sample space. All that the experiment tells us about θ is that an event has occurred, the probability of which is $f(x, \theta)$, which is a function of θ. For deciding between two values θ_1, θ_2 of θ, all that the experiment gives us is the ratio $f(x, \theta_1)/f(x, \theta_2)$ of the probabilities for the two values of θ, the likelihood ratio. We may call two sample points x_1, x_2 equivalent if all likelihood ratios are the same for x_1 as for x_2, that is, if for all θ_1, θ_2,

$$\frac{f(x_1, \theta_1)}{f(x_1, \theta_2)} = \frac{f(x_2, \theta_1)}{f(x_2, \theta_2)}$$

that is, if

$$f(x_1, \theta) = kf(x_2, \theta) ,$$

where k is the same for all θ.

Two likelihood functions related in this way, by simple proportionality, may be said to be equivalent. The likelihood principle is that all the information about θ given by the experiment is contained in the likelihood function, and that equivalent likelihood functions give the same information. For a discrete sample space, where the observed sample has a nonzero probability, this seems sound; but when we come to a continuous distribution, we are dealing with an observation which has zero probability, and the principle seems not so intuitively appealing. However, the extension to the continuous distribution seems reasonable when we regard the continuous distribution as the limit of discrete distributions. Also, this extension is supported by the Neyman-Pearson theorem. This says that for testing the simple hypothesis H_1 that a random variable has a distribution with probability density f_1 against the alternative hypothesis H_2 that the density function is f_2, the most powerful test of given size is based on the likelihood ratio $f_2(x)/f_1(x)$. There is a critical value c; sample points which reject H_1 have $f_2(x)/f_1(x) \geqq c$, and those which do not reject H_1 have $f_2(x)/f_1(x) \leqq c$.

Suppose that a statistician has to make n tests of this kind, each of a simple hypothesis H_1 against a simple alternative H_2, the experiment and the hypotheses in one test having no connection with those in another. Suppose also that for the average size of his tests (probability of rejecting H_1 when true), he wishes to maximize the average power (probability of rejecting H_1 when H_2 is true). To do this he must use the same critical value c for the likelihood ratio in all the tests. It is sufficient to prove this for $n = 2$.

Denote the first sample space by A. Let f_1, f_2 be the densities with respect to the measure μ under H_1, H_2 respectively. Any test of H_1 is specified by a critical function φ whose value $\varphi(x)$ at the sample point x is the probability of rejecting H_1 when x is observed. Denote the second sample space by B, and regard A and B as disjoint. Let g_1, g_2 be the densities, ν the measure, and ψ the critical function in B. We have to choose φ and ψ so that for given

$$\int_A \varphi f_1 \, d\mu + \int_B \psi g_1 \, d\nu ,$$

the value of

$$\int_A \varphi f_2 \, d\mu + \int_B \psi g_2 \, d\nu$$

is a maximum.

Let $C = A \cup B$. In the space C define the functions h_1, h_2, w, by

$$h_1 = f_1,\ h_2 = f_2,\ w = \varphi \text{ at points in } A \ ,$$
$$h_1 = g_1,\ h_2 = g_2,\ w = \psi \text{ at points in } B \ .$$

For any set E in C, which is the union of a measurable set R in A and a measurable set S in B, define

$$\tau(E) = \mu(R) + \nu(S) \ .$$

The sum of the sizes of the two tests may be written as

$$\int_C w h_1 \, d\tau \ ,$$

and the total power with respect to the two H_2 is

$$\int_C w h_2 \, d\tau \ .$$

For a given value of the first integral we have to maximize the second. By the Neyman-Pearson theorem this requires

$$w(z) = \begin{cases} 1 \text{ when } h_2(z) > ch_1(z) \ , \\ 0 \text{ when } h_2(z) < ch_1(z) \ , \end{cases}$$

where c is a constant. It follows that

$$\varphi(x) = \begin{cases} 1 \text{ when } f_2(x) > cf_1(x) \ , \\ 0 \text{ when } f_2(x) < cf_1(x) \ , \end{cases}$$

$$\psi(y) = \begin{cases} 1 \text{ when } g_2(y) > cg_1(y), \\ 0 \text{ when } g_2(y) < cg_1(y). \end{cases}$$

This proves the statement.

This result suggests that, to some extent, a particular numerical value of a likelihood ratio means the same thing, whatever the experiment. This is obvious when we have *a priori* probabilities for H_1 and H_2; for if the prior odds of H_2 against H_1 are p_2/p_1, the posterior odds are $(p_2/p_1)(f_2/f_1)$, and all that comes from the experiment is the ratio f_2/f_1.

Approximation of Improper Prior Measures
by Prior Probability Measures

By CHARLES STEIN

Stanford University

I. Introduction

It is known that, ordinarily, any admissible decision procedure for a statistical decision problem is, in a fairly strong sense, a limit of Bayes procedures, and in many cases such a limit must be a formal Bayes procedure with respect to a prior measure which may be improper (unbounded). This paper is, for the most part, a non-rigorous attempt at finding, in a reasonably explicit form, conditions for such a formal Bayes procedure to be admissible. In addition, a small amount of effort is devoted to the question of approximation of improper prior measures by prior probability measures without regard to the question of admissibility.

In section 2 we introduce a distance between probability measures, to be applied to posterior distributions, first suggested by KAKUTANI [3]. We then define the separation of one prior probability measure, $\Pi^{(1)}$ from a second, $\Pi^{(2)}$ to be the expected value, under $\Pi^{(1)}$, of the squared separation between the posterior distributions corresponding to $\Pi^{(1)}$ and $\Pi^{(2)}$. Roughly speaking, this measures how far off you will be on the average if you compute posterior distribution using $\Pi^{(2)}$ as the prior distribution when $\Pi^{(1)}$ is the true prior distribution. When this separation is small we shall say that $\Pi^{(2)}$ is well approximated by $\Pi^{(1)}$. The separation, and the notion of good approximation, are, in general, highly asymmetric. The definition can be extended to the case where $\Pi^{(2)}$ is an improper (unbounded) prior measure subject only to (2.9). A result of STONE [13] concerning the approximate consistency of confidence intervals based on Student's t with a Bayes approach is reformulated, without proof, in this terminology.

In section 3, a rough argument is given to indicate that, for sufficiently smooth decision problems, the excess risk incurred by using $\Pi^{(2)}$ as the prior distribution when $\Pi^{(1)}$ is the true prior distribution is bounded by the separation of $\Pi^{(1)}$ from $\Pi^{(2)}$. It looks as if this approximation is reasonably sharp if the decision problem is sufficiently complicated, for example, if all parameters are required to be estimated with roughly equal emphasis. In section 4, an approximation is computed for the separation between $\Pi^{(1)}$ and $\Pi^{(2)}$ in the case where $\Pi^{(2)}$ is an improper prior measure and $\Pi^{(1)}$ is absolutely continuous with respect to $\Pi^{(2)}$ with a density that is smooth (on a large scale). A remainder is given explicitly, but not in a convenient form. In section 5, it is shown that, when the approximations of sections 2 and 4 are valid, the author's necessary and sufficient condition for admissibility [12] yields a more readily usable sufficient condition that is related to a generalized exterior Dirichlet problem. In section 6, this reduced problem is studied, very incompletely.

This work suffers from two major defects. First, as I have already mentioned, the main results must be considered to be conjectures rather than theorems, since the arguments in sections 3 and 5 are not rigorous. The lack of rigor in section 3 should not be difficult to overcome, but a rigorous treatment of the argument in section 5 will require a careful examination of the remainder in section 4. Perhaps more important is the fact that the admissibility of a statistical procedure is of little direct practical interest. Nevertheless, in many special cases, the study of admissibility yields useful information. For most problems concerning multivariate normal distributions, traditional procedures exploiting the invariance of the problem under translations or linear transformations (or both) seem to be inadmissible. If the characterization of admissible procedures suggested by this paper can be made precise and somewhat more explicit, it may help us find procedures, better than the usual ones, that can seriously be recommended in practice. This should be important in those problems, constituting a large proportion of statistical practice, in which classical large sample theory is not fully applicable because of the large number of unknown parameters.

I am indebted to many people who listened patiently when I presented some of these ideas in a way that was even more chaotic than the presentation in this paper. In particular, the basic idea is, in part, due to J. Kiefer.

2. A definition of separation between two prior distributions

Let X be a set (the sample space), $\mathscr{B}$ a σ-algebra of subsets of $\mathscr{X}$, λ a σ-finite measure on $\mathscr{B}$, $\mathscr{T}$ a set (the parameter space), $\mathscr{C}$ a σ-algebra of subsets of $\mathscr{T}$, and p a $\mathscr{B}\mathscr{C}$-measurable non-negative valued function on $\mathscr{X} \times \mathscr{T}$ such that

$$\int p\,(x \mid \theta)\, d\lambda\,(x) = 1\,, \tag{1}$$

for all $\theta \in \mathscr{T}$. We suppose we are to observe a random point X in $\mathscr{X}$ distributed, for some unknown $\theta \in \mathscr{T}$, according to the probability density $p\,(\cdot \mid \theta)$ with respect to λ, so that the probability that X is in a set $B \in \mathscr{B}$ is

$$P_\theta\,(B) = \int_B p\,(x \mid \theta)\, d\lambda\,(x)\,. \tag{2}$$

We are interested in making inferences about θ or making decisions having consequences depending on θ. For various reasons, as indicated in later sections, we are interested in the case where θ is a random point of $\mathcal{T}$, in which case we write Θ rather than θ. Then (2) is to be interpreted as the conditional probability that $X \in B$ given $\Theta = \theta$. If Θ is distributed according to the probability measure (prior distribution) Π on $\mathcal{C}$, the conditional probability that $\Theta \in C$ given $X = x$ is

$$\Pi_x(C) = \frac{\int_C p(x \mid \theta) \, d\Pi(\theta)}{\int p(x \mid \theta) \, d\Pi(\theta)}. \tag{3}$$

This is also called the posterior probability that $\Theta \in C$ (given $X = x$).

Following KAKUTANI [3], we define a squared distance $\delta(m_1, m_2)$ between two probability measures, m_1 and m_2 on $\mathcal{C}$ by

$$\delta(m_1, m_2) = \int \left[\sqrt{\frac{dm_1}{dm}} - \sqrt{\frac{dm_2}{dm}} \right]^2 dm, \tag{4}$$

where m is any measure with respect to which both m_1 and m_2 are absolutely continuous, for example $m = m_1 + m_2$, and $\frac{dm_i}{dm}$ is the Radon-Nikodym derivative of m_i with respect to m. We shall also write (4) in the abbreviated form

$$\delta(m_1, m_2) = \int [\sqrt{dm_1} - \sqrt{dm_2}]^2 = 2\left(1 - \int \sqrt{dm_1 \, dm_2}\right) \tag{5}$$
$$= 2[1 - \alpha(m_1, m_2)],$$

say. For two prior probability measures, $\Pi^{(1)}$ and $\Pi^{(2)}$ on $\mathcal{C}$, we define the separation $\delta^*(\Pi^{(1)}, \Pi^{(2)})$ between $\Pi^{(1)}$ and $\Pi^{(2)}$ by

$$\delta^*(\Pi^{(1)}, \Pi^{(2)}) = E^{(1)} \, \delta(\Pi_X^{(1)}, \Pi_X^{(2)}), \tag{6}$$

where the $\Pi_X^{(i)}$ are the posterior distributions defined by (3), and $E^{(1)}$ denotes expectation under the assumption that [with X given Θ distributed according to (2)] Θ is distributed according to $\Pi^{(1)}$. Similarly we define

$$\alpha^*(\Pi^{(1)}, \Pi^{(2)}) = E^{(1)} \, \alpha(\Pi_X^{(1)}, \Pi_X^{(2)}), \tag{7}$$

so that

$$\delta^*(\Pi^{(1)}, \Pi^{(2)}) = 2[1 - \alpha^*(\Pi^{(1)}, \Pi^{(2)})]. \tag{8}$$

We shall want to extend the definitions (6) and (7) to certain cases where $\Pi^{(2)}$ is not a probability measure. If Π is a non-zero measure on $\mathcal{C}$ for which

$$\int p(x \mid \theta) \, d\Pi(\theta) < \infty \, a.e. \cdot (\lambda), \tag{9}$$

we shall call Π a prior measure, improper if $\Pi(\mathcal{T}) = \infty$. If Π is a prior measure, formula (3) defines a probability measure Π_x for all x except for a set of λ-measure 0. This has been used by many people, in particular by JEFFREYS [2], and will be called the formal posterior measure given $X = x$ under Π. Then (6) and (7) can be applied to the case where $\Pi^{(2)}$ is a prior measure, possibly improper, but, of course, $\Pi^{(1)}$ must be a probability measure. Roughly speaking, $\delta^* (\Pi^{(1)}, \Pi^{(2)})$ is a measure of how far off the posterior probability computed according to $\Pi^{(2)}$ will be on the average if $\Pi^{(1)}$ is the true prior distribution. It can be proved that δ^* is a strictly convex function of its first argument, that is, for any prior measure $\Pi^{(2)}$, distinct prior probability measures $\Pi^{(1)}$ and $\Pi^{(3)}$ and $0 < \alpha < 1$,

$$\delta^* [\alpha \Pi^{(1)} + (1 - \alpha) \Pi^{(3)}, \Pi^{(2)}] <$$
$$< \alpha \delta^* (\Pi^{(1)}, \Pi^{(2)}) + (1 - \alpha) \delta^* (\Pi^{(1)}, \Pi^{(3)}) . \tag{10}$$

The proof will be omitted since it is straightforward, though tedious.

Following Kraft [5], let us compute the squared distance δ between two normal distributions in order to provide a scale for comparison, and also for later use. For $i = 1, 2$ let m_i be a univariate normal distribution with mean ξ_i and variance σ_i^2. Then

$$\alpha (m_1, m_2) = \int \sqrt{dm_1 \, dm_2}$$
$$= \frac{1}{\sqrt{2 \pi \, \sigma_1 \, \sigma_2}} \int \exp \left\{ - \frac{1}{4 \sigma_1^2} (\theta - \xi_1)^2 - \frac{1}{4 \sigma_2^2} (\theta - \xi_2)^2 \right\} d\theta \tag{11}$$
$$= \sqrt{\frac{2 \sigma_1 \sigma_2}{\sigma_1^2 + \sigma_2^2}} \exp - \frac{(\xi_1 - \xi_2)^2}{4 (\sigma_1^2 + \sigma_2^2)} ,$$

and

$$\delta (m_1, m_2) = 2 \left[1 - \sqrt{\frac{2 \sigma_1 \sigma_2}{\sigma_1^2 + \sigma_2^2}} \exp - \frac{(\xi_1 - \xi_2)^2}{4 (\sigma_1^2 + \sigma_2^2)} \right]$$
$$\approx \frac{1}{2} \left[\frac{\sigma_2}{\sigma_2} + \frac{\sigma_2}{\sigma_1} - 2 \right] + \frac{(\xi_1 - \xi_2)^2}{2 (\sigma_1^2 + \sigma_2^2)} , \tag{12}$$

when $\left| \frac{\sigma_1}{\sigma_2} - 1 \right|$ and $\frac{(\xi_1 - \xi_2)^2}{\sigma_1^2 + \sigma_2^2}$ are small.

Now let us look at an example, which brings out clearly the asymmetry of the separation δ^*. Suppose we are to observe X, normally distributed with unknown mean θ and variance 1. Let $\Pi^{(1)}$ and $\Pi^{(2)}$ be normal prior distributions for θ with mean 0 and variances σ_1^2 and σ_2^2 respectively. Then $\Pi_X^{(i)}$ is normal with mean $\frac{\sigma_i^2}{\sigma_i^2 + 1} X$ and variance $\frac{\sigma_i^2}{\sigma_i^2 + 1}$ so that, by (12),

$$\delta (\Pi_X^{(1)}, \Pi_X^{(2)}) = 2 (1 - a \, e^{- \frac{b}{2} X^2}) , \tag{13}$$

where

$$a = \sqrt{\frac{2}{\frac{\sigma_1}{\sigma_2} \sqrt{\frac{\sigma_2^2 + 1}{\sigma_1^2 + 1}} + \frac{\sigma_2}{\sigma_1} \sqrt{\frac{\sigma_1^2 + 1}{\sigma_2^2 + 1}}}} , \tag{14}$$

and

$$b = \frac{\left(\frac{\sigma_1^2}{\sigma_1^2 + 1} - \frac{\sigma_2^2}{\sigma_2^2 + 1} \right)^2}{2 \left(\frac{\sigma_1^2}{\sigma_1^2 + 1} + \frac{\sigma_2^2}{\sigma_2^2 + 1} \right)} = \frac{(\sigma_1^2 - \sigma_2^2)^2}{2 (\sigma_1^2 + 1) (\sigma_2^2 + 1) (2 \sigma_1^2 \sigma_2^2 + \sigma_1^2 + \sigma_2^2)} . \tag{15}$$

But, under $\Pi^{(1)}$, X is normally distributed with mean 0 and variance $\sigma_1^2 + 1$. Thus

$$
\begin{aligned}
\delta^*\left(\Pi^{(1)}, \Pi^{(2)}\right) &= E^{(1)}\,\delta\left(\Pi_X^{(1)}, \Pi_X^{(2)}\right) \\
&= E^{(1)}\,2\left(1 - a\,e^{-\frac{b}{2}X^2}\right) \\
&= 2\left(1 - a\,\frac{1}{\sqrt{2\pi}}\int e^{-\frac{1}{2}y^2\,[1+(\sigma_1^2+1)\,b]}\,dy\right) \\
&= 2\left(1 - \frac{a}{\sqrt{1+(\sigma_1^2+1)\,b}}\right).
\end{aligned}
\tag{16}
$$

Now suppose

$$
1 < \sigma_1 < \sqrt{\sigma_2}\,. \tag{17}
$$

Then $a \approx 1$, and

$$
(\sigma_1^2 + 1)\,b \approx \frac{1}{4\,\sigma_1^2} \approx 1\,, \tag{18}
$$

so that, by (16)

$$
\delta^*\left(\Pi^{(1)}, \Pi^{(2)}\right) < 1\,. \tag{19}
$$

However

$$
\delta^*\left(\Pi^{(2)}, \Pi^{(1)}\right) = 2\left(1 - \frac{a}{\sqrt{1+(\sigma_2^2+1)\,b}}\right), \tag{20}
$$

and

$$
(\sigma_2^2 + 1)\,b \approx \frac{\sigma_2^2}{4\,\sigma_1^4} < 1 \tag{21}
$$

so that

$$
\delta^*\left(\Pi^{(2)}, \Pi^{(1)}\right) \approx 2\,. \tag{22}
$$

It seems desirable to indicate the interaction of these notions with that of a group of transformations leaving a statistical problem invariant, although it would be inappropriate to attempt a lengthy systematic treatment here. With $(\mathscr{X}, \mathscr{B}, \lambda, \mathscr{T}, \mathscr{C}, p)$ as before, let $\mathscr{G}$ be a group of $1-1$ $(\mathscr{B}, \mathscr{B})$ measurable transformations g of $\mathscr{X}$ onto $\mathscr{X}$ such that (for each such g) there is given a $(\mathscr{C}, \mathscr{C})$ measurable transformation $\bar{g}\colon \mathscr{T} \to \mathscr{T}$ with the property that, for any θ, if X is distributed according to the density $p\,(\cdot\,|\,\theta)$ then gX is distributed according to $p\,(\cdot\,|\,\bar{g}\theta)$. We suppose this correspondence $g \to \bar{g}$ is a homomorphism of $\mathscr{G}$ onto a group $\overline{\mathscr{G}}$ of transformations of $\mathscr{T}$ [as it must be if $p\,(\cdot\,|\,\theta_1) \neq p\,(\cdot\,|\,\theta_2)$ for $\theta_1 \neq \theta_2$]. We say that $\overline{\mathscr{G}}$ operates transitively on $\mathscr{T}$ if, for any $\theta_1, \theta_2 \in \mathscr{C}$, there exists $\bar{g} \in \overline{\mathscr{G}}$ such that $\bar{g}\theta_1 = \theta_2$. We recall that a confidence procedure may be taken to be a $\mathscr{B}$-measurable function τ on $\mathscr{X}$ to $\mathscr{C}$, with the interpretation that, after observing X we assert that $\theta \in \tau\,(X)$. This procedure is said to have exact confidence coefficient α if, for all $\theta \in \mathscr{T}$,

$$
P\{\theta \in \tau\,(X)\} = \alpha\,. \tag{23}
$$

We first want to discuss, in the case where $\mathscr{G}$ operates transitively on $\mathscr{T}$, a relation between a certain formal posterior probability and confidence procedures first pointed out by PITMAN [*11*] in the special case of a group consisting of translations or scale changes or both. The conditions imposed here are unnecessarily restrictive but they cover many interesting applications.

Let us first recall some results concerning invariant measures in separable, locally compact topological groups (see for example, LOOMIS [*8*], Chapter 6). Let $\mathscr{A}$ be the σ-algebra of Borel subsets of such a group $\mathscr{G}$. Then there exists a non-zero, σ-finite measure μ on $\mathscr{A}$ that is left-invariant, that is

$$\mu\,(gA) = \mu\,(A) \tag{24}$$

for all $g \in \mathscr{G}$ and $A \in \mathscr{A}$, and any other such left invariant measure is a positive multiple of μ. We define

$$\varDelta\,(g_1) = \frac{d\mu\,(gg_1)}{d\mu\,(g)}\,, \tag{25}$$

which exists and is independent of g because of the near-uniqueness of μ. Then $\varDelta$ is a continuous homomorphism of $\mathscr{G}$ into the multiplicative group of positive real numbers. Let ν be defined by

$$\nu\,(Ag) = \mu\,(A^{-1})\,. \tag{26}$$

The ν is right-invariant, that is

$$\nu\,(Ag) = \nu\,(A)\,, \tag{27}$$

and also

$$\frac{d\nu}{d\mu}\,(g) = \varDelta\,(g^{-1})\,. \tag{28}$$

If $\mathscr{G}$ is abelian or compact $\varDelta = 1$, and in the compact case, μ may be taken to be a probability measure. The invariant measures μ and the corresponding functions $\varDelta$ are given below for some groups that arise frequently in statistical problems:

(i) If $\mathscr{G}$ is the group of all translations of a k-dimensional real linear space $\mathscr{R}^k$, μ is Lebesgue measure and $\varDelta = 1$.

(ii) If $\mathscr{G}$ is the group of all linear transformations g of $\mathscr{R}^k$, then, with g expressed as a matrix

$$d\mu\,(g) = \frac{\Pi\,d\,g_{ij}}{|\det g\,|^k}\,, \tag{29}$$

and $\varDelta = 1$.

(iii) If $\mathscr{G}$ is the group of all affine transformations $g = (h, a)$ of $\mathscr{R}^k$ operating by

$$gx = (h, a)\,x = hx + a\,, \tag{30}$$

then

$$d\mu\,(g) = \frac{\Pi\,d\,h_{ij}\,\Pi\,d\,a_i}{|\det h\,|^{k+1}}\,, \tag{31}$$

and

$$d\nu\,(g) = \frac{\Pi\,d\,h_{ij}\,\Pi\,d\,a_i}{|\det h\,|^k}\,, \tag{32}$$

so that

$$\Delta\left(g\right) = \frac{d\mu}{d\nu}\left(g\right) = \frac{1}{\mid \det h\mid} \; .$$

(33)

(iv) If $\mathscr{G}$ is the multiplicative group of all $k \times k$ lower triangular matrices g (satisfying $g_{ij} = 0$ for $j > i$), then

$$d\mu\left(g\right) = \frac{\prod\limits_{j \leq i} dg_{ij}}{\prod\mid g_{ii}\mid^{i}} \; ,$$

(34)

$$d\nu\left(g\right) = \frac{\prod\limits_{j \leq i} dg_{ij}}{\prod\mid g_{ii}\mid^{k+1-i}} \; ,$$

(35)

and

$$\Delta\left(g\right) = \frac{d\mu}{d\nu}\left(g\right) = \prod\mid g_{ii}\mid^{\,k+1-2i} \; .$$

(36)

Now let $\mathscr{V}$ be a set with a σ-algebra $\mathscr{D}$ of subsets, and suppose that $\mathscr{X} = \mathscr{G} \times \mathscr{V}$ and $\mathscr{B} = \mathscr{A}\mathscr{D}$ and $\mathscr{G}$ operates on $\mathscr{X}$ by

$$g_1 x = g_1\left(g, y\right) = \left(g_1 g, y\right) \; .$$

(37)

We suppose $\mathscr{G}$ induces a group $\overline{\mathscr{G}}$ of transformations of $\mathscr{T}$ as indicated earlier, and we also suppose that $\overline{\mathscr{G}}$ operates transitively on $\mathscr{T}$. We write $X = (G, Y)$ and define the probability measure ϱ on $\mathscr{D}$ by

$$\varrho\left(D\right) = P_\theta\left\{Y \in D\right\} .$$

(38)

Because of the transitivity of $\overline{\mathscr{G}}$, this is independent of θ. Choose $\lambda = \mu\varrho$. Now let $\mathscr{H}$ be the subgroup of $\mathscr{G}$ consisting of all $h \in \mathscr{G}$ for which $\overline{h}\,\theta_0 = \theta_0$ where θ_0 is an arbitrary point of $\mathscr{T}$ fixed for the remainder of this discussion, and suppose $\mathscr{H}$ is compact. Let Π be the measure induced in $\mathscr{T}$ by the right-invariant measure ν in $\mathscr{G}$, that is, with $y: \mathscr{G} \to \mathscr{T}$ defined by

$$\gamma g = \overline{g}\,\theta_0,$$

(39)

let Π be defined by

$$\Pi\left(C\right) = \nu\left(\gamma^{-1}_y\, C\right) \; .$$

(40)

Let $\tau: \mathscr{X} \to \mathscr{C}$ be chosen so that

$$\tau\left(gx\right) = \overline{g}\left(\tau x\right) \quad \text{for all } g \in \mathscr{G} \text{ and } x \in \mathscr{X}$$

(41)

and

$$\Pi_x\left\{\tau\left(x\right)\right\} = \alpha \quad \text{for all } x \; .$$

(42)

We shall show that then

$$P_\theta\left\{\theta \in \tau\left(X\right)\right\} = \alpha \quad \text{for all } \theta \; .$$

(43)

With $x = (g, y)$,

$$\alpha = \Pi_x\left\{\tau\left(x\right)\right\} = \frac{\int_{\theta \in \tau\,(g,\, y)} p\left(g, y \mid \theta\right) d\,\Pi\left(\theta\right)}{\int p\left(g, y \mid \theta\right) d\,\Pi\left(\theta\right)}$$

$$= \frac{\int_{g_1 \theta_0 \in \tau\,(g,\, y)} p\left(g, y \mid g_1\,\theta_0\right) d\nu\left(g_1\right)}{\int p\left(g, y \mid g_1\,\theta_0\right) d\nu\left(g_1\right)}$$

(44)

$$= \frac{\int_{\theta_0 \in \tau (g_1^{-1} g, \, v)} p\,(g_1^{-1} g, v \mid \theta_0)\, dv\,(g_1)}{\int p\,(g_1^{-1} g, v \mid \theta_0)\, dv\,(g_0)}$$

$$= \int_{\theta_0 \in \tau (g_2, \, v)} p\,(g_2, v \mid \theta)\, d\mu\,(g_2)\,,$$

for all v. It follows that

$$P_{\theta_0}\{\theta_0 \in \tau (X)\} = \int d\varrho\,(v) \int_{\theta_0 \in \tau\,(g,v)} p\,(g, v \mid \theta)\, d\mu\,(g) = \alpha\,. \tag{45}$$

But, by the transitivity assumption, any θ can be expressed as $\bar g\theta_0$ for some $g \in \mathcal{G}$, so that

$$P_\theta\{\theta \in \tau (X)\} = P_{\bar g\theta_0}\{\bar g\theta_0 \in \tau (X)\} \tag{46}$$

$$= P_{\bar g\theta_0}\{\theta_0 \in \tau (g^{-1} X)\} = P_{\theta_0}\{\theta_0 \in \tau (X)\} = \alpha\,.$$

We note that the defining conditions (41) and (42), and the conclusion (43) do not depend on the explicit representation $\mathcal{X} = \mathcal{G} \times \mathcal{V}$, so that we need only be assured of the possibility of such a representation, and need not obtain it explicitly. It is easy to weaken the conditions in certain rather trivial ways. For example it is enough that $\mathcal{X} = (\mathcal{G}/\mathcal{H}) \times \mathcal{V}$ with $\mathcal{H}$ compact, or that $\mathcal{X}$ be a countable union of such spaces. However, it would be desirable to give a proof valid under conditions not expressed in terms of such an explicit representation. This result, (43), was obtained by PITMAN [17] in the special case of a group consisting of translations or scale changes, or both.

At first sight the result is a bit puzzling because the problem of finding a Π such that (41) and (42) imply (43) is invariant under transformation on the left by $\mathcal{G}$ and $\bar{\mathcal{G}}$ so that we would expect the solution Π (assuming it exists and is essentially unique) to be invariant under transformation on the left by $\bar{\mathcal{G}}$. However we observe that, in (44) $\Pi_x\{\tau (x)\}$ is homogeneous of degree 0 in Π so that any relatively invariant Π is effectively invariant for this problem, and which of the relatively invariant measures (if any) is the solution cannot be decided by this qualitative argument. A measure Π is called relatively invariant if there is a function δ on $\mathcal{G}$ such that

$$\Pi\,(\bar g\, C) = \delta\,(g)\, \Pi\,(C) \tag{47}$$

for all $g \in \mathcal{G}$ and $C \in \mathcal{C}$.

It is of some interest to ask whether this prior measure Π induced by the right invariant measure in $\mathcal{G}$ can be approximated by prior probability measures, that is whether, for any $\epsilon > 0$ there exists a prior probability measure Π^ϵ for which

$$\delta^*\,(\Pi^\epsilon, \Pi) < \epsilon\,. \tag{48}$$

This question can be answered in a large class of interesting special cases by essentially the methods used by KUDO [6] and KIEFER [4], for the question of whether the corresponding formal Bayes procedures are minimax. The result is that Π can be approximated by prior probability measures if $\mathscr{G}$ contains a finite sequence of closed subgroups $\mathscr{G}_0 = \{1\} \subset$ $\subset \mathscr{G}_1 \subset \cdots \subset \mathscr{G}_k = \mathscr{G}$ such that each quotient group $\mathscr{G}_k/\mathscr{G}_{k-1}$ is either abelian or compact. A special case of this was considered by M. STONE [13]. He showed that if $X_1, \ldots, X_n$ (with $n \geq 2$) are independently normally distributed with unknown mean ξ and unknown variance σ^2, then the usual confidence intervals $I(X)$ for ξ, based on Student's t, are approximately sets of posterior probability equal to the confidence coefficient α, in the sense that for any $\in\; > 0$ there exists a prior probability measure $\Pi^\in$ such that

$$\Pi^\in \{x: |\prod_x{}^\in [\xi \in I(x)] - \alpha| > \in\} < \in. \tag{49}$$

Of course, here

$$I(X) = (\overline{X} - a\sqrt{S}, \overline{X} + a\sqrt{S}), \tag{50}$$

where

$$\overline{X} = \frac{1}{n} \Sigma X_i, \tag{51}$$

and

$$S = \Sigma (X_i - \overline{X})^2, \tag{52}$$

and

$$\alpha = P_{\xi,\sigma} \{\xi \in I(X)\}. \tag{53}$$

However it is not always true that the prior measure Π induced by the right-invariant measure in $\mathscr{G}$ can be approximated by prior probability measures. In particular such approximation is not in general, possible for problems of multivariate analysis invariant under the full linear group. Counter examples (again in the context of the minimax problem) are given in JAMES and STEIN [7], and LEHMANN [7]. An earlier counter example of PEISAKOFF [10] concerning the free group with two generators may also be relevant.

I had hoped to include a computation of $\inf_{\Pi^{(1)}} \delta^* (\Pi^{(1)}, \Pi)$ in a multivariate situation but I have been unable to overcome a major difficulty.

I find that the material on groups in this section has been done in much greater detail by RAJINDER BIR HORA in an unpublished Ph.D. thesis at the University of Minnesota.

3. Statistical Decision Problems

In this section, we shall look at a rough argument which indicates that, for sufficiently smooth statistical decision problems, the excess risk incurred by using the formal Bayes procedure corresponding to the prior measure Π when the true prior distribution is $\Pi^{(1)}$ does not exceed a constant multiple of $\delta^* (\Pi^{(1)}, \Pi)$ where the constant depends on the decision problem and on Π. It should not be difficult to turn this rough argument into a reasonably simple proof of a usable precise assertion, but I have not succeeded in doing so.

In addition to the structure $(\mathscr{X}, \mathscr{B}, \lambda, \mathscr{T}, \mathscr{C}, p)$ considered in section 2, we suppose we are also given a finite-dimensional real coordinate space $\mathscr{Z}$, the action space, and a non-negative valued function L, the loss function, on $\mathscr{T} \times \mathscr{Z}$, jointly measurable in both variables and twice continuously differentiable in the second variable z for each value of the first variable θ. A decision function φ is a measurable function on the sample space $\mathscr{X}$ to the action space $\mathscr{Z}$ and its risk function $\varrho\,(\cdot, \varphi)$ is defined by

$$\varrho\,(\theta, \varphi) = E_\theta\, L\,[\theta, \varphi\,(x)] = \int L\,[\theta, \varphi\,(x)]\, p\,(x \mid \theta)\, d\lambda\,(x)\,. \tag{1}$$

For a prior probability measure Π on $\mathscr{C}$, a Bayes procedure is, by definition, a decision function ψ that minimizes the expected risk under Π, that is

$$\int \varrho\,(\theta, \psi)\, d\,\Pi\,(\theta) = \inf_{\varphi} \int \varrho\,(\theta, \varphi)\, d\,\Pi\,(\theta) < \infty\,. \tag{2}$$

We shall suppose that for this ψ, $\varrho\,(\theta, \psi)$ is a bounded function of θ. As explained at the beginning of section 5, this [or even constant $\varrho\,(\theta, \psi)$] can be achieved by an appropriate change of notation provided $\varrho\,(\theta, \psi)$ is everywhere finite (and non zero). The value of the left hand side of (2) is called the Bayes risk under Π. It is known that, if a Bayes procedure ψ exists, it satisfies

$$\int L\,[\theta, \psi\,(x)]\, d\,\Pi_x\,(\theta) = \inf_{z} \int L\,(\theta, z)\, d\,\Pi_x\,(\theta)\,, \tag{3}$$

except possible on a set B of x for which

$$\int d\,\Pi\,(\theta) \int_{B} p\,(x \mid \theta)\, d\lambda\,(x) = 0\,. \tag{4}$$

Formula (3) can be used to define a formal Bayes procedure with respect to an improper prior measure Π. We recall that a decision procedure ψ is said to be admissible if there exist no decision procedure φ such that

$$\varrho\,(\theta, \varphi) \leq \varrho\,(\theta, \psi) \tag{5}$$

for all $\theta \in \mathscr{T}$ with strict inequality for some θ. As indicated in section 5, in studying the question of whether the formal Bayes procedure ψ corresponding to a given improper prior measure Π is admissible we shall be interested in an upper bound for $\int \varrho\,(\theta, \psi)\, d\,\Pi^{(1)}\,(\theta) - \inf_{\varphi} \int \varrho\,(\theta, \varphi)\, d\,\Pi^{(1)}\,(\theta)$

for certain prior probability measures $\Pi^{(1)}$. We indicate such a bound by a rough argument, considering only the case where $\dim \mathscr{Z} = 1$.

Let Π be an improper prior measure, ψ the corresponding formal Bayes procedure, $\Pi^{(1)}$ a prior probability measure and φ_1 the corresponding Bayes procedire. Denoting differentiation of L with respect to its second argument by prime, we have

$$L\left[\theta, \psi\left(x\right)\right] \approx L\left[\theta, \varphi_1\left(x\right)\right] + \left[\psi\left(x\right) - \varphi_1\left(x\right)\right] L'\left[\theta, \varphi_1\left(x\right)\right] + \tag{6}$$
$$+ \tfrac{1}{2}\left[\psi\left(x\right) - \varphi_1\left(x\right)\right]^2 L''\left[\theta, \varphi_1\left(x\right)\right],$$

so that

$$\int \left\{L\left[\theta, \psi\left(x\right)\right] - L\left[\theta, \varphi_1\left(x\right)\right]\right\} d\,\Pi_x^{(1)}\left(\theta\right) \approx$$
$$\approx \left[\psi\left(x\right) - \varphi_1\left(x\right)\right] \int L'\left[\theta, \varphi_1\left(x\right)\right] d\,\Pi_x^{(1)}\left(\theta\right) + \tag{7}$$
$$+ \tfrac{1}{2}\left[\psi\left(x\right) - \varphi_1\left(x\right)\right]^2 \int L''\left[\theta, \varphi_1\left(x\right)\right] d\,\Pi_x^{(1)}\left(\theta\right)$$
$$= \tfrac{1}{2}\left[\psi\left(x\right) - \varphi_1\left(x\right)\right]^2 \int L''\left[\theta, \varphi_1\left(x\right)\right] d\,\Pi_x^{(1)}\left(\theta\right).$$

Also

$$L'\left[\theta, \psi\left(x\right)\right] \approx L'\left[\theta, \varphi_1\left(x\right)\right] + \left[\psi\left(x\right) - \varphi_1\left(x\right)\right] L''\left[\theta, \varphi_1\left(x\right)\right]. \tag{8}$$

Integrating with respect to $\Pi_x^{(1)}$ we have

$$\psi\left(x\right) - \varphi_1\left(x\right) \approx \frac{\int L'\left[\theta, \psi\left(x\right)\right] d\,\Pi_x^{(1)}\left(\theta\right)}{\int L''\left[\theta, \psi\left(x\right)\right] d\,\Pi_x^{(1)}\left(\theta\right)}. \tag{9}$$

Substituting in (7) we find

$$\int \left\{L\left[\theta, \psi\left(x\right)\right] - L\left[\theta, \varphi_1\left(x\right)\right]\right\} d\,\Pi_x^{(1)}\left(\theta\right)$$
$$\approx \frac{1}{2}\frac{\left\{\int L'\left[\theta, \psi\left(x\right)\right] d\,\Pi_x^{(1)}\left(\theta\right)\right\}^2}{\int L''\left[\theta, \psi\left(x\right)\right] d\,\Pi_x^{(1)}\left(\theta\right)} \tag{10}$$

But

$$\left\{\int L'\left[\theta, \psi\left(x\right)\right] d\,\Pi_x^{(1)}\left(\theta\right)\right\}^2$$
$$= \left\{\int L'\left[\theta, \psi\left(x\right)\right] \left[d\,\Pi_x^{(1)}\left(\theta\right) - d\,\Pi_x\left(\theta\right)\right]\right\}^2$$
$$= \left\{\int L'\left[\theta, \psi\left(x\right)\right] \left[\sqrt{d\,\Pi_x^{(1)}\left(\theta\right)} + \sqrt{d\,\Pi_x\left(\theta\right)}\right] \left[\sqrt{d\,\Pi_x^{(1)}\left(\theta\right)} - \sqrt{d\,\Pi_x\left(\theta\right)}\right]\right\}^2$$
$$\leq \int L'^2\left[\theta, \psi\left(x\right)\right] \left[\sqrt{d\,\Pi_x^{(1)}\left(\theta\right)} + \sqrt{d\,\Pi_x\left(\theta\right)}\right]^2 \cdot \tag{11}$$
$$\cdot \int \left[\sqrt{d\,\Pi_x^{(1)}\left(\theta\right)} - \sqrt{d\,\Pi_x\left(\theta\right)}\right]^2 \leq$$
$$\leq 2 \int L'^2\left[\theta, \psi\left(x\right)\right] d\left[\Pi_x^{(1)}\left(\theta\right) + \Pi_x\left(\theta\right)\right] \delta\left(\Pi_x^{(1)}, \Pi_x\right).$$

Finally, we have, approximately

$$\int \varrho\left(\theta, \psi\right) d\,\Pi^{(1)}\left(\theta\right) - \inf \int \varrho\left(\theta, \varphi\right) d\,\Pi^{(1)}\left(\theta\right)$$
$$= \int d\,\Pi^{(1)}\left(\theta'\right) \int \varrho\left(x \mid \theta'\right) d\lambda\left(x\right) \int \left[L\left(\theta, \psi\left(x\right) - L\left(\theta, \varphi_1\left(x\right)\right)\right] d\,\Pi_x^{(1)}\left(\theta\right) \leq$$
$$\leq \int d\,\Pi^{(1)}\left(\theta'\right) \int \varrho\left(x \mid \theta'\right) d\lambda\left(x\right) \frac{\int L'^2\left[\theta, \psi\left(x\right)\right] d\left[\Pi_x^{(1)}\left(\theta\right) + \Pi_x\left(\theta\right)\right]}{\int L''\left[\theta, \psi\left(\theta\right)\right] d\,\Pi_x^{(1)}\left(\theta\right)} \tag{12}$$
$$\delta\left(\Pi_x^{(1)}, \Pi_x\right).$$

Thus if

$$\frac{\int L'^2\left[\theta, \psi\left(x\right)\right] d\left[\Pi_x^{(1)}\left(\theta\right) + \Pi_x\left(\theta\right)\right]}{\int L''\left[\theta, \psi\left(x\right)\right] d\,\Pi_x^{(1)}\left(\theta\right)}$$

is bounded function of x and the approximations leading to (10) are valid, we have

$$\int \varrho\left(\theta, \psi\right) d\,\Pi^{(1)}\left(\theta\right) - \inf \int \varrho\left(\theta, \varphi\right) d\,\Pi^{(1)}\left(\theta\right) \leq K\,\delta^*\left(\Pi^{(1)}, \Pi\right). \tag{13}$$

4. An Approximation to $\delta^*\left(\Pi^{(1)}, \Pi\right)$

We return to the structure $(\mathscr{X}, \mathscr{B}, \lambda, \mathscr{T}, \mathscr{C}, p)$ introduced at the beginning of section 2, where $\mathscr{T}$ is now a k-dimensional real coordinate space with k finite. We suppose we are given an improper prior measure Π and show that under certain conditions, the separation $\delta^* (\Pi^{(1)}, \Pi)$ is given approximately by

$$\delta^* (\Pi^{(1)}, \Pi) \approx \frac{1}{4} \int \frac{1}{q(\theta)} \sum \frac{\partial q(\theta)}{\partial \theta^i} \frac{\partial q(\theta)}{\partial \theta^j} g^{ij}(\theta)\, d\Pi(\theta) , \qquad (1)$$

where $\Pi^{(1)}$ is a prior probability measure with

$$q = \frac{d\Pi^{(1)}}{d\Pi} , \qquad (2)$$

and $g(\theta)$ is the covariance matrix, when X is distributed according to $p(\cdot, \theta)$, of the posterior mean $\hat\theta(x)$ of Θ given X under Π. We denote the ith coordinate of θ by θ^i and the (i, j) coordinate of $g(\theta)$ by $g^{ij}(\theta)$. In many cases, if Π is reasonably smooth, we can expect $g(\theta)$ to be approximately the inverse of the information matrix:

$$\{[g(\theta)]^{-1}\}_{ij} \approx E_\theta \frac{\partial \log p(X \mid \theta)}{\partial \theta^i} \frac{\partial \log p(X \mid \theta)}{\partial \theta^j} . \qquad (3)$$

Of course, a precise justification of this could come only from largesample theory, but, when we come to apply formula (1) in section 5, we shall see that we need only that the ratio of the two sides is bounded away from 0 and ∞. We give a precise form of the approximation (1) with remainder in formulas (24), (25), (17), (16), (20), and (21). However the remainder is not expressed in a usable form.

All of our work will be subject to the assumption that the prior probability measure $\Pi^{(1)}$ is absolutely continuous with respect to Π and $q = d\Pi^{(1)}/d\Pi$ is twice continuously differentiable. In addition, roughly speaking, we suppose that q and its first and second partial derivatives are nearly the same at any two points θ, θ' whose distance is small or moderate in the Riemannian metric determined by $\theta \to [g(\theta)]^{-1}$. Let

$$\varrho (\Pi_x^{(1)}, \Pi_x) = 1 - \tfrac{1}{2} \delta (\Pi_x^{(1)}, \Pi_x) = \int \sqrt{d\Pi_x^{(1)}\, d\Pi_x} , \qquad (4)$$

and

$$\varrho^* (\Pi^{(1)}, \Pi) = 1 - \tfrac{1}{2} \delta^* (\Pi^{(1)}, \Pi) = E^{(1)} \int \sqrt{d\Pi_x^{(1)}\, d\overline{\Pi}_x} \qquad (5)$$

$$= \int d\Pi^{(1)}(\theta) \int p(x \mid \theta)\, d\lambda(x) \frac{\int p(x \mid \theta') \sqrt{d\Pi^{(1)}(\theta')\, d\Pi(\theta')}}{\sqrt{\int p(x \mid \theta')\, d\Pi^{(1)}(\theta') \int p(x \mid \theta'')\, d\Pi(\theta'')}}$$

$$= \int d\lambda(x) \frac{[\int p(x \mid \theta) \sqrt{d\Pi^{(1)}(\theta)\, d\Pi(\theta)}] \sqrt{\int p(x \mid \theta)\, d\Pi^{(1)}(\theta)}}{\sqrt{\int p(x \mid \theta)\, d\Pi(\theta)}} .$$

We shall apply this with $\Pi^{(1)}$ related to Π by (2). We shall use the summation convention whereby the second term on the right hand side of (7) represents a sum over i, and the third term on the right hand side of (8) a sum over i and j.

Let

$$\hat\theta(x) = \frac{\int \theta\, p(x \mid \theta)\, d\Pi(\theta)}{\int p(x \mid \theta)\, d\Pi(\theta)} . \qquad (6)$$

Then, by Taylor's formula,

$$q(\theta) = q[\hat{\theta}(x)] + [\theta^i - \hat{\theta}^i(x)]\, q_i[\hat{\theta}(x)] +$$
$$+ \tfrac{1}{2}[\theta^i - \hat{\theta}^i(x)][\theta^j - \hat{\theta}^j(x)]\, q_{ij}[\theta^*(x)]\,, \qquad (7)$$

where $\theta^*(x)$ is a point of the line segment joining θ and $\theta(x)$ and

$$q_i(\theta) = \frac{\partial q(\theta)}{\partial \theta^i}\,,\quad q_{ij}(\theta) = \frac{\partial^2 q(\theta)}{\partial \theta^i\,\partial \theta^j}\,, \qquad (8)$$

Consequently

$$\int p(x\mid\theta)\, q(\theta)\, d\Pi(\theta)$$
$$= \int p(x\mid\theta)\left\{ q[\hat{\theta}(x)] + [\theta^i - \hat{\theta}^i(x)]\, q_i[\hat{\theta}(x)] + \right.$$
$$\left. + \tfrac{1}{2}[\theta^i - \hat{\theta}^i(x)][\theta^j - \hat{\theta}^j(x)]\, q_{ij}[\theta^*(x)]\right\} d\Pi(\theta) \qquad (9)$$
$$= q[\hat{\theta}(x)] \int p(x\mid\theta)\, d\Pi(\theta) +$$
$$+ \tfrac{1}{2}\int p(x\mid\theta)\,[\theta^i - \hat{\theta}^i(x)][\theta^j - \hat{\theta}^j(x)]\, q_{ij}[\theta^*(x)]\, d\Pi(\theta)\,,$$

by (6). Again, using Taylor's formula, we find that, with $\theta^{**}(x)$ on the line segment joining θ and $\hat{\theta}(x)$,

$$\sqrt{q(\theta)} = \sqrt{q[\hat{\theta}(x)]} + [\theta^i - \hat{\theta}^i(x)]\frac{q_i[\theta(x)]}{2\sqrt{q[\theta(x)]}}$$
$$+ \tfrac{1}{2}[\theta^i - \hat{\theta}^i(x)][\theta^j - \hat{\theta}^j(x)]\left\{ \frac{q_{ij}[\theta^{**}(x)]}{2\sqrt{q[\theta^{**}(x)]}} - \tfrac{1}{4}\frac{q_i[\theta^{**}(x)]\, q_j[\theta^{**}(x)]}{\left\{q[\theta^{**}(x)]\right\}^{\frac{3}{2}}} \right\}.$$

Thus

$$\int p(x\mid\theta)\sqrt{q(\theta)}\, d\Pi(\theta)$$
$$= \sqrt{q[\theta(x)]}\int p(x\mid\theta)\, d\Pi(\theta)$$
$$+ \tfrac{1}{2}\int p(x\mid\theta)\,[\theta^i - \hat{\theta}^i(x)][\theta^j - \hat{\theta}^j(x)] \qquad (11)$$
$$\cdot\left\{ \frac{q_{ij}[\theta^{**}(x)]}{2\sqrt{q[\theta^{**}(x)]}} - \tfrac{1}{4}\frac{q_i[\theta^{**}(x)]\, q_z[\theta^{**}(x)]}{\left\{q[\theta^{**}(x)]\right\}^{\frac{3}{2}}} \right\} d\Pi(\theta)\,.$$

Then (5) becomes

$$\varrho^*(\Pi^{(1)}, \Pi)$$
$$= \int d\lambda(x)\,\frac{[\int p(x\mid\theta)\sqrt{q(\theta)}\, d\Pi(\theta)]\,\sqrt{\int p(x\mid\theta)\, q(\theta)\, d\Pi(\theta)}}{\sqrt{\int p(x\mid\theta)\, d\Pi(\theta)}}$$
$$= \int d\lambda(x)\,\sqrt{1 + \frac{1}{2\,q[\theta(x)]}\,\frac{\int[\theta^i - \theta^i(x)][\theta^j - \theta^j(x)]\, q_{ij}[\theta^*(x)]\, p(x\mid\theta)\, d\Pi(\theta)}{\int p(x\mid\theta)\, d\Pi(\theta)}}$$
$$\cdot q[\hat{\theta}(x)]\int p[x\mid\theta]\, d\Pi(\theta) \qquad (12)$$
$$\left\{ 1 + \tfrac{1}{2}\,\frac{\int[\theta^i - \theta^i(x)][\theta^j - \theta^j(x)]\left\{\dfrac{q_{ij}[\theta^{**}(x)]}{\sqrt{2[\theta^{**}(x)]\, q[\theta(x)]}} - \tfrac{1}{4}\dfrac{q_i[\theta^{**}(x)]\, q_j[\theta^{**}(x)]}{\sqrt{q[\theta(x)]}\left\{q[\theta^{**}(x)]\right\}^{\frac{3}{2}}}\right\}}{\dfrac{p(x\mid\theta)\, d\Pi(\theta)}{\int p(x\mid\theta)\, d\Pi(\theta)}} \right\}.$$

However

$$\sqrt{1+t} = 1 + \frac{t}{2} + 0\,[\min(|t|, t^2)] \qquad (13)$$

since

$$\frac{\sqrt{1+t}-1-\dfrac{t}{2}}{t^2} \tag{14}$$

is a continuous function of t vanishing as $1/t$ at $\pm \infty$. If we apply this and the inequality,

$$| uv | \leq \tfrac{1}{2} (u^2 + v^2) \tag{15}$$

to (12), we obtain

$$\begin{aligned}
&\varrho^* (\Pi^{(1)}, \Pi) \\
&= \int d\lambda (x) \, q \, [\hat{\theta} (x)] \int p (x \mid \theta') \, d\Pi (\theta') \\
&\quad \cdot \left\{ 1 + \frac{1}{4 \, q \, [\theta (x)]} \frac{\int [\theta^i - \theta^i (x)] [\theta^j - \theta^j (x)] \, q_{ui} \, [\theta^* (x)] \, p (x \mid \theta) \, d\Pi (\theta)}{\int p (x \mid \theta) \, d\Pi (\theta)} \right. \\
&\quad + \tfrac{1}{2} \, \frac{\int [\theta^i - \theta^i (x)] [\theta^j - \theta^j (x)] \left\{ \dfrac{q_{ij} \, [\theta^{**} (x)]}{\sqrt{2 \, q \, [\theta^{**} (x)] \, q \, [\theta (x)]}} - \dfrac{1}{4} \dfrac{q_i \, [\theta^{**} (x)] \, q_j \, [\theta^{**} (x)]}{\sqrt{q \, [\theta (x)]} \, \{ q \, [\theta^{**} (x)] \}^{\frac{3}{2}}} \right\}}{\int p (x \mid \theta) \, d\Pi (\theta)} \\
&\quad \left. \cdot \, p (x \mid \theta) \, d\Pi (\theta) + R_1 (x) \right\},
\end{aligned} \tag{16}$$

$$| R_1 (x) | \leq K \{ A^2 (x) + B^2 (x) \}, \tag{17}$$

with K an absolute constant, $A (x)$ the second term in braces in (16) and $B (x)$ the third term in braces in (16). But, by (7),

$$\begin{aligned}
&\int d\lambda (x) \, q \, [\theta (x)] \int p (x \mid \theta') \, d\Pi (\theta') \\
&\quad = \int d\lambda (x) \int d\Pi (\theta) \, p (x \mid \theta) \big\{ q (\theta) - [\theta^i - \hat{\theta}^i (x)] \, q_i \, [\hat{\theta} (x)] - \\
&\quad\quad - \tfrac{1}{2} [\theta^i - \hat{\theta}^i (x)] [\theta^j - \hat{\theta}^j (x)] \, q_{ij} \, [\theta^* (x)] \big\} \\
&\quad = 1 - \tfrac{1}{2} \int d\lambda (x) \int [\theta^i - \hat{\theta}^i (x)] [\theta^j - \hat{\theta}^j (x)] \, q_{ij} \, [\theta^* (x)] \, p (x \mid \theta) \, d\Pi (x) .
\end{aligned} \tag{18}$$

Using (18) to evaluate the term in (16) arising from the 1 in braces, we find

$$\begin{aligned}
&\varrho^* (\Pi^{(1)}, \Pi) = 1 + \int d\lambda (x) \, q \, [\hat{\theta} (x)] \, R_1 (x) \int \varrho (x \mid \theta) \, d\Pi (\theta) - \\
&\quad - \tfrac{1}{8} \int d\lambda (x) \int [\theta - \hat{\theta}^i (x)] [\theta^j - \hat{\theta}^j (x)] \frac{q_i \, [\theta^{**} (x)] \, q_j \, [\theta^{**} (x)]}{\{ q \, [\theta^{**} (x)] \}^{\frac{3}{2}} \{ q \, [\theta (x)] \}} \, p (x \mid \theta) \, d\Pi (\theta) + \\
&\quad + \int d\lambda (x) \, R_2 (x)
\end{aligned} \tag{19}$$

where

$$\begin{aligned}
R_2 (x) = &\int [\theta^i - \hat{\theta}^i (x)] (\theta^j - \hat{\theta}^j (x)) \left\{ \frac{q_{ij} \, [\theta^{**} (x)]}{\sqrt{q \, [\theta^{**} (x)] \, q^{-1} \, [\theta (x)]}} - q_{ij} \, [\theta^* (x)] \right\} \\
&\cdot p (x \mid \theta) \, d\Pi (\theta) .
\end{aligned} \tag{20}$$

Then, letting

$$\begin{aligned}
R_3 (x) = &\int [\theta^i - \hat{\theta}^i (x)] [\theta^j - \hat{\theta}^j (x)] \cdot \\
&\cdot \left\{ \frac{q_i \, [\theta^{**} (x)] \, q_j \, [\theta^{**} (x)]}{\{ q \, [\theta^{**} (x)] \}^{\frac{3}{2}} \{ q \, [\theta (x)] \}^{\frac{1}{2}}} - \frac{q_i (\theta) \, q_j (\theta)}{q (\theta)} \right\} p (x \mid \theta) \, d\Pi (\theta) ,
\end{aligned} \tag{21}$$

we have

$$\delta^* \left(\Pi^{(1)}, \Pi\right) = 2 \left[1 - \varrho^* \left(\Pi^{(1)}, \Pi\right)\right]$$

$$= \tfrac{1}{4} \int d\lambda \left(x\right) \int \left[\theta^i - \hat{\theta}^i \left(x\right)\right] \left[\theta^j - \hat{\theta}^j \left(x\right)\right] \frac{q_i \left(\theta\right) q_j \left(\theta\right)}{q \left(\theta\right)} p \left(x \mid \theta\right) d\Pi \left(\theta\right) - \qquad (22)$$

$$- 2 \int d\lambda \left(x\right) \left\{ q \left[\hat{\theta} \left(x\right)\right] R_1 \left(x\right) \int p \left(x \mid \theta\right) d\Pi \left(\theta\right) + R_2 \left(x\right) + R_3 \left(x\right) \right\}.$$

But

$$\int d\lambda \left(x\right) \int \left[\theta^i - \hat{\theta}^i \left(x\right)\right] \left[\theta^j - \hat{\theta}^j \left(x\right)\right] \frac{q_i \left(\theta\right) q_j \left(\theta\right)}{q \left(\theta\right)} p \left(x \mid \theta\right) d\Pi \left(\theta\right)$$

$$= \int d\Pi \left(\theta\right) \frac{q_i \left(\theta\right) q_j \left(\theta\right)}{q \left(\theta\right)} \int d\lambda \left(x\right) p \left(x \mid \theta\right) \left[\theta^i - \hat{\theta}^i \left(x\right)\right] \left[\theta^j - \hat{\theta}^j \left(x\right)\right] \quad (23)$$

$$= \int d\Pi \left(\theta\right) \frac{q_i \left(\theta\right) q_j \left(\theta\right)}{q \left(\theta\right)} g^{ij} \left(\theta\right)$$

where $g^{ij} \left(\theta\right)$ is the covariance matrix of $\hat{\theta} \left(X\right)$ when X is distributed according to $p \left(\cdot \mid \theta\right)$. Thus we have, finally

$$\delta \left(\Pi^{(1)}, \Pi\right) = \tfrac{1}{4} \int \frac{q_i \left(\theta\right) q_j \left(\theta\right)}{q \left(\theta\right)} g^{ij} \left(\theta\right) d\Pi \left(\theta\right) - 2 \int R \left(x\right) d\lambda \left(x\right), \qquad (24)$$

where

$$R \left(x\right) = q \left[\hat{\theta} \left(x\right)\right] R_1 \left(x\right) \int p \left(x \mid \theta\right) d\Pi \left(\theta\right) + R_2 \left(x\right) + R_3 \left(x\right). \qquad (25)$$

5. Rough Derivation of a Condition for Admissibility

We consider a statistical decision problem $(\mathscr{X}, \mathscr{B}, \lambda, \mathscr{T}, \mathscr{C}, p, \mathscr{F}, L)$ as formulated in Section 2. In addition, we suppose we are given an improper prior measure Π and wish to find out whether the formal Bayes procedure ψ with respect to Π is admissible. We shall argue that, under certain conditions, we can expect ψ to be admissible if

$$\liminf_{\lambda \downarrow 0} \int \left\{ g^{ij} \left(\theta\right) \frac{\partial f \left(\theta\right)}{\partial \theta^i} \frac{\partial f \left(\theta\right)}{d\theta^j} + \lambda f^2 \left(\theta\right) \right\} d\Pi \left(\theta\right) = 0, \qquad (1)$$

where

$$g^{ij} \left(\theta\right) = E_\theta \left[\hat{\theta}^i \left(X\right) - \theta^i\right] \left[\hat{\theta}^j \left(X\right) - \theta^j\right], \qquad (2)$$

with

$$\theta^i \left(x\right) = \frac{\int \theta^i q \left(x \mid \theta\right) d\Pi \left(\theta\right)}{\int p \left(x \mid \theta\right) d\Pi \left(\theta\right)}, \qquad (3)$$

and f ranges over the set of continuously differentiable functions taking on the value 1 for all points of a given non-empty open set S with compact closure. We shall not be able to state a precise theorem, but the conditions needed are roughly the following:

(i) The loss function L is twice continuously differentiable in the action and parameter point together. Both $\mathscr{X}$ and $\mathscr{T}$ are differentiable manifolds.

(ii) The formal Bayes procedure ψ with respect to Π is unique.

(iii) The risk function of ψ is bounded.

(iv) The improper prior measure Π is absolutely continuous with respect to Lebesgue measure, say with density π.

(v) The functions $\log \pi$ and $\log g$ are uniformly continuous in this Riemannian metric.

(vi) It must be possible to choose the nearly minimizing functions f in (1) so that f^2 satisfies the conditions required in Section 4 to make the remainder in (4.24) smaller than, say, half the leading term*.

If, in addition, the decision problem is sufficiently complicated, for example if it requires the complete estimation of θ with a loss function equal to the distance between true and estimated values, we can expect (1) to be necessary as well as sufficient, although the argument in this case is even less compelling than for the sufficiency.

First let us observe that the condition (iii) that the risk function of φ_0 is bounded is a condition on the form in which the problem is presented rather than on the problem itself. I believe this was first observed (in a different context) by LINDLEY. We suppose the risk function ϱ of ψ_0 is everywhere finite and $\neq 0$. We define a new loss function L' and a new prior measure Π' by

$$L'(\theta, z) = \frac{1}{\varrho(\theta)} L(\theta, z) \tag{4}$$

and

$$d\,\Pi'(\theta) = \varrho(\theta)\, d\,\Pi(\theta) \ . \tag{5}$$

Then ψ is also the formal Bayes procedure with respect to Π' for the new problem and its risk is the constant 1. If Π' is a bounded measure, then ψ is admissible because of the uniqueness assumption (ii). If not, we want

$$\int p\,(x \mid \theta)\, d\,\Pi'(\theta) < \infty \ , \tag{6}$$

in order to be able to consider Π' an improper prior measure. Hereafter we suppose this modification has already been made, if necessary.

Next we apply a slight modification of the necessary and sufficient condition for admissibility given in the author's paper [12]. In order that ψ be almost admissible with respect to Lebesgue measure it is necessary and sufficient that, for any open set $S \subset \mathscr{I}$ with compact closure, and any $\epsilon > 0$, there exist $\delta > 0$ and a probability measure $\Pi^{(1)}$ in $\mathscr{I}$, absolutely continuous with respect to Lebesgue measure such that,

$$\Pi^{(1)}(S) \geq \delta \ , \tag{7}$$

and

$$\int \varrho\,(\theta, \psi)\, d\,\Pi^{(1)}(\theta) \leq \inf_{\varphi} \int \varrho\,(\theta, \varphi)\, d\,\Pi^{(1)}(\theta) + \epsilon\delta \ , \tag{8}$$

where

$$\varrho\,(\theta, \varphi) = \int L\,[\theta, \varphi\,(x)]\, p\,(x \mid \theta)\, d\lambda\,(x) \ , \tag{9}$$

for any decision function $\varphi \colon \mathscr{X} \to \mathscr{Z}$. In [12], the condition is stated for sets A consisting of a single point, but, by examining the proof, it is not difficult to see that the change made here is inessential.

---

* See the note added in proof at the end of the paper.

The argument given at the end of Section 2 indicates that

$$\int \varrho\,(\theta, \psi)\, d\,\Pi^{(1)}\,(\theta) - \inf \int \varrho\,(\theta, \varphi)\, d\,\Pi^{(1)}\,(\theta) \le K\,\delta^*\,(\Pi^{(1)}, \Pi),\quad (10)$$

where K is a constant. Now let us take $\Pi^{(1)}$ absolutely continuous with respect to Π, with $q = d\,\Pi^{(1)}$ and apply (4.24). We find that Π is admissible if, for any open set $S \subset \mathscr{T}$ with compact closure and any $\epsilon > 0$ there exist $\delta > 0$ and a twice continuously differentiable function q on $\mathscr{T}$ such that

$$\int_{\mathscr{T}} q\,d\,\Pi = 1\,,\tag{11}$$

$$\int_{S} q\,d\,\Pi \ge \delta\,,\tag{12}$$

$$\int \frac{q_i\,(\theta)\,q_j\,(\theta)}{q\,(\theta)}\, g^{ij}\,(\theta)\, d\,\Pi\,(\theta) \le \epsilon\,\delta\,,\tag{13}$$

and the remainder in (4.24) is negligible. Now let

$$f\,(\theta) = \sqrt{\frac{q\,(\theta)}{\delta}}\,.\tag{14}$$

Conditions (11), (12), and (13) become

$$\int_{\mathscr{Y}} f^2\, d\,\Pi = \frac{1}{\delta}\,,\tag{15}$$

$$\int_{S} f^2\, d\,\Pi \ge 1\,,\tag{16}$$

and

$$\int f_i\,(\theta)\, f_j\,(\theta)\, g^{ij}\,(\theta)\, d\,\Pi\,(\theta) \ge \epsilon\,.\tag{17}$$

A simple Lagrange multiplier argument yields the form given at the beginning of this section.

6. Partial solution of the reduced problem

In section 5 we have been led to ask for conditions on the continuous positive valued function π on $\mathscr{R}^K$ and the continuous function g on $\mathscr{R}^K$ taking positive-definite symmetric $K \times K$ matrices as values, under which, for any open set S with compact closure,

$$\lim_{\lambda \downarrow 0}\ \inf_{f \in \mathscr{A}\,(S)} \int \left\{ g^{ij}\,(x)\, \frac{\partial f\,(x)}{\partial x^i}\, \frac{\partial f\,(x)}{\partial x^j} + \lambda f^2\,(x) \right\} \pi\,(x)\, dx = 0\,,\tag{1}$$

where $\mathscr{A}\,(S)$ is the set of all continuously differentiable functions f for which $f\,(x) = 1$ for all $x \in S$, and

$$dx = dx^1\,\ldots\,dx^K\,,\tag{2}$$

and we use the summation convention, so that the first term in braces in (1) represents a summation over $i, j = 1\,\ldots\,K$. The corresponding problem for general differentiable manifolds can also arise, but we shall not try to consider it. Subject to certain conditions that have been indicated, somewhat vaguely, in Section 5, if π is an improper prior density for a given observational situation and g is the expected formal posterior covariance matrix (computed under π) when the true parameter value is x, condition (1) is sufficient for formal Bayes solutions with respect to π to be admissible for any sufficiently smooth decision problem and, if the decision problem is sufficiently complicated, (1) can also be expected to be necessary for admissibility of these formal Bayes procedures.

We shall solve this problem in two rather trivial special cases, the one-dimensional case and the spherically symmetric case, and also make some remarks about the general problem. Since the question of whether (1) is satisfied remains unchanged when π and g are replaced by π' and g' with π'/π and the characteristic roots of g' relative to g bounded away from 0 and ∞, the result in the spherically summetric case is of fairly wide applicability. We observe also that the class of problems considered is invariant under continuously differentiable homeomorphisms of $\mathscr{R}^K$, g transforming as a symmetric contravariant tensor of the second rank and π as a scalar density. More explicitly, let ψ be a $1 - 1$ continuously differentiable function of $\mathscr{R}^K$ onto $\mathscr{R}^K$ with continuously differentiable inverse, and, for $y \in \mathscr{R}^K$ let

$$F(y) = f[\psi^{-1}(y)], \tag{3}$$

$$\Pi(y) = \pi[\psi^{-1}(y)] \det\left(\frac{\partial[\psi^{-1}(y)]^j}{\partial y^i}\right), \tag{4}$$

and

$$G^{ij}(y) = g^{kl}[\psi^{-1}(y)] \frac{\partial\psi^i}{\partial x^k}[\psi^{-1}(y)] \frac{\partial\psi^j}{\partial x^l}[\psi^{-1}(y)]. \tag{5}$$

Then

$$\int \left\{ g^{ij}(x) \frac{\partial f(x)}{\partial x^i} \frac{\partial f(x)}{\partial x^j} + \lambda f^2(x) \right\} \pi(x)\, dx$$

$$= \int \left\{ G^{ij}(y) \frac{\partial F(y)}{\partial y^i} \frac{\partial F(y)}{\partial y^i} + \lambda F^2(y) \right\} \Pi(y)\, dy. \tag{6}$$

Of course, these transformations (3) to (5) are also appropriate to f, π, g as they arise from the statistical problem [when $g^{-1}(x)$, the inverse of $g(x)$, is the information matrix]. It seems likely that a really satisfactory solution of our problem will exploit its tensorial character. We shall see that g and π seem to enter mainly (but not entirely) through their product, so that, in a way, the geometry of this problem is not that of a Riemannian manifold, but rather that associated with the contravariant tensor density πg.

Now let us look at the one-dimensional case, where a complete solution is almost trivial. Condition (1) reduces to

$$\lim_{\lambda \downarrow 0} \inf_{\substack{f(x)=1 \\ x \in [-1,1]}} \int_{-\infty}^{\infty} \left\{ g(x) \left[\frac{df(x)}{dx}\right]^2 + \lambda f^2(x) \right\} \pi(x)\, dx = 0, \tag{7}$$

which is equivalent to

$$\lim_{\lambda\downarrow 0}\inf_{f(1)=1}\int_1^\infty \left\{g(x)\left[\frac{df(x)}{dx}\right]^2 + \lambda f^2(x)\right\}\pi(x)\,dx = 0\,,\tag{8}$$

together with the corresponding condition on $(-\infty, 0]$. Since the two problems are completely similar, we consider only (8). If

$$\int_1^\infty \pi(x)\,dx < \infty\,,\tag{9}$$

condition (8) is trivially satisfied with $f(x)\equiv 1$. If

$$\int_1^\infty \pi(x)\,dx = \infty\,,\tag{10}$$

we shall see that a necessary and sufficient condition for (8) is

$$\int_1^\infty \frac{dx}{g(x)\,\pi(x)} = \infty\,.\tag{11}$$

We make a change of variable to

$$y = \int_1^x \frac{dt}{g(t)\,\pi(t)}\,,\tag{12}$$

and write y_∞ for the value of y [infinite if and only if (11) holds] corresponding to $x = \infty$. Let

$$F(y) = f(x)\,.\tag{13}$$

Then

$$\int_1^\infty \left\{g(x)\left[\frac{df(x)}{dx}\right]^2 + \lambda f^2(x)\right\}\pi(x)\,dx$$

$$= \int_1^\infty \left[\frac{dF(y)}{df}\right]^2 \frac{dx}{g(x)\,\pi(x)} + \lambda \int_1^\infty F^2(y)\,\pi(x)\,dx\tag{14}$$

$$= \int_1^{y_\infty} \left\{\left[\frac{dF(y)}{df}\right]^2 + \lambda F^2(y)\,H(y)\right\}dy$$

where

$$H(y) = g(x)\,\pi^2(x)\,.\tag{15}$$

If $y_\infty = \infty$ we can take

$$F(y) = \begin{cases} 1 - \dfrac{y}{A_\lambda} & \text{for } 0 \le y \le A_\lambda\,, \\[2mm] 0 & \text{for } y \ge A_\lambda\,, \end{cases}\tag{16}$$

where A_λ is chosen so that

$$\int_0^{A_\lambda} H(y)\,dy = \lambda^{-\frac{1}{2}}\,.\tag{17}$$

Since H is continuous,

$$\lim_{\lambda \downarrow 0} A_\lambda = \infty , \tag{18}$$

and thus

$$\int_1^\infty \left\{ \left[\frac{dF(y)}{dy} \right]^2 + \lambda F^2(y) \, H(y) \right\} dy$$
$$= \frac{1}{A_\lambda} + \lambda \int_1^{A_\lambda} \left(1 - \frac{y}{A_\lambda} \right)^2 H(y) \, dy < \frac{1}{A_\lambda} + \sqrt{\lambda} , \tag{19}$$

so that (8) is satisfied, by (14). On the other hand, if $y_\infty < \infty$ and f is chosen so as to make (14) finite, we must have

$$\lim_{y \to y_\infty} F(y) = \lim_{x \to \infty} f(x) = 0 , \tag{20}$$

so that

$$\int_1^{y_\infty} \left[\frac{dF(y)}{dy} \right]^2 dy \geq \frac{1}{y_\infty} \left[\int_1^{y_\infty} \frac{dF(y)}{dy} \, dy \right]^2 = , \frac{1}{y_\infty} \tag{21}$$

and (14) is at least $\dfrac{1}{y_\infty}$. Thus (8) cannot hold and we have proved.

Proposition 1: In order that (7) hold for given continuous positive-valued functions on the real line it is necessary and sufficient that

$$\text{(i) if } \int_0^\infty \pi(x) \, dx = \infty, \text{ then } \int_0^\infty \frac{dx}{g(x) \, \pi(x)} = \infty ,$$

and

$$\text{(ii) if } \int_{-\infty}^0 \pi(x) \, dx = \infty, \text{ then } \int_{-\infty}^0 \frac{dx}{g(x) \, \pi(x)} = \infty .$$

Next let us look at the spherically symmetric case. We shall prove

Proposition 2: Let π be a continuous positive-valued function on $\mathscr{R}^K$ of the form

$$\pi(x) = \varphi(\| x \|^2) , \tag{22}$$

where

$$\| x \|^2 = \sum_{i=1}^K (x^i)^2 , \tag{23}$$

and let the continuous function g on $\mathscr{R}^K$ to the space of positive-definite symmetric $K \times K$ matrices be given by

$$g^{ij}(x) = \alpha(\| x \|^2) \, \delta^{ij} + \beta(\| x \|^2) \, x^i \, x^j \tag{24}$$

where

$$\delta^{ij} = \begin{cases} 1 & \text{if } i = j \\ 0 & \text{if } i \neq j . \end{cases} \tag{25}$$

Then, in order that (1) hold, that is

$$0 = \lim_{\lambda \downarrow 0} \inf_{\substack{f(x)=1 \\ ||x|| \leq 1}} \int \left\{ g^{ij}(x) \frac{\partial f(x)}{\partial x^i} \frac{\partial f(x)}{\partial x^j} + \lambda f^2(x) \right\} \pi(x)\, dx \tag{26}$$

$$= \lim_{\lambda \downarrow 0} \inf_{\substack{f(x)=1 \\ ||x|| \leq 1}} \int \left\{ [\alpha(||x||^2)\, \delta^{ij} + \beta(||x||^2)\, x^i x^j] \frac{\partial f(x)}{\partial x^i} \frac{\partial f(x)}{\partial x^j} \right.$$
$$\left. + \lambda f^2(x) \right\} \varphi(||x||^2)\, dx\,,$$

it is necessary and sufficient that, if

$$\int_1^\infty \varphi(t)\, t^{\frac{k}{2}-1}\, dt = \infty\,, \tag{27}$$

then

$$\int_1^\infty \frac{dt}{[\alpha(t) + t\beta(t)]\, \varphi(t)\, t^{\frac{k}{2}}} = \infty\,. \tag{28}$$

Proof: Because the integral in (26) is a convex function of f invariant under the compact group of orthogonal transformations $T: \mathscr{R}^K \to \mathscr{R}^K$ [operating by taking f into $\overline{T}f$ defined by

$$(\overline{T}f)\, x = f(T^{-1} x)]\,, \tag{29}$$

it follows that the condition (26) is equivalent to the corresponding condition with f restricted to be invariant under orthogonal transformations, say

$$f(x) = \xi(||x||^2), \text{ with } \xi(1) = 1\,. \tag{30}$$

Thus (26) is equivalent to

$$0 = \lim_{\lambda \downarrow 0} \inf_{\xi(1)=1} \int \left\{ [\alpha(||x||^2)\, \delta^{ij} + \beta(||x||^2)\, x^i x^j] \frac{\partial \xi(||x||^2)}{\partial x^i} \frac{\partial \xi(||x||^2)}{\partial x^j} + \right.$$
$$\left. + \lambda \xi^2(||x||^2) \right\} \varphi(||x||^2)\, dx$$

$$= 4 \lim_{\lambda \downarrow 0} \inf_{\xi(1)=1} \int \left\{ [\alpha(||x||^2)\, \delta^{ij} + \beta(||x||^2)\, x^i x^j]\, \xi'^2(||x||^2)\, x^i x^j + \right.$$
$$\left. + \lambda \xi^2(||x||^2) \right\} \varphi(||x||^2)\, dx \tag{31}$$

$$= C \lim_{\lambda \downarrow 0} \inf_{\xi(1)=1} \int \left\{ [t\alpha(t) + t^2 \beta(t)]\, \xi'^2(t) + \right.$$
$$\left. + \lambda \xi^2(t) \right\} \varphi(t)\, t^{\frac{k}{2}-1}\, dt\,,$$

where C is a positive constant. By applying Proposition 1, we obtain the conclusion of Proposition 2.

A partial solution of the general problem, which may be useful in special cases, can be obtained by observing that, if the contour surfaces of f are preassigned, the problem is reduced to the one-dimensional case, which is solved by Proposition 1. Let ϱ be a continuously differentiable, positive valued function on $\mathscr{R}^K$ such that

$$\varrho(x) = 0 \text{ for all } x \in S . \tag{32}$$

and, for all real r,

$$\int_{\varrho(x)<r} \pi(x)\, dx < \infty , \tag{33}$$

and, as $h \downarrow 0$,

$$\int_{r<\varrho(x)\leq r+h} \pi(x)\, dx = 0\,(h) , \tag{34}$$

uniformly for r in any compact set.

Then

$$\lim_{\lambda\downarrow 0}\ \inf_{f\in a(S)}\ \int \left\{ g^{ij}(x)\frac{\partial f(x)}{\partial x^i}\frac{\partial f(x)}{\partial x^j} + \lambda f^2(x) \right\} \pi(x)\, dx \leq$$

$$\leq \lim_{\lambda\downarrow 0}\ \inf_{\xi(0)=1}\ \int \left\{ g^{ij}(x)\frac{\partial \xi\,[\varrho(x)]}{\partial x^i}\frac{\partial \xi\,[\varrho(x)]}{\partial x^j} + \lambda \xi^2\,[\varrho(x)] \right\} \pi(x)\, dx \tag{35}$$

$$= \lim_{\lambda\downarrow 0}\ \inf_{\xi(0)=1}\ \int_0^\infty \left\{ \xi'^2(r)\frac{d}{dr}\int_{\varrho(x)<r} g^{ij}(x)\frac{\partial\varrho(x)}{\partial x^i}\frac{\partial\varrho(x)}{\partial x^j} \pi(x)\, dx + \right.$$

$$\left. + \lambda\xi^2(r)\frac{d}{dr}\int_{\varrho(x)<r} \pi(x)\, dx \right\} dr .$$

By Proposition 1, a necessary and sufficient condition for the right hand side of (35) to be 0 is that either

$$\int_0^\infty \frac{dr}{\dfrac{d}{dr}\displaystyle\int_{\varrho(x)<r} g^{ij}(x)\frac{\partial\varrho(x)}{\partial x^i}\frac{\partial\varrho(x)}{\partial x^j}\pi(x)\, dx} = \infty \tag{36}$$

or π is integrable (a trivial case we exclude in the following discussion). Thus for (1) to hold it is sufficient that, for some choice of ϱ, satisfying (32) — (34), condition (36) is satisfied. On the other hand, if (1) holds, we can nearly obtain equality in (35) by determining ϱ from the contour surfaces of a suitably chosen, nearly minimizing f for small λ, so that, again by Proposition 1, the condition stated below (36) is also necessary. Thus we have

Proposition 3: Under the conditions stated at the beginning of this section, with π not integrable, in order that (1) hold it is necessary and sufficient that there exist a continuously differentiable positive valued function ϱ on $\mathscr{R}^K$ such that (32) — (34) and (36) are satisfied.

Proposition 3 can be specialized to obtain more explicit sufficient conditions. First we observe that g and π enter (36) only in the combination $g\pi$, and although π enters separately in (33) a sufficient condition [for (33)], compactness of the sets $\{x : \varrho(x) \leq r\}$ can be given without reference to π. Now $g\pi$ is not a tensor, but rather a contravariant tensor density. This means that (3) — (6) imply

$$G^{ij}(y)\,\Pi(y) \tag{37}$$

$$= g^{kl}\,[\psi^{-1}(y)]\,\pi\,[\psi^{-1}(y)]\,\frac{\partial \psi^i}{\partial x^k}\,[\psi^{-1}(y)]\,\frac{\partial \psi^i}{\partial \psi^l}\,[\psi^{-1}(y)]\,\det\left[\frac{\partial\,[\psi^{-1}(y)]^{j'}}{\partial y^i}\right],$$

which differs from the formula (5) for the transformation of a contravariant tensor by the final factor of a determinant. In the case $K \geq 3$, the geometry determined by such a tensor density $g\pi$ is equivalent to the (Riemannian) geometry determined by the covariant tensor h defined by

$$h = (g\pi)^{-1}\det(g\pi)^{\frac{1}{K-2}}\,. \tag{38}$$

It should be possible to obtain more explicit necessary conditions and sufficient conditions in terms of this geometry. For example, it is sufficient that (36) hold with $\varrho(x)$ equal to the distance from a point in S when this distance is large. The case $K = 2$ must be treated separately.

I am indebted to R. FINN, K. LOEWNER, and A. NOVIKOFF for some helpful conversations in connection with the material of this section. In particular, FINN brought to my attention a paper of MEYERS and SERRIN [9] dealing with a similar, but perhaps more difficult, problem.

Note added in proof: On rereading this paper, I am again struck by the fact that it is very badly written. However, I feel that the basic idea is important enough to make the paper more helpful than harmful, provided the reader understands that the must work things out for himself, using this paper only as a guide. In particular, neglect of the factor displayed above (3.13) causes some trouble. However, I believe the results are basically right, at least in the important case where the problem is invariant under a group operating transitively on the parameter space.

References

[1] JAMES, W., and C. STEIN: Estimation with quadratic loss. Proc. Fourth Berkeley Symposium on Math. Stat. and Prob. p. 361. Berkeley and Los Angeles: Univ. of Calif. Press 1961.

[2] JEFFREYS, H.: Theory of Probability. Third Edition. Oxford: Clarendon Press 1961.

[3] KAKUTANI, S.: On equivalence of infinite product measures. Ann. Math. 49, 214 (1948).

[4] KIEFER, J.: Invariance, minimax sequential estimation, and continuous time processes. Ann. Math. 28, 573 (1957).

[5] KRAFT, C.: Some conditions for consistency and uniform consistency of statistical procedures. Univ. Calif. Publ. in Statistics. 2, No. 6, 125 (1955).

[6] KUDO, H.: On minimax invariant estimators of the transformation parameter. Nat. Sci. Rep., Ochanomizu Univ. 6, 31 (1955).

[7] LEHMANN, E.: Testing Statistical Hypotheses. New York: Wiley 1959.

[8] LOOMIS, L.: An Introduction to Abstract Harmonic Analysis. New York: Van Nostrand 1953.
[9] MEYERS, N., and J. SERRIN: The exterior Dirichlet problem for second order partial differential equations. J. Math. Mech. **9**, 513 (1960).
[10] PEISAKOFF, M.: Transformation of parameters, Unpublished Ph. D. Thesis, Princeton 1951.
[11] PITMAN, E. J. G.: Location and scale parameters. Biometrika **30**, 391 (1939).
[12] STEIN, C.: A necessary and sufficient condition for admissibility. Ann. Math. Statist. **26**, 518 (1955).
[13] STONE, M.: The posterior t distribution. Ann. Math. Statist. **34**, 568 (1963).

Stationary Gaussian Processes Satisfying the Strong Mixing Condition and Best Predictable Functionals

By **A. M. Yaglom**

Academy of Sciences of the USSR

1. Wide sense strong mixing condition

The well-known problem of least squares prediction of the stationary stochastic process $x(t)$ is the problem of finding the functional $\tilde{x}(\tau)$ of the values $x(t)$, $t \leq 0$, which is the least squares approximation of the "future" value of the process $x(\tau)$, $\tau > 0$. In addition to the functional $\tilde{x}(\tau)$ it is important to know the mean square error of the prediction

$$\sigma^2(\tau) = E \mid x(\tau) - \tilde{x}(\tau) \mid^2, \tag{1}$$

or, what is equivalent, the correlation coefficient

$$\varrho(\tau) = \frac{E x(\tau)\, \tilde{x}(\tau)}{\left\{ E\,[x(\tau)]^2\, E\,[\tilde{x}(\tau)]^2 \right\}^{\frac{1}{2}}} = \sqrt{1 - \frac{\sigma^2(\tau)}{E\,[x(\tau)]^2}} \tag{2}$$

between $x(\tau)$ and $\tilde{x}(\tau)$ [here and later we can consider without loss of generality only the processes $x(t)$ with $Ex(t) = 0$]. If the process $x(t)$ is Gaussian, the least squares approximation $\tilde{x}(\tau)$ is linear; therefore, we can say that the problem of linear least squares prediction of the stationary process $x(t)$ is the wide sense version of the general problem of least squares prediction (see Doob [1], Chapter II, Section 3).

Let us now use the fact that the Hilbert space $\mathscr{H}_x$ generated by all random variables $x(t)$ (with the norm $\| x \| = \left\{ E \mid x \mid^2 \right\}^{1/2}$) is isomorphic to the Hilbert space $\mathscr{L}$ of all functions $\varphi(\lambda)$ (where $-\pi \leq \lambda \leq \pi$ for discrete parameter processes and $-\infty < \lambda < \infty$ for continuous parameter processes) which are quadratically integrable with respect to the spectral measure $dF(\lambda)$ of the process $x(t)$, and that in this isomorphism the function $e^{it\lambda}$ corresponds to the random variable $x(t)$. Let $\mathscr{L}_0^-$ be the closed linear subspace of $\mathscr{L}$ generated by the functions $e^{it\lambda}$, $t \leq 0$ (in many cases the subspace $\mathscr{L}_0^-$ can be described simply as the space of boundary values of functions which are analytic outside the unit circle for the case of discrete parameter and analytic in the lower half-plane of the complex variables plane for the case of continuous parameter). Then the problem of least squares linear prediction may be formulated analytically as finding the function $\Phi_\tau(\lambda) \in \mathcal{L}_0^-$ which maximizes the expression

$$\varrho(\tau) = \frac{\left| \int\limits_\Lambda e^{i\tau\lambda}\, \overline{\Phi_\tau(\lambda)}\, dF(\lambda) \right|}{\left\{ \int\limits_\Lambda |\Phi_\tau(\lambda)|^2\, dF(\lambda) \cdot \int\limits_\Lambda dF(\lambda) \right\}^{\frac{1}{2}}}, \tag{3}$$

where the region of the integration Λ is the interval $-\pi \leq \lambda \leq \pi$ in the case of a discrete parameter t and the whole real line $-\infty < t < \infty$ in the case of a continuous parameter t. The least squares approximation $\tilde{x}(\tau)$ will be given by the equation

$$\tilde{x}(\tau) = \int\limits_\Lambda \Phi_\tau(\lambda)\, d\mathcal{J}(\lambda), \tag{4}$$

where $\Phi_\tau(\lambda)$ is one of the functions maximizing the right side of (3), that is, $\Phi_\tau(\lambda)$ in (4) can differ from $\Phi_\tau(\lambda)$ in (3) only by a constant factor which can easily be found. In (4), $d\mathcal{J}(\lambda)$ is the random spectral measure of the process $x(t)$ involved in its spectral representation

$$x(t) = \int\limits_\Lambda e^{it\lambda}\, d\mathcal{J}(\lambda). \tag{5}$$

The value $\varrho(\tau)$ defined by (3) turns out to be equal to the correlation coefficient (2), so that finding the least upper bound of the expressions (3) for $\Phi_\tau(\lambda) \in \mathcal{L}_0^-$ gives immediately the mean square error of the optimum prediction.

It is natural to expect that as a rule $\sigma^2(\tau)$ will tend to $E[x(\tau)]^2 = R(0)$ as $\tau \to \infty$ and consequently $\varrho(\tau)$ will tend to zero as $\tau \to \infty$. Processes satisfying this condition are called *regular* (or *purely indeterministic*). The conditions necessary and sufficient for such regularity are given by the following well-known main theorem in the theory of linear least squares prediction of stationary processes:

Theorem 1 (Kolmogorov, Krein). *The stationary random process $x(t)$ will be regular if and only if its spectral function $F(\lambda)$ is absolutely continuous and the spectral density $f(\lambda) = F'(\lambda)$ satisfies*

$$\int\limits_{-\pi}^{\pi} \log f(\lambda)\, d\lambda > -\infty \tag{6}$$

(for a discrete parameter) or

$$\int\limits_{-\infty}^{\infty} \frac{\log f(\lambda)}{1 + \lambda^2}\, d\lambda > -\infty \tag{6'}$$

(for a continuous parameter).

The proof of the theorem can be found, for example, in Doob [7]. According to this theorem, for regularity of the process $x(t)$ it is only necessary that the spectral density exist and vanish no more than at isolated points without "sticking too close" to the λ-axis at these points.

Note now that KOLMOGOROV [2] and KREIN [3] gave the general formulas expressing $\sigma^2(\tau)$ [or $\varrho(\tau)$] for all $\tau > 0$ through the spectral function $F(\lambda)$ of the process $x(t)$. However, the explicit expression for the functional $\tilde{x}(\tau)$ giving the optimum prediction can be found only in some special cases, the most important of which is the case of spectral density rational in $e^{i\lambda}$ or in λ, depending on whether the parameter t is discrete or continuous (see WIENER [4] and YAGLOM [5]).

In the study of limit theorems for stochastic processes it is often necessary to replace the regularity condition by some more restrictive condition. The most important among these conditions is probably that introduced by ROSENBLATT [6], which is called the *strong mixing condition*. For the general formulation of this condition it is necessary to consider σ-algebras $\mathfrak{M}_0^-$ and $\mathfrak{M}_\tau^+$ of random events generated by the events of the form $x(t) < a$, $t \leq 0$ and, correspondingly, $x(t) < a$, $t \geq \tau$. Let us denote

$$\alpha(\tau) = \underset{A \in \mathfrak{M}_0^-,\, B \in \mathfrak{M}_\tau^+}{\text{l.u.b.}} \, |\, P(AB) - P(A)\, P(B)\,|\,. \tag{7}$$

The strong mixing condition is the condition

$$\alpha(\tau) \to 0 \text{ as } \tau \to \infty\,. \tag{8}$$

Let us consider now the closed linear subspace $\mathscr{H}_0^-$ and $\mathscr{H}_\tau^+$ of the Hilbert space $\mathscr{H}_x$, generated by the random variables $x(t)$, $t \leq 0$ and, correspondingly, $x(t)$, $t \geq \tau$. Let $\varrho_1(\tau)$ be the maximum correlation coefficient between the elements of $\mathscr{H}_0^-$ and $\mathscr{H}_\tau^+$, that is,

$$\varrho_1(\tau) = \underset{U_0^- \in \mathscr{H}_0^-,\, V_\tau^+ \in \mathscr{H}_\tau^+}{\text{l.u.b.}} \, \varrho(U_0^-, V_\tau^+)\,, \tag{9}$$

where $\varrho(U, V)$ is the correlation coefficient between the random variables U and V. It is easy to see that $\varrho_1(\tau) \geq \alpha(\tau)$ always. For the special case of Gaussian stationary random processes $x(t)$ it was shown by KOLMOGOROV and ROZANOV [7] that the inequality $\varrho_1(\tau) \leq 2\pi\alpha(\tau)$ also holds. Consequently, for Gaussian processes the strong mixing condition is equivalent to the condition

$$\varrho_1(\tau) \to 0 \text{ as } \tau \to \infty. \tag{10}$$

Thus, condition (10) is the wide sense version of the general strong mixing condition (8). From now on, we shall consider only this wide sense version, which will be called simply the strong mixing condition.

Going from the Hilbert space $\mathscr{H}_x$ to the isomorphic Hilbert space $\mathscr{L}$ we can write

$$\varrho_1(\tau) = \underset{\Phi_\tau^-(\lambda) \in \mathscr{L}_0^-,\, \Psi_\tau^+(\lambda) \in \mathscr{L}_0^+}{\text{l.u.b.}} \frac{|\int_\Lambda e^{i\tau\lambda}\overline{\Phi_\tau^-(\lambda)}\,\Psi_\tau^+(\lambda)\,dF(\lambda)|}{\left\{\int_\Lambda |\Phi_\tau^-(\lambda)|^2\,dF(\lambda) \cdot \int_\Lambda |\Psi_\tau^+(\lambda)|^2\,dF(\lambda)\right\}^{\frac{1}{2}}}\,, \tag{11}$$

where $\mathscr{L}_0^+$ is the closed linear subspace of $\mathscr{L}$ generated by the functions $e^{it\lambda}$, $t \geq 0$. It is obvious that $\varrho_1(\tau) \geq \varrho(\tau)$ and, consequently, the strong mixing condition is more restrictive than the regularity condition. Therefore, the strong mixing condition is invalid if the spectral function $F(\lambda)$ is not absolutely continuous, or if the spectral density vanishes so strongly that condition (6) [or (6′)] does not hold. On the other hand, Kolmogorov and Rozanov [7] showed that the strong mixing condition will certainly hold if the time t is discrete and the spectral density $f(\lambda)$ is continuous and nonvanishing, or if the time t is continuous and the density $f(\lambda)$ is uniformly continuous, nonvanishing and decreases at infinity as some power function. However, the problem of necessary and sufficient conditions for strong mixing of Gaussian processes turned out to be more difficult than the corresponding problem of conditions of regularity and still remains unsolved. Nevertheless, recently Ibragimov [8], [9] obtained some interesting necessary conditions. The most important of these conditions can be stated as the following theorem.

Theorem 2 (Ibragimov). *The strong mixing condition* (10) *will be invalid if at least one of the following three conditions holds:*

1. *the spectral density $f(\lambda)$ has jump discontinuities;*
2. *the spectral density $f(\lambda)$ has a pole λ_0 such that*

$$\lim_{\lambda \to \lambda_0} |\lambda - \lambda_0|^\delta f(\lambda) \neq 0$$

for some $\delta > 0$;

3. *the spectral density has a zero λ_0 such that*

$$\lim_{\lambda \to \lambda_0} \frac{\log f(\lambda)}{\log |\lambda - \lambda_0|}$$

is different from an even integer.

The theorem explains the Rosenblatt example [10] of stationary Gaussian processes with spectral density proportional to $|\lambda|^{-\alpha}$, $1/2 < \alpha < 1$, for small $|\lambda|$, and consequently not satisfying the strong mixing condition.

Proceeding from Ibragimov's general results one can expect that the strong mixing condition will hold (at least for discrete t) if the spectral density $f(\lambda)$ would be sufficiently smooth and would have only very weak poles and zeroes close enough to the simplest algebraic zeroes of an even order. However, the final solution of the problem of necessary and sufficient conditions of strong mixing for Gaussian processes requires overcoming significant analytical difficulties.

2. The theory of canonical correlation for Gaussian stationary processes
A new characteristic property of rational spectral densities

In Ibragimov's paper [9] one finds conditions on $f(\lambda)$ which are necessary and sufficient for the maximum correlation coefficient $\varrho_1(\tau)$

to decrease at infinity as a power, or exponentially, or faster than any exponential function (for the discrete parameter case). However, results of this kind are much weaker than those concerning the problem of the mean square error $\sigma^2(\tau)$ [or correlation coefficient $\varrho(\tau)$] of the linear last squares prediction. In fact, the theory of linear least squares prediction enables us to find $\varrho(\tau)$ for every $\tau > 0$. It is natural, therefore, to put a question on the values $\varrho_1(\tau)$ for finite τ.

Up to now, the values $\varrho_1(\tau)$ for finite τ were apparently obtained only for the case when the time t is discrete, $\tau = 1$, and the spectral density $f(\lambda)$ is rational in $e^{i\lambda}$. It follows from IBRAGIMOV's results that the processes with rational spectral density always satisfy the strong mixing condition, $\varrho_1(\tau)$ decreasing exponentially for such processes. For this special case HELSON and SZEGO [11] showed recently that here one can write explicitly an algebraic equation of finite degree having $\varrho_1(1)$ as its largest root. However, it is easy to show that for the rational spectral density case and for discrete and for continuous parameter t one can obtain $\varrho_1(\tau)$ for every $\tau > 0$ as the largest root of some algebraic equation, all other roots of which have clear statistical meaning. To understand this meaning one need only remember some principal facts of the general theory of *canonical normal correlation*, which was developed in the thirties independently by HOTELLING [12] and by OBOUKHOV [13], [14] for random vectors (see also ANDERSON [15], Chapter 12). Later this theory was extended to the case of random processes by GELFAND and YAGLOM [16] (see also HANNAN [17]). The theory of canonical correlation enables us also to find a new characteristic property of stationary processes with rational spectral density, which may be of interest independently of the problem on the expression for $\varrho_1(\tau)$.

According to the general theory of canonical normal correlation of random vectors, for any two normally distributed vectors $\underset{\sim}{u} = (u_1, u_2, \ldots, u_n)$ and $\underset{\sim}{v} = (v_1, v_2, \ldots, v_m)$ one can find a transformation of coordinates in the spaces of these vectors such that all the components of the compound vector

$$(U_1, V_2, \ldots, U_n, V_1, V_2, \ldots, V_m) \tag{12}$$

(where U_i and V_j are the components of $\underset{\sim}{u}$ and $\underset{\sim}{v}$ in the new coordinate systems) will be pairwise uncorrelated with the exception only of pairs (U_i, V_i), $i = 1, \ldots, l$ where $l \leq \min(n, m)$. The general method of obtaining the canonical variables (U_i, V_i), $i = 1, \ldots, l$, and canonical correlation coefficients $\varrho_i = \varrho(U_i, V_i)$ $i = 1, \ldots, l$, can be described in purely geometrical terms as follows (see [16]). Let us consider the multidimensional space $\mathscr{H}_{u,v}$ of all linear combinations $w = \sum_1^n \alpha_i u_i + \sum_1^m \beta_j v_j$ [with the usual scalar product $(w_1, w_2) = E w_1 \overline{w}_2$) and two multidimensional "planes" $\mathscr{H}_u$ and $\mathscr{H}_v$ of this space consisting of vectors of the form $\sum_1^n \alpha_i u_i$ and, correspondingly, $\sum_1^m \beta_j v_j$. Let $\mathscr{P}_1$ be the matrix of projection in $\mathscr{H}_{u,\,v}$ on $\mathscr{H}_u$ and $\mathscr{P}_2$ be the matrix of projection on $\mathscr{H}_v$. Then the matrices $\mathscr{B}_1 = \mathscr{P}_1 \mathscr{P}_2$ and $\mathscr{B}_2 = \mathscr{P}_2 \mathscr{P}_1$ will determine linear transformations in the subspaces $\mathscr{H}_u$ and $\mathscr{H}_v$ (these transformations can also be determined by matrices $\widetilde{\mathscr{B}}_1 = \mathscr{P}_1 \mathscr{P}_2 \mathscr{P}_1$ and $\widetilde{\mathscr{B}}_2 = \mathscr{P}_2 \mathscr{P}_1 \mathscr{P}_2$ operating in $\mathscr{H}_{u,\,v}$). It is easy to see that the projection of every eigenvector of $\mathscr{B}_1$ on the space $\mathscr{H}_v$ will be an eigenvector of $\mathscr{B}_2$ with the same eigenvalue and vice versa. Consequently, the nonzero eigenvalues of $\mathscr{B}_1$ and of $\mathscr{B}_2$ will coincide with each other [so that the number l of such eigenvalues is $\leq \min (n, m)$], and the eigenvectors of $\mathscr{B}_1$ and of $\mathscr{B}_2$ are obtained from one another with the help of projections $\mathscr{P}_1$ and $\mathscr{P}_2$. These eigenvectors will be the canonical variables (12) and the corresponding eigenvalues ϱ_i, $i = 1, 2, \ldots, l$ will coincide with the canonical correlation coefficients $\varrho\,(U_i, V_i)$. From the geometrical point of view, the numbers ϱ_i, $i = 1, 2, \ldots, l$ will represent the whole collection of isometric invariants of the two subspaces $\mathscr{H}_u$ and $\mathscr{H}_v$. In this connection, it is natural to define the angles between the multidimensional planes $\mathscr{H}_u$ and $\mathscr{H}_v$ as the angles $\Theta_i = \cos^{-1} \varrho_i$ (see, for example, SHIROKOV [18]).

It is obvious that the method described above can be applied to the general case of two linear subspaces $\mathscr{H}_1$ and $\mathscr{H}_2$ of a Hilbert space $\mathscr{H}$, if the matrices $\mathscr{P}_1$ and $\mathscr{P}_2$ are replaced by the projection operators in $\mathscr{H}$. Then $\mathscr{B}_1$ and $\mathscr{B}_2$ would be nonnegative self-adjoint operators in $\mathscr{H}_1$ and $\mathscr{H}_2$ with norm not greater than unity. These operators will also have common spectrum which, in general, can be continuous (refer to [17]). If $\mathscr{H}_1 = \mathscr{H}_x$ and $\mathscr{H}_2 = \mathscr{H}_y$ where $x\,(t), t \in T$, and $y\,(s), s \in S$ are two infinite collections of Gaussian random variables (that is, two Gaussian random processes on arbitrary parameter sets T and S), then the least upper bound of the spectrum of these operators will coincide with the maximum correlation coefficient for the processes $x\,(t)$ and $y\,(s)$. When the spectrum of the operators $\mathscr{B}_1$ and $\mathscr{B}_2$ is purely discrete, the theory of normal canonical correlation of random vectors can be simply generalized to Gaussian random processes $x\,(t)$ and $y\,(s)$. Namely, in this case we can find two sequences $U_1, V_2, \ldots$ and $V_1, V_2, \ldots$ of linear functionals of the values $x\,(t), t \in T$, and, correspondingly, $y\,(s), s \in S$, with the following properties: the elements of the first sequence form a basis in the space $\mathscr{H}_x$, the elements of the second sequence form a basis in the space $\mathscr{H}_y$, and all the elements of the compound sequence $U_1, U_2, \ldots,$ $V_1, V_2, \ldots$ are pairwise uncorrelated with the exception only of pairs (U_i, V_i), $i = 1, \ldots, l$ where l is equal to some integer or to infinity.

The problem on the value $\varrho_1\,(\tau)$ considered in Section 1 concerns the case when $x\,(t), t \leq 0$, and $y\,(s) = x\,(s), s \geq \tau$, are parts of the same Gaussian stationary process separated by "empty" interval of length

$\tau > 0$. In this case the projection operator $\mathscr{P}_1$ transforms a variable $y \in \mathscr{H}_\tau^+$ into its least squares approximation in the space $\mathscr{H}_0^-$, that is, into its linear least square prediction. Therefore, in this case we can explicitly write the operator $\mathscr{P}_1$ (after going from the Hilbert space $\mathscr{H}_x$ to the isomorphic space $\mathscr{L}$) with the help of the general theory of linear least square prediction; the general expression of the operator $\mathscr{P}_2$ in $\mathscr{L}$ can also be similarly obtained. However, in general these operators are so complicated that they can hardly be helpful for the problem on the spectrum of the operators $\mathscr{B}_1 = \mathscr{P}_1 \mathscr{P}_2$ and $\mathscr{B}_2 = \mathscr{P}_2 \mathscr{P}_1$.

Let us now suppose that the process $x(t)$ has spectral density $f(\lambda)$, which is a rational function of λ, that is, the density $f(\lambda)$ has the form

$$f(\lambda) = B \frac{\left| \prod_{k=1}^{M} (\lambda - \beta_k) \right|^2}{\left| \prod_{l=1}^{N} (\lambda - \alpha_l) \right|^2}, \qquad -\infty < \lambda < \infty, \tag{13}$$

where $B > 0$, $M < N$ and imaginary parts of all β_k are nonnegative, imaginary parts of all α_l are positive (to be definite, from now on we shall consider only the case of a continuous parameter t). In this case the situation becomes immediately much simpler. In fact, it is well known that for processes with spectral density (13), the projection of the function $e^{it\lambda} \in \mathscr{L}$ (where $\tau > 0$) on the subspace $\mathscr{L}_0^-$ (which corresponds to the projection of $x(\tau)$ on $\mathscr{H}_0^-$) has the form

$$\Phi_\tau(\lambda) = \frac{\gamma_\tau(\lambda)}{\prod_{k=1}^{M} (\lambda - \beta_k)}, \tag{14}$$

where $\gamma_\tau(\lambda)$ is a polynomial of degree not greater than $N-1$ (see [5]). It can easily be deduced from this that for all $\tau > 0$ the projection of the subspace $\mathscr{L}_\tau^+$ on the subspace $\mathscr{L}_0^-$ coincides with the N-dimensional linear manifold $\mathscr{L}^{(N)} \subset \mathscr{L}_0^-$ generated by the functions $\lambda^j / \prod_{k=1}^{M} (\lambda - \beta_k)$, $j = 0, 1, \ldots, N-1$. Consequently, the operator $\mathscr{B}_1 = \mathscr{P}_1 \mathscr{P}_2$ will be identically zero on the orthogonal complement of the N-dimensional subspace $\mathscr{L}^{(N)}$, and therefore it cannot have more than N nonzero eigenvalues ϱ_i. In other words, for stationary Gaussian processes $x(t)$ with rational spectral density $f(\lambda)$ of the form (13), there cannot exist more than N canonical linear functionals $U_1, \ldots, U_N$ of the values $x(t), t \leq 0$, and corresponding to them canonical linear functionals $V_1, \ldots, V_N$ of the values $x(t), t \geq \tau$. It can be obtained also from the theory of linear least square prediction for processes with rational spectral density that the number of pairs (U_i, V_i) of canonical functionals will be exactly equal to N. The correlation coefficients $\varrho_i = \varrho(U_i, V_i)$ between the functionals U_i and V_i completely describe the statistical dependence of the "future" of the process $x(t)$ upon its "past." If we put the pairs (U_i, V_i) in order of decreasing correlation coefficients ϱ_i, then $\varrho_1 = \varrho(U_1, V_1)$ will coincide with the maximum correlation coefficient $\varrho_1(\tau)$ of Section 1.

Similar results can be obtained for a more general stationary process $x(t)$ with spectral function $F(\lambda)$ having derivative $F'(\lambda) = f(\lambda)$ of the form (13), and besides this a finite number (let us say K) of discrete jump discontinuities. In this case, the projection of the function $e^{i\tau\lambda} \in \mathscr{L}$, $\tau > 0$ on the space $\mathscr{L}_0^-$ will be almost everywhere of the form (14), but at discontinuity points of $F(\lambda)$ this projection must have the value $e^{i\tau\lambda}$. Hence, the projection of the subspace $\mathscr{L}_\tau^+$ on the subspace $\mathscr{L}_0^-$ forms an $(N + K)$-dimensional linear manifold. Consequently, the number of pairs (U_i, V_i) of canonical variables with canonical correlation coefficients $\varrho_1 > 0$ will not be greater than $N + K$ (in fact, it will be exactly equal to $N + K$).

The last result has a simple converse. Let us suppose that the stationary process $x(t)$ has the property that the number of pairs of canonical variables (U_i, V_i) with $\varrho_i > 0$ for the parts of the process $x(t)$, $t \leq 0$, and $x(t)$, $t \geq \tau$, will be finite (not greater than Q), for every $\tau > 0$ (and will be equal to Q for sufficiently small τ). In this case the projection of the subspace $\mathscr{H}_\tau^+ \subset \mathscr{H}_x$ on the subspace $\mathscr{H}_0^-$ for every $\tau > 0$ will not be more than Q-dimensional and the projection of $\mathscr{H}_0^+$ on $\mathscr{H}_0^-$ will form a Q-dimensional linear manifold $\mathscr{H}^{(Q)} \subset \mathscr{H}_0^-$. Let the vectors $W_0 = x(0)/\{E[x(0)]^2\}^{1/2}, W_1, \ldots, W_{Q-1}$ form an orthonormal basis in the manifold $\mathscr{H}^{(Q)}$. Let us denote

$$\psi_k(\tau) = Ex(\tau)\,\overline{W}_k = (x(\tau), W_k),\ \chi_k(s) = EW_k\,\overline{x(-s)} = (W_k, x(-s)).$$

Then evidently both the functions $\psi_0(\tau), \ldots, \psi_{Q-1}(\tau)$ and the functions $\chi_0(s), \ldots, \chi_{Q-1}(s)$ will be linearly independent and

$$R(\tau + s) = Ex(\tau)\,\overline{x(-s)} = \frac{R(\tau)\,R(s)}{R(0)} + \sum_{k=1}^{Q-1} \psi_k(\tau)\,\chi_k(s), \qquad (15)$$
$$\text{for } \tau \geq 0,\ s \geq 0,$$

where $R(\tau)$ is the covariance function of the process $x(t)$. It can easily be deduced from the functional equation (15) that the function $R(\tau)$ must be of the form

$$R(\tau) = \sum_{j=1}^{n} C_j(\tau)\,e^{i\alpha_j\tau}, \qquad (16)$$

where $C_j(\tau)$ is a polynomial of degree m_j and $\Sigma_j(m_j + 1) = Q$. In fact, let us suppose $\tau = \tau_0, 2\tau_0, \ldots, (Q-1)\tau_0$ in (15), where τ_0 is a fixed number, and exclude the functions $\chi_1(s), \ldots, \chi_{Q-1}(s)$ from the obtained $Q - 1$ equations; then we find that the continuous function $R(s)$ satisfies a linear difference equation with constant coefficients of order Q whose solutions are of the form (16). From (16) and from the nonnegative definiteness of the function $R(\tau)$ it follows immediately that its Fourier-Stieltjes transform $F(\lambda)$ will have derivative of the form (13), will have no singular component, and can have no more than a finite number K of jump discontinuities, where $N + K = Q$.

Thus, we obtained the following theorem.

Theorem 3. *Let $x(t)$ be a stationary random process with continuous parameter. Then the parts $\{x(t), t \leq 0\}$ and $\{x(t), t \geq \tau\}$ of the process will have a finite number Q of pairs of canonical variables (U_i, V_i) with canonical correlation coefficients $\varrho_i = \varrho(U_i, V_i) > 0$ if and only if the spectral function $F(\lambda)$ of the process $x(t)$ is the sum of the integral of a rational function of the form (13) and of a monotone nondecreasing jump function increasing only in a finite number K of jump discontinuities, where $N + K = Q$.*

The theorem is evidently similar to the well-known theorem of DOOB [19] according to which the processes involved in the theorem can be characterized also as the component processes of finite-dimensional [namely $(N + K)$-dimensional] stationary Gaussian Markov processes.

3. Explicit expressions for maximum correlation coefficient and the best predictable functional for the stationary process with rational spectral density

The work of GELFAND and YAGLOM [16] deals with the problem on canonical correlation for finite parts of two different stationary Gaussian processes with rational spectral densities. In this case the number of nonzero canonical correlation coefficients is infinite and in [16] an explicit expression for $-\Sigma_i \log(1 - \varrho_i^2)$ was obtained. In the simpler case of two infinite parts $\{x(t), t \leq 0\}$ and $\{x(t), t \geq \tau\}$ of the same Gaussian process with rational spectral density (13), the number of nonzero canonical correlation coefficients is finite, and it is easy to find here explicit expressions for all coefficients ϱ_i and for the corresponding linear functionals U_i and V_i. For this purpose one can use, for example, the general method of solving linear least squares approximation problems for stationary processes with rational spectral density developed in [5].

Let us suppose that the spectral representation of the real stationary process $x(t)$ has the form (5) (where $\varLambda$ is the line $-\infty < \lambda < \infty$) and let us denote

$$U_i = \int_{-\infty}^{\infty} \varPhi_i^-(\lambda)\, d\mathscr{J}(\lambda), \quad V_i = \int_{-\infty}^{\infty} e^{i\tau\lambda} \varPhi_i^+(\lambda)\, d\mathscr{J}(\lambda). \qquad (17)$$

Let us assume that $EU_i^2 = EV_i^2 = 1$, that is,

$$\int_{-\infty}^{\infty} |\varPhi_i^-(\lambda)|^2 f(\lambda)\, d\lambda = 1, \qquad \int_{-\infty}^{\infty} |\varPhi_i^+(\lambda)|^2 f(\lambda)\, d\lambda = 1 \qquad (18)$$

(possible since the functionals U_i and V_i are defined only up to a constant factor). In this case, evidently,

$$E\,(V_i - \varrho_i U_i)\,x\,(-t) = 0, \qquad t \geq 0, \tag{19}$$

that is,

$$\int_{-\infty}^{\infty} [e^{it\lambda}\,\Phi_i^+\,(\lambda) - \varrho_i\,\Phi_i^-\,(\lambda)]\,e^{it\lambda} f\,(\lambda)\,d\lambda = 0, \qquad t \geq 0\,, \tag{20}$$

where $\varrho_i = \varrho_i\,(\tau)$ is the correlation coefficient between the random variables U_i and V_i. Similarly

$$E\,(U_i - \varrho_i V_i)\,x\,(\tau + t) = 0, \qquad t \geq 0, \tag{21}$$

that is,

$$\int_{-\infty}^{\infty} [e^{-it\lambda}\,\Phi_i^-\,(\lambda) - \varrho_i\,\Phi_i^+\,(\lambda)]\,e^{-it\lambda} f\,(\lambda)\,d\lambda = 0, \qquad t \geq 0\,. \tag{22}$$

Condition (20) will be satisfied if $[e^{it\lambda}\Phi_i^+\,(\lambda) - \varrho_i\Phi_i^-\,(\lambda)]\,f\,(\lambda)$ is the boundary value of a function analytic in the upper half-plane of the complex variable λ. Similarly, condition (22) will be satisfied if the function $[e^{-it\lambda}\,\Phi_i^-\,(\lambda) - \varrho_i\Phi_i^+\,(\lambda)]\,f\,(\lambda)$ is the boundary value of a function analytic in the lower half-plane. Besides this, from the fact that the functional U_i belongs to the subspace $\mathscr{H}_\tau^-$ and the functional V_i belongs to the subspace $\mathscr{H}_\tau^+$, it follows that the function $\Phi_i^+\,(\lambda)$ is the boundary value of a function analytic in the upper half-plane and the function $\Phi_i^-\,(\lambda)$ is the boundary value of a function analytic in the lower half-plane (see [5]). Easily, all the stated conditions can be satisfied if we suppose that

$$\Phi_i^+\,(\lambda) = \frac{\gamma_i^+\,(\lambda)}{\displaystyle\prod_{k=1}^{M}\,(\lambda - \bar{\beta}_k)}\,, \qquad \Phi_i^-\,(\lambda) = \frac{\gamma_i^-\,(\lambda)}{\displaystyle\prod_{k=1}^{M}\,(\lambda - \beta_k)} \tag{23}$$

where $\gamma_i^+\,(\lambda)$ and $\gamma_i^-\,(\lambda)$ are polynomials of degree not greater than $N - 1$. In fact, the functions $\Phi_i^+\,(\lambda), \Phi_i^-\,(\lambda)$ will then from the very beginning have the necessary analytical properties, and to satisfy conditions (20) and (22) it is sufficient to determine the coefficients of the polynomials $\gamma_i^+\,(\lambda)$ and $\gamma_i^-\,(\lambda)$ from the equations

$$e^{it\lambda} \prod_{k=1}^{M}\,(\lambda - \beta_k)\,\gamma_i^+\,(\lambda) - \varrho_i \prod_{k=1}^{M}\,(\lambda - \bar{\beta}_k)\,\gamma_i^-\,(\lambda) = 0\,, \tag{24}$$

$$\text{for } \lambda = \alpha_1,\,\ldots,\,\alpha_N$$

and

$$-\varrho_i \prod_{k=1}^{M}\,(\lambda - \beta_k)\,\gamma_i^+\,(\lambda) - e^{-it\lambda} \prod_{k=1}^{M}\,(\lambda - \bar{\beta}_k)\,\gamma_i^-\,(\lambda) = 0\,, \tag{25}$$

$$\text{for } \lambda = \bar{\alpha}_1,\,\ldots,\,\bar{\alpha}_N$$

(in the case when not all the roots α_j are distinct these equations undergo evident changes).

Thus, we obtain a system of $2N$ homogeneous linear equations for the $2N$ unknown coefficients of the polynomials $\gamma_i^+(\lambda)$ and $\gamma_i^-(\lambda)$. The system will have nontrivial solutions only if its determinant is equal to zero. Therefore, the possible nonzero canonical correlation coefficients ϱ_i can be found from the determinantal equation

$$\begin{vmatrix} e^{i\tau\alpha_1}\prod_k(\alpha_1-\beta_k), \ldots, & \alpha_1^{N-1}e^{i\tau\alpha_1}\prod_k(\alpha_1-\beta_k), & -\varrho\prod_k(\alpha_1-\bar\beta_k), \ldots, & -\varrho\alpha_1^{N-1}\prod_k(\alpha_1-\bar\beta_k) \\ \cdots & \cdots & \cdots & \cdots \\ e^{i\tau\alpha_N}\prod_k(\alpha_N-\beta_k), \ldots, & \alpha_N^{N-1}e^{i\tau\alpha_N}\prod_k(\alpha_N-\beta_k), & -\varrho\prod_k(\alpha_N-\bar\beta_k), \ldots, & -\varrho\alpha_N^{N-1}\prod_k(\alpha_N-\bar\beta_k) \\ -\varrho\prod_k(\bar\alpha_1-\beta_k), \ldots, & -\varrho\bar\alpha_1^{N-1}\prod_k(\bar\alpha_1-\beta_k), & e^{-i\tau\bar\alpha_1}\prod_k(\bar\alpha_1-\bar\beta_k), \ldots, & \bar\alpha_1^{N-1}e^{-i\tau\bar\alpha_1}\prod_k(\bar\alpha_1-\bar\beta_k) \\ \cdots & \cdots & \cdots & \cdots \\ -\varrho\prod_k(\bar\alpha_N-\beta_k), \ldots, & -\varrho\bar\alpha_N^{N-1}\prod_k(\bar\alpha_N-\beta_k), & e^{-i\tau\bar\alpha_N}\prod_k(\bar\alpha_N-\bar\beta_k), \ldots, & \bar\alpha_N^{N-1}e^{-i\tau\bar\alpha_N}\prod_k(\bar\alpha_N-\bar\beta_k) \end{vmatrix} = 0. \tag{26}$$

It is easily seen that all the coefficients if this algebraic equation of degree $2N$ will be real and that a negative root $-\varrho_i$ corresponds to every positive root ϱ_i of this equation. Let us now put the positive roots $\varrho_1, \ldots, \varrho_N$ of equation (26) in decreasing order; then the first root $\varrho_1 = \varrho_1(\iota)$ will determine the maximum correlation coefficient of Section 1. Putting any of the roots $\varrho_1, \ldots, \varrho_N$ into equations (24) and (25), we shall obtain a homogeneous system of $2N$ linear equations with zero determinant for $2N$ coefficients of the polynomials $\gamma_i^+(\lambda)$ and $\gamma_i^-(\lambda)$. The system determines the unknown coefficients up to a constant factor, and the factor itself can be determined up to a factor ± 1 from conditions (18). The corresponding functionals (17) will be the canonical functionals for our stationary random process.

The results can be stated as follows.

Theorem 4. *The maximum correlation coefficient between the "future" and the "past" of the stationary random process $x(t)$ with rational spectral density (13) is equal to the largest positive root of the determinantal equation (26). All other positive roots of the equation determine the other nonzero canonical correlation coefficients between the "future" and the "past" of the process. The canonical variables U_i and V_i corresponding to the correlation coefficient ϱ_i can be determined up to the factor ± 1 from equations (17), (18), (23), (24) and (25).*

References

[1] Doob, J. L.: Stochastic Processes. New York: Wiley 1953.
[2] Kolmogorov, A. N.: Interpolation und Extrapolation von stationären zufälligen Folgen. Izv. Akad. Nauk SSSR. Ser. Mat. 5, 3 (1941).
[3] Krein, M. G.: On a problem of extrapolation of A. N. Kolmogoroff. Dokl. Akad. Nauk SSSR 46, 306 (1945).

[4] WIENER, N.: Extrapolation, Interpolation, and Smoothing of Stationary Time Series. New York: Wiley 1949.

[5] YAGLOM, A. M.: An Introduction to the Theory of Stationary Random Functions. New York: Prentice-Hall 1962.

[6] ROSENBLATT, M.: A central limit theorem and a strong mixing condition. Proc. Nat. Acad. Sci. USA. **42**, 43 (1956).

[7] KOLMOGOROV, A. N., and YU. A. ROZANOV: On a strong mixing condition for a stationary random Gaussian process. Teor. Veroyatnost. i Primenen. **5**, 222 (1960).

[8] IBRAGIMOV, I. A.: Spectral functions of certain classes of Gaussian stationary processes. Dokl. Akad. Nauk SSSR. **137**, 1046 (1961).

[9] — Stationary Gaussian sequences that satisfy the strong mixing condition. Dokl. Akad. Nauk SSSR. **147**, 1282 (1962).

[10] ROSENBLATT, M.: Independence and dependence. Proc. Fourth Berkeley Symp. on Math. Stat. and Prob. **II**, 431—443. Berkeley and Los Angeles: University of California Press 1961.

[11] HELSON, H., and G. SZEGÖ: A problem in prediction theory. Ann. Matem. Pura ed Appl. **51**, 107 (1960).

[12] HOTELLING, H.: Relation between two sets of variates. Biometrica. **28**, 321 (1936).

[13] OBOUKHOV, A. M.: Normal correlation of vectors. Izv. Akad. Nauk SSSR. Ser. Mat. 3, 339 (1938).

[14] — Theory of correlation of vectors. Uchen. Zap. Mosk. Gosud. Univ., Matem. **45**, 73 (1940).

[15] ANDERSON, T. W.: An Introduction to Multivariate Statistical Analysis. New York: Wiley 1959.

[16] GELFAND, I. M., and A. M. YAGLOM: Calculation of the amount of information about a random function contained in another such function. Uspehi Mat. Nauk. **12**, No. 1 (73), 3 (1957).

[17] HANNAN, E. J.: The general theory of canonical correlation and its relation to functional analysis. J. Austral. Math. Soc. 2, 229 (1961).

[18] SHIROKOV, P. A.: Tensor Calculus. Moskva-Leningrad 1934.

[19] DOOB, J. L.: The elementary Gaussian processes. Ann. Math. Statist. **15**, 229 (1944).

Strong Limit Theorems for Stochastic Processes
and Orthogonality Conditions for Probability Measures

By **A. M. Yaglom**

Academy of Sciences of the USSR

1. Introduction

Let $x(t)$, $0 \leq t \leq T$, be the Wiener process, that is, a real Gaussian stochastic process with

$$Ex(t) = 0, \quad Ex(t)\, x(s) = R(t, s) = \min\{t, s\} \tag{1}$$

and let N_n for $n = 1, 2, \ldots$ be the sequence of increasing integers. Consider the following functional of $x(t)$:

$$U_n\,[x(t)] - \sum_{k=1}^{N_n} \left[x\left(\frac{kT}{N_n}\right) - x\left(\frac{(k-1)\,T}{N_n}\right)\right]^2. \tag{2}$$

In 1940 Lévy [1] discovered the following interesting result concerning this functional.

Lévy's theorem: *If $N_n = 2^n$ then with probability one*

$$\lim_{n \to \infty} U_n\,[x(t)] = T. \tag{3}$$

The restriction on the sequence $\{N_n\}$ can be considerably weakened (see for example, Doob [2], Kozin [3]), but we shall not dwell on it.

Later Lévy's theorem was independently discovered by Cameron and Martin [4]; for many years it was considered a theorem of an important specific property of the Wiener process (see, e.g., Doob [2]). It can be proved by two different methods: 1. by means of direct estimation of the mathematical expectation and variance of $U_n\,[x(t)]$ and application of the Tchebychev inequality and Borel-Cantelli lemma (Lévy, Cameron and Martin) and 2. with the help of the theory of martingales (Doob). For the general Wiener process with $R(t, s) = \sigma^2 \min\{t, s\}$ the equality (3) takes the form

$$\lim_{n \to \infty} U_n\,[x(t)] = \sigma^2\, T. \tag{4}$$

Thus, with the part of realization of the general Wiener process on the interval of an arbitrary length T we can find the exact value σ^2 with probability one. It follows that two of Wiener's probability measures with different variances σ^2 in the space of functions on the interval $[0, T]$

are mutually orthogonal. This conclusion is the main result of CAMERON and MARTIN [4].

Recently it became clear that Lévy's theorem is a very special case of a wide class of strong limit theorems for stochastic processes. These theorems allow us to find a great number of sufficient orthogonality conditions for various probability measures. Now we shall give a review of these general strong limit theorems.

2. The generalized Lévy's theorem for Gaussian processes

The Wiener process is a Gaussian-Markov process with independent increments. At the present time the Lévy type theorems may be formulated for many kinds of stochastic processes which have, at least, one of these three properties (GAUSSIAN, MARKOV, independent increments); in a number of cases it is probably sufficient to assume that some weakened version of one of these three properties is valid. We begin with the case of the Gaussian processes. As for the other process we shall make only some short remarks in Section 3. It should also be noted that we deal only with strong limit theorems proved with the help of the Tchebychev and Borel-Cantelli lemmas. The question as to what extent the other methods (for example, the methods of martingales theory) can be used to prove strong limit theorems requires special investigation.

The first generalization of Lévy's theorem on a relatively wide class of Gaussian random processes was the theorem proved by BAXTER [5] in 1956. The theorem deals with the Gaussian processes $x\,(t)$, $0 \leqq t \leqq T$, with mathematical expectation $Ex\,(t) = m\,(t)$ having a bounded first derivate $m'\,(t)$ and covariance function $R\,(t, s) = Ex\,(t)\,x\,(s) - m\,(t)\,m\,(s)$ with uniformly bounded second derivatives in $0 \leqq t, s \leqq T, t \neq s$. It follows from the stated conditions that the function $R\,(t, s)$ has the right and left first derivatives on the diagonal $t = s$, that is, the following limits exist:

$$D^+\,(t) = \lim_{s \to t+} \frac{R\,(t, t) - R\,(s, t)}{t - s}\,,\ D-\,(t) = \lim_{s \to t-} \frac{R\,(t, t) - R\,(s, t)}{t - s}\,. \tag{5}$$

Thus, under the stated conditions the covariance function $R\,(t, s)$ decrease linearly near the diagonal $t = s$. Now note that if $\lim_{n \to \infty} U_n\,[x\,(t) - m\,(t)]$ exists then $\lim_{n \to \infty} U_n\,[x\,(t)]$ also exists and is equal to the first limit because of the boundness of $m'\,(t)$. Thus, without loss of generality one can suppose that $m\,(t) \equiv 0$ (further, for simplicity we shall consider only the processes satisfying the last condition). If $m\,(t) \equiv 0$ then

$$E\left\{x\left(\frac{kT}{N_n}\right) - x\left[\frac{(k-1)\,T}{N_n}\right]\right\}^2 = E\,[\varDelta_k^{(n)}\,x]^2 \tag{6}$$

$$= R\left(\frac{kT}{N_n}, \frac{kT}{N_n}\right) - 2\,R\left[\frac{kT}{N_n}, \frac{(k-1)\,T}{N_n}\right] + R\left[\frac{(k-1)\,T}{N_n}, \frac{(k-1)\,T}{N_n}\right],$$

so that for the process $x\,(t)$

$$\lim_{n\to\infty} E\,U_n\,[x\,(t)] = \lim_{N_n\to\infty} \sum_{k=1}^{N_n} \left[D^-\left(\frac{kT}{N_n}\right) - D^+\left(\frac{kT}{N_n}\right) \right] \cdot \frac{T}{N_n}$$

$$= \int_0^T [D^-\,(t) - D^+\,(t)]\,dt\,. \tag{7}$$

Let now $N_n = 2^n$; then as it is easily seen

$$D\,U_n\,[x\,(t)] = E\,U_n^2 - (E\,U_n)^2 = 2 \sum_{k,\,j=1}^{2^n} [E\,\varDelta_k^{(n)}\,x\,\varDelta_j^{(n)}\,x]^2 = O\left(\frac{1}{2^n}\right)\,. \tag{8}$$

According to Tchebychev's inequality and the Borel-Cantelli lemma we can derive the following result from (7) and (8).

Baxter's theorem. *If* $N^n = 2^n$ *and* $x\,(t)$, $0 \le t \le T$, *is a Gaussian process satisfying the stated assumptions, then with probability one*

$$\lim_{n\to\infty} U_n\,[x\,(t)] = \int_0^T [D^-(t) - D^+(t)]\,dt\,. \tag{9}$$

Baxter's theorem is an obvious generalization of Lévy's theorem. It follows from it that if $\int_0^T [D_1^-\,(t) - D_1^+\,(t)]\,dt \neq \int_0^T [D_2^-\,(t) - D_2^+\,(t)]\,dt$ then the Gaussian measures μ_1 and μ_2 of the two processes $x_1\,(t)$ and $x_2\,(t)$ satisfying the conditions of Baxter's theorem with the functions $D_1^-\,(t)$, $D_1^+\,(t)$, and correspondingly $D_2^-\,(t)$, $D_2^+\,(t)$ in (5) would be orthogonal. For the stationary processes $x_1\,(t)$ and $x_2\,(t)$ with rational spectral densities $f_1\,(\lambda)$ and $f_2\,(\lambda)$, this fact denotes that the measures μ_1 and μ_2 would be orthogonal if $\lim_{|\lambda|\to\infty} f_1\,(\lambda)/f_2\,(\lambda) \neq 1$ (SLEPIAN [6]); orthogonality conditions of the same simplicity can be obtained for the case when both Gaussian processes $x_1\,(t)$ and $x_2\,(t)$ are the Markov processes (VARBERG [7]).

The most restricting condition in the statement of Baxter's theorem is the condition of the existence of limits (5). However, actually a similar theorem can be obtained for the processes satisfying much weaker conditions. For example, GLADYSHEV [8] considered the case when the following limits exist:

$$D_\alpha^+\,(t) = \lim_{s\to t+} \frac{R\,(t,\,t) - R\,(s,\,t)}{(t-s)^\alpha}\,,$$

$$D_\alpha^-\,(t) = \lim_{s\to t-} \frac{R\,(t,\,t) - R\,(s,\,t)}{(t-s)^\alpha}\,, 0 < \alpha < 2\,, \tag{10}$$

and, respectively, the product $|\,t - s\,|^{\,2-\alpha}\,|\,\partial^2 R\,(t,\,s)/\partial t\,\partial s\,|$ is considered bounded. In this case it is easily seen that

$$E\,[\varDelta_k^{(n)}\,x]^2 \approx \left[D_\alpha^-\left(\frac{kT}{N_n}\right) - D_\alpha^+\left(\frac{kT}{N_n}\right) \right] \cdot \left(\frac{T}{N_n}\right)^\alpha\,, \tag{11}$$

and thus,

$$\lim_{n \to \infty} \left(\frac{T}{N_n}\right)^{1-\alpha} E \, U_n \, [x \, (t)] = \int_0^T [D_\alpha^- \, (t) - D_\alpha^+ \, (t)] \, dt \, . \qquad (12)$$

If $N_n = 2^n$, then for the variance of $U_n \, [x \, (t)]$ the estimation

$$D \, U_n \, [x \, (t)] = O \, (\max\{2^{-n\,(2\alpha-1)}, \, n \, 2^{-2n}\}) \qquad (13)$$

can be obtained (cf. ALEKSEEV [9], who considered only the processes having spectral densities). With the help of (12) and (13) the following result can be proved.

Gladyshev's theorem. *If $x \, (t)$ is the Gaussian stochastic process satisfying the stated conditions and $N_n = 2^n$, then with probability one*

$$\lim_{n \to \infty} \left(\frac{T}{2^n}\right)^{1-\alpha} U_n \, [x \, (t)] = \int_0^T [D_\alpha^- \, (t) - D_\alpha^+ \, (t)] \, dt \, . \qquad (14)$$

If $\alpha = 1$ then Gladyshev's theorem coincides with Baxter's theorem. It can be used, for example, to obtain the orthogonality condition for the probability measures of two Gaussian stationary processes with the spectral densities decreasing as $|\lambda|^{-\beta}$ when $|\lambda| \to \infty$ if β is not an odd integer. However, if β is an odd integer then Gladyshev's theorem is insufficient. In the case $\beta = 3$ it is necessary to suppose that the following limits exist:

$$\widetilde{D}_2^+ \, (t) = \lim_{s \to t+} \frac{R(t,t) - R(s,t)}{(t-s)^2 \cdot |\log|t-s||} \, , \quad \widetilde{D}_2^- \, (t) = \lim_{s \to t-} \frac{R(t,t) - R(s,t)}{(t-s)^2 \cdot |\log|t-s||} \, . \qquad (15)$$

The last case was also considered by GLADYSHEV. He proved, in fact, that under certain conditions (including the condition of stationarity) if $N_n = 2^n$ and the limits (15) exist, then with probability one

$$\lim_{n \to \infty} \frac{2^n}{Tn \log 2} \, U_n[x \, (t)] = \int_0^T [\widetilde{D}_2^- \, (t) - \widetilde{D}_2^+ \, (t)] \, dt \, . \qquad (16)$$

Equations (14) and (16) suggest that for a very wide class of Gaussian processes on the internal $[0, T]$ the following generalized Lévy theorem holds.

If the sequence of the integers N_n for $n = 1, 2, \ldots$ increases rapidly enough, then with probability one

$$\lim_{n \to \infty} \frac{\sum_{k=1}^{N_n} [\Delta_k^{(n)} x]^2}{\sum_{k=1}^{N_n} E \, [\Delta_k^{(n)} x]^2} = 1 \, . \qquad (17)$$

However, in such a form the theorem has never been proved. For the stationary Gaussian process the theorem can be written in a much simpler form; here *with probability one*

$$\lim_{n \to \infty} \frac{\sum_{k=1}^{N_n} [\Delta_k^{(n)} x]^2}{2 N_n [R(0) - R(T/N_n)]} = \lim_{n \to \infty} \frac{1}{2 \, N_n \, [R(0) - R(T/N_n)]} \, U_n \, [x \, (t)] = 1 \qquad (18)$$

where $R(t) = Ex(s + t)x(s)$. The last equation under wide conditions has recently been proved by ROZANOV, who suggested also that equations (17) and (18) express some special case of the general strong law of large numbers for sequences of series of dependent random variables. If it is right than the general theorem formulated above must have some unknown much wider generalization. Unfortunately, even the formulation of the strong law of large numbers for sequences of series of dependent variables meets some serious difficulties which are to be overcome in the future.

3. Strong limits theorems of the Lévy-Baxter type for non-Gaussian processes

Let us now discuss the question of the formulation of strong limits theorems for some classes of non-Gaussian stochastic processes.

The situation is quite simple in the case of processes with independent increments. In fact, in this case the random variables $[\Delta_k^{(n)} x]^2$ for $k = 1, 2, \ldots, N_n$, are mutually independent and we can apply the usual methods for proving the strong law of large numbers for independent variables to obtain equations (17) and (18). Therefore, it is not surprising that as early as 1957 KOZIN [3] showed that a theorem similar to BAXTER's is valid for general processes with stationary independent increments satisfying the condition $\lim E [x(t) - x(s)]^4 / |t - s| = 0$. For such processes it is obvious that $E[x(t) - x(s)]^2 = h \cdot |t - s|$, where h is nonnegative constant and KOZIN proved that here

$$\lim_{n \to \infty} U_n [x(t)] = hT \tag{19}$$

with probability one if $N_n = 2^n$ (the last condition can be essentially weakened). It is natural to think that KOZIN's result may also be generalized for many processes with nonstationary independent increments.

Let us now consider the Markov processes. For the diffusion type Markov processes $x(t)$ with the well-behaved infinitesimal moments of the first two orders $m(t, x)$ and $\sigma^2(t, x)$ determined by the equations

$$m(t,x) = \lim_{h \to 0} \frac{E[x(t+h) - x(t) \mid x(t) = x]}{h},$$

$$\sigma^2(t,x) = \lim_{h \to 0} \frac{E[(x(t+h) - x(t))^2 \mid x(t) = x]}{h},$$

$$\tag{20}$$

the increment $dx(t)$ of the process for a short period satisfies the symbolic Langevin equation

$$dx(t) = m(t,x)\, dt + \sigma^2(t,x)\, dy(t)\,, \tag{21}$$

where $y(t)$ is the Wiener process (for a precise meaning of this equation see [2]). Proceeding from (21) it is easy to see that there exists a rapidly enough increasing sequence $N_n, n = 1, 2, \ldots$, such that with probability one

$$\lim_{n \to \infty} U_n \left[x(t) \right] = \int_0^T \sigma^2 \left(t, x(t) \right) dt . \tag{22}$$

It should be noted that in this case the right side of the equation will not generally be a constant but a random variable. It depends on the fact that the mean square of the increment of the general Markov process at the moment t is a function of $x(t)$. Equation (22) can evidently be used to obtain certain orthogonality conditions for Markov probability measures (compare YAGLOM [10]). Results of the form (22) can also be stated for some non-Markov processes satisfying the Ito integral equation which is the precise form of the Langevin equation (such processes are called Ito processes in GIRSANOV [11]).

Finally let us note that the theorems stated in Section 2 for Gaussian processes can also be proved for many non-Gaussian processes. In fact, in all the proofs of strong limit theorems in Section 2 only the expressions for the second and the fourth moments of the process are used. Therefore, it was not necessary to suppose that the processes were Gaussian; it was sufficient to suppose only that the fourth cumulants of the process were equal to zero. This remark is of no real importance, as non-Gaussian processes with vanishing fourth cumulants are extremely artificial. However, it is clear that for the application of the Borel-Cantelli lemma it is not necessary that the fourth cumulants be exactly equal to zero. Instead it is sufficient to require that these cumulants satisfy some inequalities of as yet not precisely known form. The inequalities will determine exactly for which random variables $\Delta_k^{(n)} x$ the general equations (17) and (18) hold and are of interest to probability theory.

4. More elaborate strong limit theorems for Gaussian processes

In the following, we shall again consider only the Gaussian processes, although it should be noted that the general remarks in Section 3 can be applied to all strong limit theorems to be discussed below. Let us suppose for simplicity, that all the processes considered are those with stationary increments, though this condition is also not obligatory. Up to now we considered only the limit theorems dealing with $\lim_{n \to \infty} U_n \left[x(t) \right]$. Are there any other strong limit theorems for stochastic processes?

Even now we can say that the theorems of the Lévy-Baxter's type are only rather special examples of strong limit theorems. First of all, there are more precise theorems about the limiting behavior of the functional $U_n \left[x(t) \right]$. Such theorems were recently published, for example, by ALEKSEEV [9] (see also YAGLOM [12]). According to the theorems of BAXTER and GLADYSHEV in Section 2 for the Gaussian process $x(t)$ with stationary increments and with

$$E\,[x\,(t) - x\,(0)]^2 = B\,(t) = C\,|\,t\,|^\alpha + o\,(\,|\,t\,|^\alpha)\ \text{as}\ |\,t\,| \to 0 \qquad (23)$$

where $0 < \alpha < 2$, and for $N_n = 2^n$

$$\lim_{n \to \infty} 2^{(\alpha-1)n}\,U_n\,[x\,(t)] = \lim_{n \to \infty} 2^{(\alpha-1)n}\,E\,U_n\,[x\,(t)] = CT^\alpha \qquad (24)$$

with probability one. It follows from this that two Gaussian measures of stochastic processes $x_1\,(t)$ and $x_2\,(t)$ satisfying (23) are orthogonal if the corresponding values of α and/or C are different. Let now

$$B\,(t) = C_1\,|\,t\,|^{\alpha_1} + C_2\,|\,t\,|^{\alpha_2} + o\,(\,|\,t\,|^{\alpha_2})\,, \qquad (25)$$

where $0 < \alpha_1 < \alpha_2 < 2$, $\alpha_2 - \alpha_1 < 1/2$. Then it is possible to show that

$$E\left\{U_n[x\,(t)] - N_n \cdot C_1\,(T/N_n)^{\alpha_1}\right\} = N_n \cdot C_2\,(T/N_n)^{\alpha_2} + o\,(N_n^{1-\alpha_2})\,. \qquad 26)$$

Besides, it follows from (13) that

$$D\left\{2^{n(\alpha_2-1)}\,U_n\,[x\,(t)]\right\} = O\left\{\max[2^{-n(1+2\alpha_1-2\alpha_2)},\ n2^{-2n(2-\alpha_2)}]\right\}\,. \qquad (27)$$

Applying the Tchebychev inequality and the Borel-Cantelli lemma one can obtain the following result (it was, in fact, precisely proved in [9] only for the processes having spectral densities).

Alekseev's theorem. *If $x\,(t)$, $0 \le t \le T$, is a Gaussian process with stationary increments satisfying (25) and if $N_n = 2^n$, then with probability one*

$$\lim_{n \to \infty}\left\{2^{n(\alpha_2-1)}\,U_n[x\,(t)] - 2^{n(\alpha_2-\alpha_1)}\,C_1\,T^{\alpha_1}\right\} = C_2\,T^{\alpha_2}\,. \qquad (28)$$

It follows from the theorem that the Gaussian measures of two processes satisfying the stated conditions may be orthogonal even if the values α_1 and C_1 for these processes are equal; it is sufficient that the values α_2 and/or C_2 be different.

Let us now consider the strong limit theorems dealing with the functionals different from $U_n\,[x\,(t)]$. It is natural to begin with the general quadratic functionals of the increments of $x\,(t)$. By suggestion of the author such functionals were studied recently by ALEKSEEV [13, 14]. They can be written in the following form:

$$q_n[x\,(t)] = (\Delta^{(n)}\,x)'\,Q^{(n)}\,\Delta^{(n)}\,x \qquad (29)$$

where $Q^{(1)}$, $Q^{(2)}$, ... is the sequence of quadratic matrices of orders $N_1, N_2, \ldots$ and $\Delta^{(n)}\,x = \left\{\Delta_1^{(n)}\,x, \ldots, \Delta_{N_n}^{(n)}\,x\right\}$ is the vector of the increments of $x\,(t)$ on the intervals of the length T/N_n. It is clear that all the estimations for such functionals are more complicated than for $U_n\,[x\,(t)]$, but in the case when $Q^{(n)}$ is the covariance matrix of some auxilliary process with stationary increments, we can use the fact that $E\,q_n\,[x\,(t)]$ is equal to the trace of the product of two covariance matrices and that there exist some expressions helpful for evaluation of such traces. Using this method ALEKSEEV [13, 14] obtained some interesting new strong limit theorems. The theorems are rather complicated and we shall present here only the general form of one of his results (all of which are proved in [13, 14] in the assumption of existence of spectral densities simplifying the evaluations of first and second moments). Let

$$B(t) = C_1 \,|\, t \,|^{\alpha_1} + C_2 \,|\, t \,|^{\alpha_2} + o\,(\,|\, t \,|^{\alpha_2}) \,, \tag{30}$$

$$0 < \alpha_1 < \alpha_2 < 2, \; \alpha_2 - \alpha_1 < 1/2$$

and $Q^{(n)}$ is the covariance matrix of the increment vector of the process $z\,(t)$ with

$$B_z(t) = E[z(t) - z(0)]^2 = |\, t \,|^\beta, \; \beta + \alpha_2 < 3 \,. \tag{31}$$

If $N_n = 3^n$ then for some normalizing constants A_n with probability one

$$\lim_{n \to \infty} A_n \cdot q_n[x(t)] = D_1(C_1, \alpha_1) \,, \tag{32}$$

where $D_1(C_1, \alpha_1) = \lim_{n \to \infty} A_n \cdot Eq_n$ is a numerical function of C_1 and α_1. Besides, with probability one

$$\lim_{n \to \infty} A'_n \left\{ q_n[x(t)] - A_n^{-1} \cdot D_1(C_1, \alpha_1) \right\} = D_2(C_2, \alpha_2) \,, \tag{33}$$

where A'_n are the new normalizing constants $(A'_n \to \infty$ as $n \to \infty)$ and $D_2(C_2, \alpha_2)$ is another numerical function. It follows that two Gaussian measures of the processes with the function $B(t)$ of the form (30) will be orthogonal if, at least, one of the parameters $(\alpha_1, C_1, \alpha_2, C_2)$ is different for two processes. A similar result was obtained by ALEKSEEV for the case when $\alpha_2 = \alpha_1 + 1/2$. It should be noted that in the case when $\alpha_2 - \alpha_1 > 1/2$ such theorems cannot exist, because according to the results of [13] in this case the Gaussian measures with different values of α_2 and C_2 would be equivalent.

Quite recently new strong limit theorems of interest were obtained by YU. A. ROZANOV, who considered some other quadratic functionals of the process $x\,(t)$. In particular, he considered the functional of the form

$$U_n[x(t), \tau] = \sum_{k=1}^{N_n} \left[x\left(\frac{kT'}{N_n}\right) - x\left(\frac{(k-1)\,T'}{N_n}\right) \right] \left[x\left(\tau + \frac{kT'}{N_n}\right) - \right.$$
$$\left. - x\left(\tau + \frac{(k-1)\,T'}{N_n}\right) \right], \tag{34}$$

where $T' = T - \tau$ and showed that under wide conditions it follows from the existence of right and left derivatives of the function $B(t)$ for all t that for some sequence $N_1, N_2, \ldots$ with probability one

$$\lim_{n \to \infty} U_n[x(t), \tau] = \frac{T'}{2} [B'(\tau - 0) - B'(\tau + 0)]. \tag{35}$$

From this result there follows, for example, the result of HAJEK [15] about the orthogonality of the Gaussian measures of the stationary processes $x_1\,(t)$ and $x_2\,(t)$ on the interval $[0, T]$ with the covariance functions $R_1(\tau) = e^{-|\tau|}$ and $R_2(\tau) = \max\{1 - |\tau|, 0\}$ if $T > 1$ (although it is known that the measures would be equivalent if $T \leq 1$; compare [*15, 16*]).

Until now we considered only the strong limit theorem for quadratic functionals of the realization of the process. However, it is natural to expect that there also exist many theorems of this kind for nonquadratic functionals. We may, for example, replace the square of the increments $\Delta_k^{(n)} x$ in equations (17), and (18) by some other function of the increments or of the values of the process; there are also many other nonlinear functionals which are suitable in this respect. The recent results of ALEKSEEV and ROZANOV give the impression that for deriving the orthogonality condition for Gaussian measures these more elaborate strong limit theorems may be unnecessary: it is likely that any two orthogonal Gaussian measures may be distinguished with the help of some quadratic functional. However, the general strong limit theorems for nonquadratic functionals may be of interest from a purely mathematical point of view. Besides this, the nonquadratic functional may be an important tool for testing the statistical hypotheses for non-Gaussian stochastic processes.

Thus, one can conclude that many interesting results recently originated from Lévy's theorem on the Wiener process are but a small part of the whole field of the strong limit theorems for stochastic processes and that this field is still far from being settled.

Aknowledgment. The author is indebted to YU. A. ROZANOV for valuable discussions connected with the content of the paper.

References

[*1*] LÉVY, P.: Le mouvement Brownien plan, Amer. J. Math. **62**, 487 (1940).

[*2*] DOOB, J. L.: Stochastic processes. New York: Wiley 1953.

[*3*] KOZIN, F.: A limit theorem for processes with stationary independent increments. Proc. Amer. Math. Soc. **8**, 960 (1957).

[*4*] CAMERON, R. H., and W. T. MARTIN: The behavior of measure and measurability under the change of scale in Wiener space. Bull. Amer. Math. Soc. **53**, 130 (1947).

[*5*] BAXTER, G.: A strong limit theorem for Gaussian processes. Proc. Amer. Math. Soc. **7**, 522 (1956).

[*6*] SLEPIAN, D.: Some comments on the detection of Gaussian signals in Gaussian noise. IRE Trans. on Inform. Theory. **4**, 65 (1958).

[*7*] VARBERG, D. E.: On equivalence of Gaussian measures. Pacif. J. Math. **11**, 751 (1961).

[*8*] GLADYSHEV, E. G.: A new limit theorem for random processes with Gaussian increments. Teor. Veroyatnost. i Primenen. **6**, 57 (1961).

[*9*] ALEKSEEV, V. G.: On perpendicularity conditions for Gaussian measures corresponding to two stochastic processes. Teor. Veroyatnost. i Primenen. **8**, 304 (1963).

[*10*] YAGLOM, A. M.: Supplement to the Russian edition of: U. Grenander. Stochastic processes and statistical inference. Moscow: Publ. House of Foreign Literature 1961.

[11] GIRSANOV, I. V.: On transforming a certain class of stochastic processes by absolutely continuous substitution of measures. Teor. Veroyatnost. i Primenen. **5**, 314 (1960).
[12] YAGLOM, A. M.: On the equivalence and perpendicularity of two Gaussian probability measures in function space, in the book Time series analysis (M. Rosenblatt, ed.) p. 327. New York: Wiley 1963.
[13] ALEKSEEV, V. G.: On the orthogonality and equivalence conditions for Gaussian measures in function space. Dokl. Akad. Nauk SSSR **147**, 751 (1962).
[14] — New theorems on almost-sure properties of realizations of Gaussian stochastic processes. Litovsky Matemat. Sbornik, Vilnius. **3**, No. 2, 5 (1963).
[15] HAJEK, J.: On linear statistical problems in stochastic processes. Czech. Math. J. **12** (87), 404 (1962).
[16] ROZANOV, YU. A.: On the equivalence of probability measures corresponding to Gaussian stationary processes. Teor. Veroyatnost. i Primenen. **8**, 241 (1963).